PHYSICAL ACTIVITY

HUMAN GROWTH AND DEVELOPMENT

Contributors

FORREST H. ADAMS

ERLING ASMUSSEN

HELEN M. ECKERT

V. REGGIE EDGERTON

ALAN G. INGHAM

GERALD S. KENYON

ROBERT L. LARSON

JOHN W. LOY

BARRY D. MCPHERSON

ROBERT M. MALINA

JANA PARÍZKOVÁ

G. LAWRENCE RARICK

ROBERT N. SINGER

PHYSICAL ACTIVITY

Human Growth and Development

Edited by *G. LAWRENCE RARICK*

Department of Physical Education
University of California
Berkeley, California

ACADEMIC PRESS   *New York and London*   *1973*

ACADEMIC PRESS, INC.
111 Fifth Avenue, New York, New York 10003

United Kingdom Edition published by
ACADEMIC PRESS, INC. (LONDON) LTD.
24/28 Oval Road, London NW1

LIBRARY OF CONGRESS CATALOG CARD NUMBER: 72-77339

PRINTED IN THE UNITED STATES OF AMERICA

Contents

v

Chapter   4.   **Factors Affecting the Working Capacity of Children and Adolescents**

Forrest H. Adams

Chapter   5.   **Body Composition and Exercise during Growth and Development**

Jana Parízková

Chapter 10. **Motor Performance of Mentally Retarded Children**

G. Lawrence Rarick

Chapter 11. **Play, Games, and Sport in the Psychosociological Development of Children and Youth**

John W. Loy and Alan G. Ingham

Chapter 12. **Becoming Involved in Physical Activity and Sport: A Process of Socialization**

Gerald S. Kenyon and Barry D. McPherson

List of Contributors

Numbers in parentheses indicate the pages on which the authors' contributions begin.

FORREST H. ADAMS, Division of Cardiology, Department of Pediatrics, UCLA School of Medicine, Los Angeles, California (80)

ERLING ASMUSSEN, Laboratory for Theory of Gymnastics, August Krogh Institute, University of Copenhagen, Copenhagen, Denmark (60)

HELEN M. ECKERT, Department of Physical Education, University of Califonia, Berkeley, California (154)

V. REGGIE EDGERTON, Department of Kinesiology, University of California, Los Angeles, California (1)

ALAN G. INGHAM, Department of Sociology, University of Massachusetts, Amherst, Massachusetts (257)

GERALD S. KENYON, Faculty of Human Kinetics nd Leisure Studies, University of Waterloo, Waterloo, Ontario (303)

ROBERT L. LARSON, Department of Athletic Medicine, University of Oregon, Eugene, Oregon (32)

JOHN W. LOY, Department of Physical Education and Department of Sociology, University of Massachusetts, Amherst, Massachusetts (257)

BARRY D. MCPHERSON, University of Waterloo, Waterloo, Ontario (303)

ROBERT M. MALINA, Department of Anthropology, University of Texas, Austin, Texas (125, 333)

JANA PARÍZKOVÁ, Physical Culture Institute, Ujezd, Czechoslovakia (97)

G. LAWRENCE RARICK, Department of Physical Education, University of California, Berkeley, California (125, 201, 225, 364)

ROBERT N. SINGER, Department of Physical Education, Florida State University, Tallahassee, Florida (176)

Preface

Physical activity is generally held to be an important factor in the growth and development of children and adolescents. Yet its significance has not been given adequate attention in volumes on growth and development or in textbooks on the physiology of exercise. This is not surprising for published research on physical activity and human growth has increased dramatically in recent years. Hence, these sources have been able to provide only limited coverage of important advances in knowledge in an area that is of concern to all who are interested in improving physical activity programs for the young.

This volume is entirely devoted to physical activity as it relates to the growth, development, and health of children. Its approach is of necessity multidisciplinary, for any attempt to deal with the complex nature of human growth and development must tap the resources of many fields of knowledge. The orientation of the volume reflects this, for it brings into focus the research of biological and behavioral scientists that is related to the physical activity needs and problems of children and youth.

The volume consists of fourteen topical chapters, each a self-contained unit although all have a common theme—physical activity and human growth. Responsibility for the selection of the topics rests solely with the editor. The reader must judge for himself the effectiveness of this selection. It is believed that collectively the chapters cover in a systematic and meaningful way the most significant research findings in this diversified and emerging field of inquiry.

To encompass the accumulated knowledge on physical activity and growth in a book of limited size is no small task when one considers the scope of the published research in this area. Investigations range from systematic observations of structural and functional adaptations at the cellular level in response to the stress of exercise to the gross morphological

xiii

changes following long periods of physical training. Of equal significance are the numerous studies of the working capacity of boys and girls. These investigations have added new dimensions to our understanding of the physiological adjustments that the growing body makes to the demands of exercise. More recently, investigators in the psychology and sociology of sports have advanced our understanding of individual and group behavior in a variety of sports situations. Most certainly our knowledge of the benefits and hazards of sports participation by young people has increased markedly in recent years. It would seem self-evident that there has been a need for some time to bring together in a single source an account of what we know about the motor behavior of children and the effects of physical activity upon human growth as well as the direction future research should take if we are to deal effectively with the problems in this area.

The book has been written primarily for upper division and graduate students in physical education, but should be of value to physicians and to those in the fields of growth and development and exercise physiology. It provides a reliable, ready reference source to selected aspects of physical activity and human growth.

The authors' contributions are gratefully acknowledged. They were selected in part for their general knowledge of physical education, but more particularly for their competence and research productivity in their own areas of specialization. They have done a highly commendable job in extracting from the literature pertinent information on their respective topics and organizing and presenting it in a way that will be meaningful to the reader.

G. LAWRENCE RARICK

Exercise and the Growth and Development of Muscle Tissue

V. Reggie Edgerton

I. General Muscle Development

Most structural units of a human are responsive to its routine functional load. That is, the size of the particular unit varies directly with its use. Skeletal muscles and the heart can increase in size if subjected to greater than routine loads. The increased size theoretically could be accounted for by an increase in the volume of existing cells or an increase in number of cells or both. In regard to muscle tissue, both of these possibilities have been subjected to experimentation for almost a century. The nature of this experimentation has varied from a study of muscle weight to rate of RNA synthesis and DNA content.

A. LONGITUDINAL GROWTH OF MUSCLE

Relatively little attention has been given to longitudinal growth of extra-fusal muscle fibers in normal or exercised animals. Longitudinal growth reportedly occurs at the tendo-musculo-junction by the addition of sarcomeres (MacKay and Harrop, 1969). It seems obvious that muscle fibers have to elongate as an animal grows. It has been shown in mice that sarcomere length is proportional to muscle length. A change in sarcomere length has been observed and appears to result from less myofilamental overlap rather than elongation of actin or myosin filaments (Rowe and Goldspink, 1968). It was suggested that the greater overlap at an early age is beneficial in that it is compatible with efficiency in maintaining isometric tensions as is typical of its normal behavior at that time. As long as a muscle remains of sufficient length to be stretched optimally, it should not be a critical factor in muscle development as affected by exercise. Interestingly, it has been suggested that the increasing tension on a muscle resulting from bone growth may be a stimulus to a muscle fiber to increase in size (Stewart, 1968).

B. DIAMETRICAL GROWTH OF MUSCLE

An increase of girth of fiber size as normal growth continues is apparently due to an increase in the number of myofibrils as well as possibly other cellular components. It is known that myosin, actin, and tropomyosin are synthesized on large polyribosome aggregates in the sarcoplasm between myofibrils. These proteins apparently polymerize into filaments which then attach in some way to Z-band material (Heywood *et al.,* 1967). It appears that hypertrophy induced by exercise employs mechanisms similar to those active during normal growth, that is, the formation and addition of new myofibrils (Goldspink, 1970). When myofibrils were partially split, the parent myofibril was usually about twice as large as the two new myofibril branches. The addition of new myofibrils roughly coincides with an increase in the percent total nitrogen of the myofibrillar fraction while a proportionate decrease occurs in mitochondrial and sarcoplasmic fractions (Perry, 1970). Perry also suggested some qualitative changes in myofibrillar proteins during this rapid growth period. While there is no good evidence at this time that actin and tropomyosin change with age, myosin differs during early growth in terms of its Ca^{2+}-activated adenosinetriphosphatase (ATPase) activity and amino acid composition as well as other specific chemical characteristics. The amino acid sequence of myosin seems to be related to the functional fiber types as well as age (Perry, 1970). Changes in contractile properties with age may be only a manifestation of qualitative changes of existing myofilaments.

II. Overload Hypertrophy of Muscle

When one speaks of growth and development of muscle tissue in relation to physical activity, hypertrophy first comes to mind. However, before hypertrophy can be meaningfully studied certain considerations must be dealt with. Principally, identification of hypertrophy necessitates a workable definition. That is, hypertrophy can be used in terms of an increase in absolute muscle weight or muscle weight relative to body weight. Having established a definition of hypertrophy, then it can be identified or detected. Perhaps the more critical, but least used, approach is to compare regression lines of muscle weight to body weight in control and experimental muscles. Statistical tests are available which can determine the probability of the two slopes being similar (Steel and Torri, 1960). This analysis is more informative than the conventional procedures in that it can be easily determined whether a change in muscle weight was proportional to normal whole body growth or whether the rate of muscle growth proportionately exceeded body growth. For example, hypertrophy of muscle could be incorrectly indicated in a group of animals if muscle to body weight ratios were used as the criterion and if control subjects tended to gain predominantly fat. This may very well be the case in some experiments in which control animals are compared to exercised animals. A more direct approach, although more laborious, is measuring the size of individual fibers. Oddly, changes in muscle weight have not always corresponded to fiber size measures. However, as will be apparent later in the chapter, these results may not be as contradictory as they seem. Taking into consideration the variables that are often overlooked, it should be of no surprise that data from various papers and even data of a single report seem incompatible in terms of recognizing muscle hypertrophy.

Cellular hypertrophy is quite simple, being only an enlargement of a cell relative to some previous point in time. Thus, normal growth of a fiber is hypertrophy. But hypertrophy of a muscle fiber in response to exercise would be enlargement of the fiber beyond that expected simply by growth.

A. EXERCISE-INDUCED GROWTH

The effect of exercise on general body growth has been assessed numerous times in laboratory animals but with varying results. Because total skeletal muscle makes up a very large component of the whole body by weight, general body growth is a reflection of muscle growth. Young Fischer rats allowed to exercise at will grew faster than littermates confined to small quarters (Ring *et al.,* 1970). However, Bloor *et al.* (1970) found a retardation of overall general growth proportionate to severity of exercise, attributing the retarded growth (body, liver, kidney, adrenal, and spleen), in part, to a decrease in the number of cells. Cardiac hypertrophy was evident in these animals exemplifying a significant training effect. Similar results were reported for animals raised in a hypoxic environment. In the heart and liver fewer cells and greater cytoplasmic mass were evident. In old rats, exercise elicited catabolic responses sufficient to induce a loss of body and organ weights (Bloor et al., 1970).

Gordon et al. (1967a) reported that rats trained with highly repetitive and low-resistant-type exercise routines gained less body weight than did controls, and muscle weights were generally smaller. But rats trained with forceful exercise were claimed to have had a greater gain in muscle weight relative to body weight than inactive rats (Gordon et al., 1967b). Other reports suggest the same specificity of response (Goldspink, 1964; Holmes and Rasch, 1958) in that a greater rate of gain in body weight has been found in forcefully exercised animals than in controls. This is not consistent with findings by Gordon et al. (1967b) who found a slower rate of body weight gain but a high muscle to body weight ratio in rats subjected to high resistance type of overload for at least 7 weeks, as opposed to controls. A significantly lower rate of gain in body weight with proportional changes in muscle weight has been observed repeatedly by this author in rats and guinea pigs exposed to daily endurance-type exercise programs (Edgerton, 1970). A slower rate of gain of body weight has not been evident in adult nonhuman primates (lesser bush baby) exposed to daily endurance treadmill exercises. If the differences in the species and muscles assessed and length of training period are significant, then it must also be realized that generalizations in regard to the results of all of these studies dealing with hypertrophy in various laboratory animals are severely limited.

In addition to the variables of resistance and repetition of an exercise, isometric and isotonic movements may be specific in their muscular effects. It has been proposed that isometric contractions may induce an increase in muscle strength without hypertrophy. Ikai and Fukunaga (1970) reported a 92% increase in strength with only a 23% increase in cross-sectional area of arm flexors after 100 days of 6 days/week, three times a day of 10 sec isometric contractions. They also found a 30% increase in strength in the untrained flexors. Although the effects of high-resistant, low-repetitive exercise on endurance of a muscle is generally thought to be minimal, significant changes in endurance (proportional to strength increases) have been demonstrated with such a program (Stull and Clarke, 1970). Others have shown no evidence of any endurance effect with a high-resistant exercise program (Asmussen, 1968).

In any consideration of the cellular components that could account for fiber enlargement, cellular components related to energy metabolism, as well as those directly involved in the process of muscle shortening, should be considered when studying growth and development of skeletal muscle. In man an increase in muscle mass and strength obviously results from high resistance overload training with very few repetitions. This muscular overload places maximal demands on the contractile elements and minimal demands on the energy production elements of the cell. But hypertrophy may result from a highly repetitive and low resistance exercise routine thus overloading the energy producing components of the cell. For example, mitochondria make up a significant volume of some muscle fibers and this organelle is known to be responsive to endurance training programs. Evidence of these specific responses to different types of exercise is becoming increasingly apparent.

Growth and development of muscle fibers of laboratory animals and exercised humans have been studied rather extensively. In rats, as pointed out earlier, Gordon *et al.* (1967b) have reported an increase in contractile proteins with no change in sarcoplasmic proteins in muscle of rats trained with a forceful-type training program. They found significantly greater mean fiber sizes in the white fibers of forcefully exercised rats, but not for red fibers. However, some appropriate statistical parameters were absent in this report. But perhaps even more critical, an adequate description of the method of determining the fiber type was lacking. Obviously, these ideas must be tested more strenuously and the experiments verified before the proposed hypothesis is acceptable.

Goldspink (1964) studied the biceps brachii of eight mice that were forced to retrieve their food (25 days) from a container suspended on a pulley and counterbalanced with a weight. He found that the larger peak of a bimodal distribution of fiber sizes became even more prominent after forceful exercise, thus demonstrating selective hypertrophy of the smaller fibers. This was true for two different levels of food intake. It was not clear, however, why mean body weight was higher in the exercised group while mean muscle weight tended to be smaller, although not significantly, and mean fiber size increased (at the higher level of food intake). It was suggested that changes in the extracellular components might explain this apparent discrepancy although no evidence to support the supposition was given.

Protein changes in contractile elements and energy systems which have been investigated may be potential contributors to cellular hypertrophy. Helander (1961) found an increase in myofilamental nitrogen in guinea pigs exposed to an endurance exercise program for several months. There was no appreciable change in sarcoplasmic nitrogen nor muscle weight in the exercised or chronically restricted animals. Stroma nitrogen (connective tissue) tended to decrease in the trained guinea pig calf muscles. Helander (1961) also found a significant increase in myofilamental nitrogen and decrease in sarcoplasmic nitrogen in rabbits with restricted activity. However, since these animals were older than controls as well as limited in regard to voluntary movement, these changes in nitrogen could be the result of age and/or activity.

Adult guinea pigs that were trained 5 days a week for 4 months on a motor driven treadmill showed an increase in the percentage of red fibers in the gastrocnemius (Barnard et al., 1970a). Red fibers were shown to be greater in size (Nelson et al., 1971). Selective enlargement of red fibers and an increase in the number of red fibers have also been demonstrated in small primates trained for 6 months on a treadmill (Edgerton et al., 1972). Consequently, it would appear that a highly repetitive-type exercise can induce mild degrees of selective hypertrophy, although the maximum tension that the muscle is capable of pulling following a maximal electrical stimulation is not altered. (Barnard et al., 1970b). The greater size of the red fibers in the trained animals may have resulted from the elevated mitochondrial content instead of a change in number of myofibrils since the trained guinea pigs also demonstrated enhanced mitochondrial yield and ability of muscle to oxidize several substrates (Barnard et al., 1970b).

The normally larger white fibers becoming "redder" with training is also a logical explanation for the apparent difference in fiber size. But in either case these changes would not logically be associated with maximal tension capabilities but with an augmented potential to yield energy for maintaining tension over an extended length of time which has been demonstrated in guinea pigs and primates (Edgerton et al.,1972).

B. SURGICALLY INDUCED GROWTH

Surgical methods have also been commonly used to investigate fundamental aspects of muscle growth. For example, partial removal or tenotomy of synergistic muscles of the mouse for 8–15 weeks resulted in a bimodal frequency distribution (controls were unimodal) in muscle fiber size in the mouse soleus and anterior tibialis. These results were interpreted as a demonstration of selective enlargement of some fibers as opposed to a slight change in all fibers (Rowe and Goldspink, 1968). These results and interpretations again are consistent with the hypothesis of selective hypertrophy. There was no attempt to relate these findings to the specific types of skeletal muscle fiber types. Hamosh et al. (1967) showed a 30% increase in rat soleus muscle weight within 4 days after tenotomy of the gastrocnemius and plantaris. This was accompanied by an increase in the RNA content of the microsomal fraction (fragmented sarcoplasmic reticulum) of the muscle although the specific activity of the RNA in promoting amino acid incorporation remained unchanged. Other authors have reported a fourfold increase in RNA polymerase (enzyme that catalyzes RNA synthesis) of the soleus within 48 hours after overload by tenotomy (Figs. 1 and 2). Nuclear synthesis of RNA was also augmented within 24 hr (Sobel and Kaufman, 1969). In short, overload by tenotomy of synergists causes an elevated rate of RNA synthesis by activating RNA polymerase. This elevated protein synthesis may be a manifestation of higher RNA content without any change in the protein synthesizing activity of existing RNA. But the physiological significance of the apparent acute hypertrophy immediately after

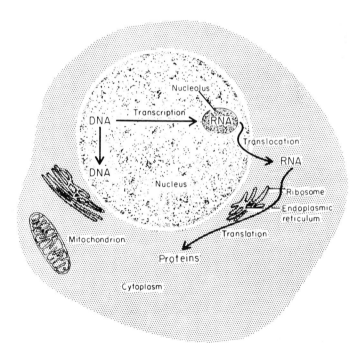

FIG. 1. Most of the DNA of a cell is located in the nucleus (some is also found in the mitochondria). Mitosis of a nucleus involves the duplication of the genetic material DNA. DNA can also transcribe its genetic program to RNA which differs from DNA only slightly chemically. The newly synthesized RNA then diffuses into the cytoplasm (translocation) and determines which and how many proteins (e.g., enzymes) will be synthesized in the cytoplasm.

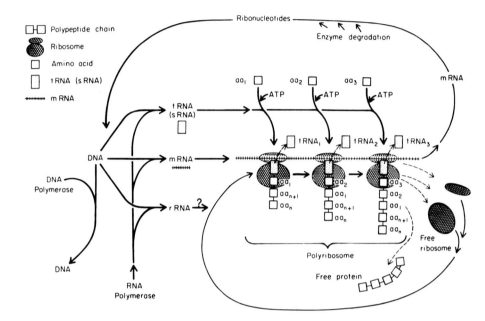

FIG. 2. A diagrammatic summary of how a specific type of protein is synthesized. Messenger RNA is the actual program or code which is read by tRNA, and each amino acid binds to a specific tRNA. In other words, mRNA determines through selection of a specific tRNA which amino acid will be incorporated onto the polypeptide chain at each point thus regulating the kind of protein (amino acid sequence) synthesized. The role of rRNA is not as well understood. In this particular diagram three polypeptides are being synthesized simultaneously, each being only at a different stage of completion. Each polypeptide will have an identical amino acid sequence since the same mRNA is being read.

tenotomy (1–4 days) is questionable with respect to exercise-induced hypertrophy since this pattern of growth is unlike that observed following any chronic overload as experienced during a normal exercise training program. The neurological ramifications of tenotomy or removal of a muscle cannot be ignored. Unfortunately, the difficulty in making inferences is further compounded in that most of the potentially significant data on overload were derived from experiments involving tenotomy of synergists and short postoperative times. Immediate signs of elevated muscle weights following surgery are common, but within a few weeks these differences disappear, in fact, even reverse in some muscles (unpublished observations). This may be a manifestation of the fact that it is most difficult to control the actual postoperative overload imposed on synergistic muscles owing to the rapid proliferation of connective tissue, which increases the utility of the tenotomized muscles.

Tomanek and Woo (1970) overloaded the plantaris in rats by denervating the gastrocnemius plus running them on a treadmill. Essentially they found that hypertrophy was not limited by age, and in examination of a more rarely considered parameter they found an equivalent concentration of elastin and collagen in sedentary old and young rats, while a concomitant increase in elastin and collagen was evident after muscular overload in young, but not old, rats.

Surgically overloaded rat heart showed no change in the mean diameter of thick myofilaments, and the authors consequently hypothesized that fiber enlargement must result from an increasing number of thick myofilaments (Carney and Brown, 1964). Richter and Kellner (1963) compared hypertrophied and normal human hearts and found no difference in the nature of the myofibrils' geometric arrangement in enlarged fibers, but the number of myofilaments increased through formation of new myofibrils. More than a 50% increase in number of myofibrils was observed in the enlarged hearts. Using autoradiography of the rat diaphragm, Morkin (1970) found that new myofibrillar proteins were added to the periphery of existing fibrils. But this same peripheral location of myofibril labeling with ^{3}H-leucine could be true whether new filaments were being added to existing fibrils or whether totally new fibrils were being formed.

C. PROTEIN SYNTHESIS

Muscle growth is necessarily accompanied by protein synthesis since its contractile elements, metabolic enzymes, and membranes are essential components of the fiber. Thus, a measure of protein synthesis assists in determining the rate of growth and development of a fiber. Since amino acids are the building blocks of proteins, a measure of the rate of incorporation of specific amino acids into proteins is a measure of protein synthesis. Incorporation is determined by using radioactive labeled amino acids such as ^{14}C-leucine and ^{14}C-lysine.

Growth of skeletal muscle can be associated with enhanced amino acid transport. Goldberg and Goodman (1969a) found α-aminoisobutyric acid (AIB) in higher concentration in hypertrophying muscles than contralateral controls in normal, diabetic, and hypophysectomized rats. This was evident within 4 hr after tenotomy of the synergists. Inhibiting protein synthesis with actinomycin D did not eliminate the elevated rates of AIB transport. Inactivity, on the other hand, caused a reduction of AIB transport (Goldberg and Goodman, 1969b). Thus, it seems that overload of a muscle may be enough to modify permeability characteristics of the sarcolemma which in turn may induce in some way, perhaps by simply making amino acids more available, elevated protein synthesis. Compensatory hypertrophy is also accompanied by elevated levels of RNA and RNA synthesis and orotic acid (an essential precursor in the synthesis of RNA) incorporation within two postoperative days (tenotomy) (Goldberg, 1967a, b, 1968). The proteins affected in this type of overload do not seem to be selective. That is, it appears that the increased incorporation of amino acids into muscle following overload is a generalized protein response and not related only to contractile or sarcoplasmic proteins.

Schreiber *et al.* (1967) found twice as much labeled leucine and lysine incorporated into protein by microsomes from guinea pig overloaded heart in a cell-free system than controls. Augmented microsomal protein synthesis began as early as 1 hr after the onset of stress. Protein synthesis may be initiated by increased cytoplasmic amino acid levels resulting from the increased protein degradation seen with muscle overload (Hatt *et al.,* 1965; Schreiber *et al.,* 1966; Tomita, 1966; Bozner and Meessen, 1969). A greater availability of microsomal RNA in overloaded muscles may also be the explanation for elevated rates of protein synthesis since there was a threefold increase in RNA specific activity of microsomes from overloaded hearts. This increase could be messenger or ribosomal RNA, or both. However, elevated protein synthesis could be accounted for simply by a change in efficiency of polysomes rather than a change in messenger RNA activity (Earl and Korner, 1966; Garren *et al.,* 1967).

Elevated ^{14}C-leucine incorporation into hypertrophied heart mitochondria of rats was observed within 3 days of aortic constriction, but control values were attained by the end of the second half of the first week (Shahab and Wollenberger, 1970). After 3 hr of elevated aortic pressure in the guinea pig, increased protein synthesis in overloaded hearts was found to be associated with the contractile protein myosin, but no myoglobin (oxygen storage) or collagen (Schreiber *et al.,* 1970). In control hearts, on the other hand, ^{3}H-leucine was incorporated equally into myosin and nonmyosin cardiac functions. Myoglobin, among other proteins, is obviously affected by long-term cardiac overload (Holloszy, 1967). Cattle acclimatized to high altitude were found to have a 40% increase in the number of mitochondria (no change in size), O_2 uptake, and cytochrome oxidase [an enzyme employed in the final stages of oxygen utilization per mitochondria (Ou and Tenney, 1970)].

Evidence of hypertrophied cardiac muscle fibers is numerous. The role that fiber hyperplasia plays in cardiac enlargement is probably insignificant. It appears that proteins associated with the contractile apparatus (myosin) and the energy-yielding system (myoglobin) are augmented within at least a few hours of the onset of overload. This is done without any significant change in muscle fiber DNA. Changes in RNA activity, possibly stimulated by high levels of cytoplasmic amino acids, or an increased availability of DNA for replication may account for the elevated levels of protein synthesis. The heterogeneity of the amino acid pool could also be critical and this pool may be the intracellular amino acid attached to tRNA (Fig. 2) (Stringfellow and Brachfeld, 1970).

D. DEVELOPMENT OF MUSCLE SPINDLE

Intrafusal fibers of rat muscles increase in number from two at birth to three at day three and to four by day six and through adult life. This development was found not to be dependent on mitotic activity since neither x-rays nor colchicine (arrests mitosis) varied the normal development (Bravo-Rey *et al.*, 1969). Maynard *et al.* (1971) studied the effect of overload by

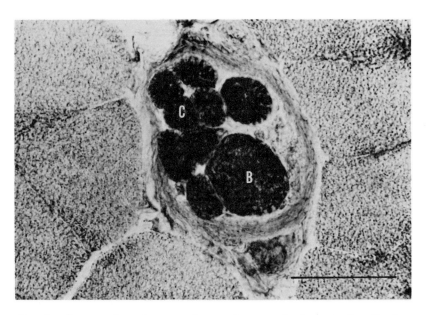

FIG. 3. Cross section of a normal cat soleus muscle demonstrating the larger extrafusal (EF) and smaller intrafusal (IF) muscle fibers. Note the relative intensity of reduced nicotinamide-adenine dinucleotide diaphorase (NADH-D) activity in the EF and IF fibers. A distinct difference in IF diameters are apparent. The larger one is a "bag" IF fiber (B) while the smaller ones are probably "nuclear chain" (C) IF fibers. The NADH-D activity of the IF fibers from hypertrophied muscles tend to be greater than in control muscles. Photograph and data were supplied by Maier (1972) (bar, 50 μ).

exercise and denervation of synergist muscles on muscle spindle morphology. No differences in mean cross-sectional area or length of intrafusal fibers were found; however, some exercise effect on intrafusal fiber size was suggested since nuclear bag fibers tended to fall on a unimodal distribution in the trained animals whereas a bimodal distribution was evident for controls and denervated muscles.

Hypertrophy of the cat gastrocnemius induced by surgical incapacitation of the synergistic muscle resulted in an increase in the firing pattern of Ia afferents with controlled muscle stretch. A similar change was observed in atrophied muscles. In this same study the intrafusal fibers also generally had greater mitochondrial enzymic activity suggesting greater contractile activity (Maier, 1972) (Fig. 3).

III. Hyperplasia

In addition to hypertrophy, hyperplasia may play a role in the induced growth of muscle tissue, although the question of hyperplasia of skeletal muscle fibers after chronic muscular overload has drawn an unwarranted

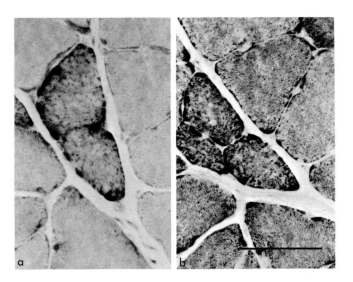

FIG. 4. Serial sections of a rat soleus muscle stained for mitochondrial (a) α-glycerophosphate dehydrogenase (α-GPD) and (b) NADH-D. The fiber in the center of the photograph appears to be sectioned at a point where the fiber is almost split. The section stained for NADH-D has apparently split into three fibers. The composite size of the three smaller fibers (b) approximates the size of the fiber in (a). This observation was made more frequently in chronically exercised rats than in sedentary controls (from Edgerton *et al.*, 1970a) (bar, 50 μ).

amount of attention. Within the last few years the issue seems to have been resolved with remaining questions of primarily academic interest.

A. DIRECT EVIDENCE IN MUSCLE FIBERS

Longitudinal muscle fiber splitting has been a generally acceptable phenomenon by clinicians in regard to human pathologic muscle. However, analogous evidence in muscular overload experiments has been considered to be insufficient. Edgerton (1970) clearly demonstrated a high incidence of longitudinal splitting in the soleus muscle of control and chronically exercised rats while none was found in the plantaris or gastrocnemius. Splitting was observed more frequently in the soleus of the exercised rats (Fig. 4). Hall-Craggs and Lawrence (1970) have also recently reported longitudinal splitting in the normal and overload (tenotomy) rat soleus muscle. It was suggested that the splitting followed damage to fibers thereby providing a way of rejecting degenerating portions of a fiber. Reitsma (1970) reported similar findings in very severely overloaded (removal of synergistic muscle and intensive treadmill exercise) frog and rat muscles. He found splitting in the rat plantaris as well as the soleus although more frequently in the soleus.

From the studies described above, it is obvious that in laboratory animals at least, fibers do split longitudinally and that it is selective to the type of muscle, being more prominent in the rat soleus which consists of predominantly slow-twitch oxidative fibers (Edgerton and Simpson, 1969; Barnard et al., 1971). It is also apparent that the splitting occurs more frequently in overloaded muscles. Fiber splitting has been seen in pathologic and normal human muscle, and, in a sense, is suggestive in normal chronically overloaded human muscle. It was suggested by Susheela (1964) that the splitting was a means of maintaining muscle bulk. In a microscopic comparison of muscle from muscular and relatively nonmuscular men who died from accidental deaths, a significantly greater number of fibers per muscle and larger muscle fibers were found in the muscular subjects (Etemadi and Hasseini, 1968). And finally, unless an intracellular longitudinal separation extends from one end of the fiber to the other, in a strict sense, it cannot be stated that hyperplasia of muscle fibers even occurred. But in spite of all of these findings its contribution to augmenting muscle function in efforts to compensate for increased demands that might be expected normally on a muscle is probably negligible.

B. ORGANELLE PROLIFERATION

As explained by Goss (1966), one need not limit the use of the term "hyperplasia" to whole fibers. It is more convenient and as correct to use the term in reference to organs, cells, and organelles of cells such as mitochondria, ribosomes, and nuclei. Cheek's (1968) working definition of hyperplasia was an increase in number of nuclei. Nuclei number was determined by dividing the total DNA by 6.2 since the amount of DNA per nucleus remains relatively constant. He assumed that the number of nuclei was equal to the number of cells. However, most authors would agree that a muscle fiber and muscle cell are the same and that a muscle fiber is multinucleated, with the number of nuclei varying from fiber to fiber.

With these basic assumptions, Cheek (1968) has demonstrated a substantial increase (up to twentyfold) in number of nuclei (therefore assumed to be muscle cells) in skeletal muscle of rats and humans postnatally. Thus, it can be said that cellular proliferation may have occurred but not muscle cell proliferation. Besides the problem of multinucleation, the following points are noteworthy in the interpretation of these results.

The number of muscle nuclei do not seem to divide by mitosis once the fiber is formed. A report of such happening has actually been retracted (Briskey et al., 1970). Satellite cells appear to be incorporated into mature fibers and may account for at least some of the apparent increase in muscle nuclei. Since satellite cells probably cannot synthesize myofilaments, it is not likely an important factor in providing information for regulation of fiber growth. An increase in nuclei can also be accounted for by proliferation of connective tissue cells or perhaps even fat cells. Thus, there are several reasonable explanations of how an increase in whole muscle DNA can be observed without an increase in "cell" number. A functional increase in the number of nuclei within muscle fibers is highly questionable and obviously one cannot relate nuclei number with cell number in fiber hyperplasia.

Because of the relative importance and prevalence of heart hypertrophy and since the heart lends itself well to studies dealing with muscle hypertrophy, it has been subjected to intensive study. The question of hyperplasia again has been of concern. Grove et al. (1969a,b) found increases in the number of nonmuscular nuclei of hypertrophied rat hearts. Compatible with the proliferation of the nonmuscular component of the heart, Buccino et al. (1969) showed an increase in collagen content in cat hearts hypertrophied by pulmonary constriction. These authors found that the concentration of nuclei was closely related to the concentration of DNA in ventricular papillary muscle although only 10–15% of the nuclei was of muscle origin. Others have reported increased DNA in enlarged hearts without proportionate increases in the number of nuclei (Grimm et al., 1970). Elevated heart weights have been associated with increased connective tissue thymidine, an essential component of DNA and nuclei, and an increased incorporation of hydroxyproline, an amino acid obtained from collagen and gelatin, both of which occur at about 11 days after aortic constriction (Grove et al.,

1969b). However, the DNA increase was significantly less after 2–4 months of overload than just after overload was introduced. Mitotic figures signifying DNA replication were about 10 times greater in number but were almost exclusively outside the muscle fibers in the enlarged heart. Even though an increase in total heart DNA was found, hypertrophy of the muscle was sufficient to cause an actual decrease in DNA concentration.

Thus, increased levels of DNA in muscle apparently is the result of augmented usage. But it appears that this DNA is, primarily at least, a result of general connective tissue cells and satellite cells rather than muscle fibers. Satellite cells are located adjacent to but outside muscle fibers and may be dormant phagocytic cells. They do not moderate protein synthesis of muscle fibers per se. One might even speculate that their phagocytic potential may be obviated in response to chronic muscular overload. But in such a case one would expect muscle atrophy not hypertrophy.

IV. Neuroendocrine Mechanisms of Exercise-Induced Growth

It would appear that the stimulus for exercise-induced growth of skeletal muscles could be mediated hormonally or neurally. Each of the changes discussed above can in some way be traced back to hormonal or neural influences. Both of these possibilities are weighed.

A. HORMONAL

1. Androgens

Male androgenic hormone can stimulate RNA polymerase, nucleolar RNA synthesis, DNA polymerase (enzyme assisting in synthesis of DNA), and DNA synthesis in target tissues (Fig. 2) (O'Malley, 1969). Testosterone increases rate of incorporation of precursors into nucleic acids and proteins of the male rat levator ani muscle. This effect is abolished 3 days after denervation of adult but not immature rats (Buresová and Gutmann, 1970). Thus, it appears that testosterone regulates to some degree muscular protein synthesis apparently through some control involving nuclear DNA and RNA.

Some of the changes induced by androgens may be related to permeability alterations. Testosterone can facilitate the penetration of the pentose, xylose, but not the hexose, L-glucose into the levator ani. The fact that the effects of insulin and electrical stimulation on xylose penetration are not additive to one another nor to the testosterone effect suggests a similar mechanism by both hormones. However, the testosterone effect on muscle glycogen was additive to an insulin effect. As was the case with proteosynthesis, the neural dependence of testosterone on augmentation of xylose penetration was abolished with denervation (Buse and Buse, 1961). Testosterone has been shown to be an important factor in the elevated levels of muscle glycogen seen after training in the male guinea pig (Gillespie and Edgerton, 1970).

The myotrophic effect of androgens is widely recognized without appreciation of its antigonadotrophic effects. Decreased gonadal ventral prostate and seminal vesicle weight has been shown, although most of the androgenic hormones have a biphasic effect on these organs. That is, one dosage may induce enlargement of the gonads while another induces atrophy. An androgen which induces solely a myotrophic effect has not been found (Boris *et al.*, 1970).

More generally, nonsteroidal anabolic agents have been used with success in improving appetite and body weight. Increased nitrogen retention within 2–3 weeks has been reported along with a sense of well-being. The anabolic effect seemed to be as great as that observed with steroidal anabolic agents (Albanese *et al.*, 1970).

Anabolic steroid treatment of competitive swimmers and weight lifters for 6 weeks (10 mg oxandrolone/day) induced elevated levels of lactate dehydrogenase (LDH), LDH isozymes I and V, creatine phosphokinase, alkaline phosphatase, urea nitrogen, and ions and protein metabolism. Strength performance increased in college athlete weight lifters, but swimmers did not become faster (O'Shea and Winkler, 1970). No toxic side effects such as edema, impaired hepatic function, or electrolytic imbalance were experienced except during the third week of treatment when muscles tended to cramp during heavy training. Improvement in performance after anabolic steroids has also been reported by Steinback (1968). On the other hand, no change in strength, motor performance, or physical working capacity was detected by Fowler *et al.* (1965).

Although the positive effect of testosterone on skeletal muscle size and glycogen has been clearly demonstrated, depressed androgen secretion rates with stress have been observed (Rose, 1969; Eleftheriou and Pattison, 1967; Eleftheriou and Church, 1968). The apparent discrepancy could be rationalized by saying that the normal exercise is not of sufficient stress to attain depression of androgen secretion. It is also possible that muscles become more sensitive to androgens during training or that during postexercise periods, androgen secretion may even be elevated.

2. Growth Hormone

Obviously postnatal growth is dependent upon pituitary hormones, namely, growth hormone (GH), although compensatory muscular hypertrophy caused by muscular overload can be induced in hypophysectomized animals (Goldberg, 1969a). It was concluded that the mechanism involved in muscle growth induced by GH and muscular work were independent. Growth hormone increases synthesis of ribosomes and mRNA as well as facilitates ribosomal attachment to mRNA in muscle (Fig. 2) (O'Malley, 1969; Martin and Wool, 1968). Thus, like androgens, GH has an anabolic effect on muscle. It is also accompanied by increased fat mobilization and is responsive to a number of situations that can be likened to those experienced during exercise. This lipoid effect is apparently important for bone growth as well (Campbell and Rastogi, 1969).

Plasma GH is elevated by hypoglycemia or intracellular deprivation of glucose, conditions which result from prolonged fasting or muscular exercise. In contrast, glucose infusion suppresses human GH secretion (Roth *et al.*, 1963). Small increases in GH in humans after an 8-km walk have been reported (Glick *et al.*, 1964), while high concentrations were observed in squash players 2–3 hr after the game (Hunter and Greenwood, 1964). In subjects that increased their energy requirement fivefold, a steadily rising concentration of plasma free fatty acids (FFA), a decreased RQ, and a marked rise in GH were observed. The peak GH level was seen after about 1 hr of exercise. In one subject, oral glucose administration prevented any of these factors from changing in response to exercise. The authors suggested that GH was released by some "triggering" mechanism (Hunter *et al.*, 1965). The GH response to exercise was immediate and present in obese as well as normal subjects. These authors found no correlation between FFA and GH. This may be related to the observations that higher concentrations of FFA were observed in the obese in response to exercise than in nonobese subjects. It was felt by the author that the elevated release of GH was due to the general stress response more than to the energy expenditure required by the exercise (Schwarz *et al.*, 1969). It may also be significant that GH in females seems to be more responsive than in males. This could turn out to be an important biological difference between sexes in the mechanism employed in muscular compensation since the gonadal hormone testosterone obviously facilitates muscular adaptation to overload in the males while its female counterpart, estrogen, is much less androgenic. Although GH increases in "fit" and "unfit" subjects, after exercise a more moderate and short-lasting rise was found in the "fit" (Sutton *et al.*, 1968).

Sperling *et al.* (1970) found that infusion of the amino acid arginine induced rises in the rate of GH release in preadolescents, adolescents, and adults in both females and males with few exceptions. Interestingly these exceptions were in adult females, and it has been shown in rats that some contraceptive steroids significantly reduced GH in the anterior pituitary (Liu and Lin, 1970). Elevated plasma amino acid levels seem to be a fundamental stimulus to GH secretion in light of the conditions mentioned earlier which are known to induce and suppress a GH response.

Goldberg and Goodman (1969a) found that hypertrophy of the rat soleus and plantaris induced by tenotomy of synergists was similar in hypophysectomized and normal animals suggesting that pituitary GH is not essential for muscle growth. Thus, muscle growth seems to be distinguishable into GH-dependent and work-induced-dependent factors with GH being particularly critical during developmental growth (Goldberg, 1967a; Bigland and Jehring, 1952; Greenbaum and Young, 1953). On the other hand, GH can support growth of denervated muscle sufficiently to maintain a weight equal to its contralateral control, obviously overcoming the loss of the "neurotrophic" influence. One might read these studies as meaning that GH and a neurotrophic influence act synergistically and similarly, but independently as mechanisms of muscle maintenance and growth.

In summary, GH obviously plays a major but nonspecific anabolic role in work-induced muscle hypertrophy. It is also apparent that GH is not the only regulatory factor in work-induced muscular growth. The mechanism through which GH produces its effect apparently involves regulation of ribosomes, RNA synthesis, and amino acid incorporation into proteins for the muscle.

3. Insulin

The rapid binding of insulin to the external cell membrane of the target organ in such a way as to increase membrane permeability to glucose (O'Malley, 1969) is a more commonly recognized effect of this hormone. But the localization of insulin throughout muscle fibers suggests some additional intracellular active site other than the membrane. This idea is compatible with the observation that insulin induces a ribosomal response (Stein and Cross, 1959).

The insulin-sensitive site in rat skeletal muscle fibers has also been localized to a specific ribosomal subunit (Martin and Wool, 1968) causing reaggregation of existing ribosomes and increased protein synthetic capacity. Thus, RNA synthesis is not necessary for insulin stimulation of protein synthetic activity in muscle. In other words, insulin seems to activate existing organelles of muscle fibers in order to mediate elevated protein synthesis as opposed to synthesis of new organelles (ribosomes).

Insulin elevates CO_2 production in skeletal muscles as well as hexokinase, pyruvate kinase, phosphofructokinase, and therefore glycolysis (Ozand and Narahara, 1964; Machiya *et al.*, 1969). Similarly, glucose uptake, lactate, and CO_2 formation is increased in rhesus monkey muscle with insulin (Bocek and Beatty, 1969). Insulin also increases glycogen transferase in rat muscle. Phosphorylase is not affected. It has no significant effect on glucose uptake in liver or kidney in rats.

Low muscle glycogen is frequently found in subjects with insufficient insulin secretions such as in mellitus diabetics. This condition can be altered quickly in man and rats with insulin treatment suggesting some dependency of muscle glycogen on circulating insulin levels (Roch-Norlund et al., 1970; Liebson et al., 1968).

Insulin may be capable of indirectly inducing a growth effect on muscle through GH since insulin can induce hypoglycemia which in turn can cause elevated secretion of GH (Roth et al., 1963 ; Krulich and McCann, 1966a,b; Katz et al., 1967). The effects of GH were discussed previously (Section III,A,2). Pituitary gland preparations from insulin-treated rats caused increased tibial cartilage widths, but preparations from saline-treated rats show no change in cartilage width (Hazelwood and Galaznik, 1970) suggesting that the pituitary of insulin-treated rats contains and perhaps can secrete more GH. Further interaction of insulin and GH is evident in that GH suppresses the insulin action on carbohydrate metabolism and possibly thereby facilitating an insulin-induced augmentation of protein synthesis (Campbell and Rastogi, 1969).

Even though insulin has the capacity to augment muscular growth and development, it does not seem likely that it serves a critical role in muscular growth as induced by exercise. Goldberg (1968) has shown that work-induced hypertrophy can occur without insulin. Glucose uptake can be facilitated without hormonal influence (Szabo et al., 1969; Rasio et al., 1966). In fact, the effects of exercise and insulin on glucose transport work through two distinctly different mechanisms (Goldstein et al., 1953).

Insulin and glucose in the blood may be lowered with a single bout of exercise (Devlin, 1963; Pruett, 1970). Cochran et al. (1966) and Rasio et al. (1966) found no increase in insulin with exercise. Physical training of obese subjects actually caused a marked depression in insulin values (Bjorntorp et al., 1970). A desirable interaction of insulin and exercise is indicated in diabetics since exercise seems to potentiate the hypoglycemic effect of exogenous insulin. But it has also been shown that lowered insulin levels with severe exercise is independent of blood glucose concentrations (Pruett, 1970). Thus, it follows that some factor, independent of insulin and pancreatic glucagon, which promotes glucose production, is released during exercise (Vranic and Wrenshall, 1969). Perhaps it is pertinent that Buse and Buse (1961) found that an optimal response to insulin as shown by cell membrane transport of the pentose sugar, D-xylose, in muscle was neural dependent.

Insulin is known to be a significant lipogenic hormone. This can be logically explained by the elevated glucose made available to cells by insulin thereby inhibiting catabolism of fats in order to avoid the excessive energy production that would result from the oxidation of too many substrates simultaneously. The great lipid mobilization in diabetics after exercise as opposed to normal individuals could be a manifestation of the greater reliability of the diabetics on the utilization of lipids at rest resulting into an adaptation of perhaps the enzymes which facilitate lipid utilization during exercise as well. After insulin therapy, this difference no longer appears (Carlstrom, 1969). At the onset of recovery from exercise, insulin increases (Pruett, 1970; Wright, 1968), and a lowering of insulin-like activity occurs with training (Devlin, 1963). On the other hand, bed rest impairs peripheral glucose utilization, this effect being independent of plasma insulin (Lipman *et al.,* 1970).

4. *Glucocorticoids*

Glucocorticoids do play some role in protein metabolism. For example, they regulate RNA metabolism probably through RNA polymerase and make DNA generally more readily available for transcription, (Elson *et al.,* 1965). This may be analogous to gene derepression. Its effect seems to be greater on mitochondrial RNA than nuclear RNA synthesis (Mansour and Nass, 1970). But larger doses may suppress mitochondrial enzymes. This suggests that the alteration in protein synthesis is closely related to the energy-producing functions of the tissue. Glucocorticoids seem to stimulate gluconeogenesis, increase liver glycogen, liver glycogen synthetase activity, and other enzymes related to carbohydrate metabolism (elevated plasmal lactate), and nitrogen excretion. But other studies have shown no change in glycogen snythetase after glucocorticoids (Hombrook *et al.,* 1966; Steiner *et al.,* 1961). They seem essential for mobilization of fatty acids from fat depots (Forbath *et al.,* 1969; Binder, 1969). The increase in glycogen synthetase activity seems to be independent of glucose-6-phosphate and apparently does not depend on protein synthesis (DeWulf and Hers, 1967).

Even though glucocorticoids can enhance the synthesis of some proteins it is well known that they promote muscle atrophy. It seems that the contractile elements which make up the bulk of a muscle is susceptible to glucocorticoid enhanced protein degradation resulting in muscle atrophy. Cortisone tends to selectively deplete amino acids from white as opposed to red muscles. In rats, denervation similarly increases the sensitity of the soleus and plantaris to cortisol while an increased work load of white muscles seems to make this muscle resistant to the wasting effect of cortisol (Goldberg,

1969). Preferential fiber atrophy of presumably fast twitch fibers was observed by Walsh *et al.* (1971) after hydrocortisone treatment in rats while ACTH injections caused mild atrophy of slow twitch fibers. Long-term administration of ACTH or glucocorticoids caused a disruption of the normal fiber morphology.

Plasma corticoids are known to increase during muscular exertion, this increase being less dramatic in trained than sedentary men and rats (Bellet *et al.,* 1969; Frenkl *et al.,* 1969; Viru and Akke, 1969). Corticoids are increased in adrenals of guinea pigs exposed to 3–5 min swim sets. Blood corticoids are elevated only after exhaustion due to swimming while no change in adrenal corticoids is found at this stage. Low blood levels could result from insufficient stimulation of adrenocorticoid activity. Pituitary involvement may be a critical factor in maintenance of plasma corticoids since they can be maintained with corticotropin injections which serve to stimulate the adrenals. Low excretion of 17-hydroxycorticoids after long-lasting exercises can be avoided in athletes by corticotropin (Viru and Akke, 1969). However, Knigge (1958) demonstrated a lack of depletion of pituitary corticotropin reserves after extensive stress. Thirty minutes of moderate exercise in man causes an increase in 11-hydroxycorticosteroid excretion (Bellet, *et al.,* 1969), but plasma levels of cortisol later decreased probably because of an increased tissue consumption, not lack of corticotropin from the anterior pituitary.

In the use of corticoids therapeutically, it should be realized that a number of potential side effects exist. Some of the known effects are bradycardia, peptic ulcer, osteoporosis, diabetes, decreased resistance to disease, myopathies, and in children, perhaps, growth retardation (Quaade, 1969). Glucocorticoids can completely suppress ACTH secretion (15–20 mg cortisol/day). Adrenal atrophy is related to length of therapy and total dosage. It has also been shown that the adrenal response to ACTH decreases after prolonged cortisol therapy. All patients who had taken glucocorticoids for more than 3–5 months had pituitary insufficiencies.

5. Prostaglandins

Prostaglandins (PG), a relatively new and little understood hormone, was first found in seminal fluid but is now known to be released from phrenic nerve-diaphragm preparation (Laity, 1969), the spinal cord, and cerebral cortex synaptosomes (Diassi and Horovitz, 1970; Hoffer *et al.,* 1969). In fact, release of PG may occur in all tissues where membranes are activated (Laity, 1969). Prostaglandins antagonize catecholamine-induced lipolysis. It may do this by preventing activation of adenyl cyclase, the enzyme which catalyzes the formation of cyclic AMP which activates lipolysis. It has also been demonstrated that prostaglandins can alter neuronal discharge rate in cat brain stems. Prostaglandins also antagonize the catecholamine effect on Purkinje cells by increasing rather than decreasing spontaneous discharge (Hoffer *et al.,* 1969), lowering blood pressure, stimulating smooth muscle, affecting sperm transport; inducing uterine movements and gastric secretions (Diassi and Horovitz, 1970) are other currently known actions of PG most of which apparently are relatively unrelated to muscle growth and development.

The role of PG in muscle growth and development can only be prematurely speculated. Our current understanding of PG suggests two channels through which muscular change could be effected. First, but least likely, is its inhibitory effect on catecholamine depression of lipid utilization via cAMP depression. Interestingly, depressed cAMP should also minimize glycogen degradation (lack of phosphorylase activation) and maximize glycogen synthesis (increase in glycogen synthetase activity). A second channel through which muscular change could be effected would be through neuronal activation. For example, one could speculate a neuronally mediated effect of PG by elevated neuronal spontaneous discharges which in turn could induce some trophic effect (Section III,B).

6. *Other Hormones*

There are other hormones that no doubt play important roles in the maintenance of muscle tissues metabolically and perhaps less so structurally. It is generally thought that these hormones do not have significant androgenic effects on the muscle, that is, their net effect is not growth and development of the muscle from a structural point of view. Most of these hormones are integral regulating components of the metabolic machinery of the muscle, however.

Glucagon, epinephrine, and thyroxine assists in metabolic regulation of muscles as does insulin. Briefly, glucagon mobilizes glucose from the liver thereby elevating plasma glucose and the availability of glucose to muscle. Epinephrine and norepinephrine are well-known stimulates of glycogenolysis in muscle as well as liver thus making glucose available as an energy source. Thyroxine generally augments metabolic rate having an activating effect on many key cellular metabolic enzymes. Parathormone, which is released from the parathyroid gland under normal conditions, probably has little effect on growth and development of muscle as it relates to physical activity. It is quite important in bone development, however.

The role of the female hormones estrogen and progesterone in anabolism of skeletal muscle and its relation to physical activity have received little attention. Estrogen has some androgenic effects on muscle although its potency is not near that of testosterone. Progesterone has a slight antiandrogenic effect. The effect of testosterone on muscle growth and metabolism (Section III,A,1) is dramatic on both the energetic and contractile components of the muscle.

B. NEURAL

Numerous studies have demonstrated the neural dependence of biochemical and physiological properties of skeletal muscle. For example, the quantity and quality of proteins as well as substrates synthesized and utilized vary markedly from fiber to fiber (Fig. 5). Amino acid incorporation into myosin, aldolase and lactate dehydrogenase of slow twitch muscle is greater than fast twitch white (Dreyfus, 1967). Penetration of nonmetabolizable amino acid ^{14}C-AIB is much faster in slow than fast white muscle. This differential transport of amino acids is thought to be the reason for the greater incorporation into the various proteins listed above (Dreyfus, 1967). Other authors have reported essentially identical findings (Goldberg, 1967b) in normal and hypophysectomized rats, suggesting lack of dependency of this

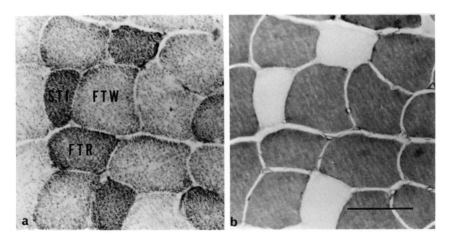

FIG. 5. Serial cross sections of the lesser bush baby *(Galago senegalenesis)* plantaris muscle stained for (a) NADH-D activity and (b) myosin adenosine triphosphate (ATPase) activity. All fibers lightly stained with NADH-D are stained darkly with ATPase [fast twitch white (FTW) fibers] as are all NADH-D dark fibers with the darker subsarcolemmal or peripheral staining relative to the core of the fiber [fast twitch red (FTR) fibers]. The relatively dark NADH-D fibers which are stained homogeneously throughout the diameter of the fiber are called slow twitch intermediates. The fibers that stain darkly with ATPase have relatively fast contractile properties while the lightly stained fibers are relatively slow in contracting and relaxing (Edgerton and Simpson, 1969; Barnard *et al.,* 1971) (bar, 50 μ).

phenomenon on GH. Ribonucleic acid content in "redder" muscles was also found to be higher than in "white" muscles. Catabolism is also greater in "red" muscle which means that the turnover rate of proteins in "red" muscle is greater than in "white" muscle.

Small motoneurons incorporate more amino acids into proteins/unit volume than do larger neurons (Peterson, 1966). Slow twitch intermediate fibers apparently are innervated by small motoneurons and have low stimulating thresholds. It also appears that neuron size varies inversely with the activity of a number of metabolic enzymes. These studies simply show that smaller neurons are the most active physiologically as might be expected if they innervate the more active muscle fibers.

Motoneurons with distinctive metabolic enzyme profiles have been tentatively identified by Campa and Engel (1971). However, they were able to differentiate only Renshaw cells, interneurons, and γ-neurons from all α-motoneurons histochemically. Since different types of motoneurons were not differentiated, they are not related to the types of muscle fibers (Edgerton and Simpson, 1969) and motor units previously described (Levine et al., 1971) (Fig. 5).

It has been claimed that the differential metabolic properties seen histochemically in various muscle fibers result from a diluting effect. That is, white fibers appear to have less metabolic activity since they are large, and red fibers apparently have high metabolic activity because they are small (Goldspink, 1969). This supposition is overwhelmingly discredited by unlimited data.

A large number of parameters have been investigated in regard to their response to cross-innervation. All of these studies simply demonstrate the control of a number of enzymes and substrates in muscle exerted by motoneurons. When a nerve trunk which normally innervates a fast twitch muscle (flexor hallucis longus) and a nerve which normally innervates a slow twitch muscle (soleus) are severed and the distal end of each nerve is sutured to the foreign proximal end of the nerve trunk, most biochemical and physiological properties in the muscle will be reversed so that the fast twitch muscle becomes slow twitch and slow becomes fast twitch muscle (Buller et al., 1960; Robbins et al., 1969).

It has been hypothesized that this neural control of properties of slow and fast muscle results from either the pattern of frequency of impulses of a motor unit or from some "neurotrophic substance" which flows down the nerve to the muscle fiber. The two possibilities are not necessarily exclusive in that it would appear that an increase in frequency of impulses also leads to an increase in axoplasmic flow and probably distal passage of a neurotrophic substance(s) (Astaf'eva, 1966).

Since the frequency of use and disuse of skeletal muscle can alter the nature of its growth and development, and since some motor units are used more frequently than others (Edgerton *et al.,* 1970b) while cross-innervation experiments can reverse these characteristics, growth and development of normal adult muscle may be for the most part regulated neurally. Then it may also seem probable that any alteration in muscle properties by varying use of a muscle is largely mediated neurally (Edgerton *et al.,* 1969).

Some attempts have been made to isolate factors that may be critical in the mediation of the neural influence on muscle. Robert and Oester (1970) silenced the sciatic nerve trunks with 8–14 day anesthetic implants. In this way impulses were stopped but not necessarily the trophic effect. They demonstrated no muscle fibrillation after the nerve block as was shown with denervation. Fast muscles of rabbits and cats subjected to daily bouts of electrical stimulation contracted and relaxed more slowly then control muscles. Slow muscles became faster following tenotomy or cordotomy, but this speeding was prevented by chronic stimulation at frequencies of 5 or 10 per second (Salmon and Vrbova, 1969). Olson and Swett (1969) found similar changes in contractile properties of hypo- and hyperactivity induced by unilateral differentiation. The hypoactive fast muscles tended to be faster and the hyperactive muscles slower.

Experimental designs are needed which will differentiate the trophic from the impulse effect. The impulses must be of sufficient character to minimize the inherent differences in impulse frequency of intact motor units but maximize any qualitative differences in neurotrophic substances that might be released. This treatment would maximize the potential trophic differences assuming the elevated stimulation levels also elevated axoplasmic flow. If the difference in the natural pattern of impulses is not obliterated, more than likely the impulse and trophic effects would not be due to different impulse patterns.

What is the nature of the neurotrophic influence on muscle growth and development? It has been found that the distribution (Axelsson and Thesleff, 1959) and sensitivity (Fambrough, 1970) of acetylcholine receptors of skeletal muscles are controlled by the neuron by controlling protein synthesis since protein inhibitors (e.g., puromycin) prevented the normal localization of acetylcholine sensitivity in mature muscles.

Other investigators have attempted to approach the question of the mechanism of muscle growth and development more directly. Schiaffino and Hanzlíková (1970) demonstrated "hypertrophy" of the soleus and plantaris muscles independent of the neural intactness in rats when synergists were tenotomized. The results suggested as do others (Sola and Martin, 1953; Feng and Lu, 1965; Gutmann *et al.,* 1966; Buresová *et al.,*

1969) that perhaps muscle tension induces hypertrophy or at least helps in maintaining some form of homeostasis of protein turnover. However, there are points of caution to be considered in making inferences in regard to exercise from the results described above (Schiaffino and Hanzlíková, 1970). First, muscle weights were taken on the fifth postoperative day, at time when it seems that muscle weights are elevated to a peak after surgery while such is not evident after 5 days of physical training. This acute postoperative effect may be unlike functional hypertrophy after a more normal type of overload. Then, hypertrophy was defined as an increase in muscle weight relative to the contralateral denervated muscle. It may be that the greater tension on muscles after tenotomy of synergists will assist in maintenance of normal weights, whereas to get true hypertrophy neural and hormonal factors will be essential.

Another experiment that seems to indicate a neurotrophic effect on skeletal muscle was reported by Gutmann et al. (1955). He observed that the glycogen content in muscle after denervation was dependent on the length of the distal nerve stump. Muscles having the shorter distal nerve stumps were not able to resynthesize glycogen levels as sufficiently as muscles having longer nerve stumps. A similar conclusion was reached by Lucco and Eyzaguirre (1955) when they tested the presence of fibrillation and sensitivity to acetylcholine after denervation at various points proximo-distally. More recently, Schuh and Albuquerque (1971) showed that depolarization of muscle fibers following denervation occurs more slowly when the nerve is cut more proximally. Some observations indicate that labeled amino acids and inorganic phosphate can be transferred from the hypoglossal nerve fibers of the rabbit to the tongue muscles (Kow et al., 1967). A transfer of amino acids across the neuromuscular junction of a snail has been reported also (Kerkut et al., 1967).

Fex and Jirmanova (1969) have provided further data indicating a neural influence in muscle hypertrophy which is relatively independent of functional neuromuscular junctions, thus the pattern of frequency of impulses per se. By implanting a foreign nerve into a muscle with its normal innervation already intact, muscle weight and isometric contractile capacity increased about 30%. Remember, it is generally accepted that normally innervated muscles will not receive further neural sources (hyperinnervation). The observed changes are not likely to be accountable by hyperinnervation since the formation of a significant number of new neuromuscular synapses by the foreign nerve was not evident. Some data demonstrating an increase in muscle fiber diameter in the soleus were also given (Fex and Jirmanova, 1969; Fex, 1969). Results from the lab have failed to confirm their report (Lee and Edgeton, unpublished results). Although changes in speed of contraction were suggested in the report, the evidence was not convincing.

A more recent and intriguing study by Lentz (1971) showed greater cholinesterase activity of muscle cultures when treated with homogenates and tissue explants from ganglia, spinal cord, liver, and nerves of the newt, suggesting a trophic effect mediated by a diffusible neurochemical substance.

In summary, skeletal muscle adapts in a number of ways to overload. That is, the contractile and energy-yielding components of the muscle fibers are responsive to chronic overload. The specificity and mechanism of this adaptation is the point of current interest. The degree to which a specific type of overload can induce a selective adaptation needs further investigation. The mechanism through which these changes are induced may involve membrane transport properties, DNA, RNA, as well as other cellular functions. These functions are to a significant degree under hormonal and neural control. The possibility of neural control has been only recently seriously considered to be a potential factor in the mechanism of muscular adaptation to exercise. Recent experiments testing the neurotrophic influence may prove to be important factors in the regulation of normal and exercise-induced growth and development of muscular tissue.

References

Albanese, A. A., Lorenze, E. J., Orto, L. A., and Wein, E. H. (1970). *Nutr. Rep. Inte.* **2,** 29.

Asmussen, E. (1968). *In* "Exercise Physiology" (H. B. Falls, ed.), p. 37. Academic Press, New York.

Astaf'eva, O. G. (1966). *Tr. Saratov. Med. Inst.* **49,** 18.

Axelsson, J., and Thesleff, S. (1959). *J. Physiol (London)* **147,** 178.

Barnard, R. J., Edgerton, V. R., and Peter, J. B. (1970a). *J. Appl. Physiol.* **28,** 762.

Barnard, R. J., Edgerton, V. R., and Peter, J. B. (1970b). *J. Appl. Physiol.* **28,** 767.

Barnard, R. J., Furukawa, T., Edgerton, V. R., and Peter, J. B. (1971). *Amer. J. Physiol.* **220,** 410.

Bellet, S., Roman, L., and Barham, F. (1969). *Metabo., Clin. Exp.* **18,** 484.

Bigland, B., and Jehring, B. J. (1952). *J. Physiol., (London)* **116,** 129.

Binder, C. (1969). *Acta Med. Scand., Suppl.* **500,** 9.

Bjorntorp, P., deJounge, K., Sjöstrom, L., and Sullivan, L. (1970). *Metabo., Clin. Exp.* **19,** 631.

Bloor, M., Pasyk, S., and Leon, A. S. (1970). *Amer. J. Pathol.* **58,** 185.

Bocek, R. M., and Beatty, C. H. (1969). *Endocrinology* **85,** 615.

Boris, A., Stevenson, R. H., and Trmal, T. (1970). *Steroids* **15,** 61.

Bozner, A., and Meessen, H. (1969). *Virchows Arch., B* **3,** 248.

Bravo-Rey, M., Wamaski, J., Eldred, E., and Maier, A. (1969). *J. Neurol.* **25,** 595.

Briskey, E. J., Cassens, R. G., and Marsh, B. B., eds. (1970). "The Physiology and Biochemistry of Muscle as a Food," Vol. II, p. 585. Univ. of Wisconsin Press, Madison.

Buccino, R. A., Harris, E., Spann, J. F., Jr., and Sonnenblick, E. H. (1969). *Amer. J. Physiol.* **216,** 425.

Buller, A. J., Eccles, J. C., and Eccles, R. M. (1960). *J. Physiol. (London)* **150,** 417.

Burešová, M., and Gutmann, E. (1970). *Life Sci.* **9**, 547.

Burešová, M., Gutmann, E., and Klicpera, M. (1969). *Experientia* **25**, 144.

Buse, M. G., and Buse, J. (1961). *Diabetes* **10**, 134.

Campa, J. F., and Engel, W. K. (1971). *Science* **171**, 198.

Campbell, J., and Rastogi, K. S. (1969). *Metabo., Clin. Exp.* **18**, 930.

Carlstrom, S. (1969). *Acta Med. Scand.* **186**, 429.

Carney, J. A., and Brown, A. L., Jr. (1964). *Amer. J. Pathol.* **44**, 521.

Cheek, D. B. (1968). "Human Growth." Lea & Febinger, Philadelphia, Pennsylvania.

Cochran, B., Jr., Marbach, E. P., Poucher, R., Steinberg, T., and Swinup, G. (1966). *Diabetes* **15**, 838.

Devlin, J. G. (1963). *Ir. J. Med. Sci.* **6**, 423.

DeWulf, H., and Hers, H. G. (1967). *Eur. J. Biochem.* **2**, 57.

Diassi, P. A., and Horovitz, Z. P. (1970). *Annu. Rev. Pharmacol.* **10**, 57.

Dreyfus, J. C. (1967). *Rev. Fr. Etud. Clin. Biol.* **12**, 343.

Earl, D. C. N., and Korner, A. (1966). *Arch. Biochem. Biophys.* **115**, 437.

Edgerton, V. R. (1971). Unpublished data.

Edgerton, V. R. (1970). *Amer. J. Anat.* **127**, 81.

Edgerton, V. R., and Simpson, D. R. (1969). *J. Histochem. Cytochem.* **17**, 828.

Edgerton, V. R., Gerchman, L., and Carrow, R. (1969). *Exp. Neurol.* **24**, 110.

Edgerton, V. R., Barnard, R. J., Peter, J. B., Simpson, D., and Gillespie, C. A. (1970a). *Exp. Neurol.* **27**, 46.

Edgerton, V. R., Simpson, D. R., Barnard, R. J., and Peter, J. B. (1970b). *Nature (London)* **225**, 866.

Edgerton, V. R., Barnard, R. J., Peter, J. B., Simpson, D., and Gillespie, C. A. (1972). *Exp. Neurol.* (In press).

Eleftheriou, B. E., and Church, R. L. (1968). *J. Endocrinol.* **42**, 347.

Eleftheriou, B. E., and Pattison, M. L. (1967). *J. Endocrinol.* **39**, 613.

Elson, D., Luck, J. M., and Boyer, P. D. (1965). *Annu. Rev. Biochem.* **34**, 447.

Etemadi, A. A., and Hasseini, F. (1968). *Anat. Rec.* **162**, 269.

Fambrough, D. M. (1970). *Science* **168**, 372.

Feng, T. P., and Lu, D. X. (1965). *Sci. Sinica* **12**, 1772.

Fex, S. (1969). *Physiol. Bohemoslov.* **18**, 205.

Fex, S., and Jirmanova, I. (1969). *Acta Physiol. Scand.* **76**, 257.

Forbath, N., Hall, J. D., and Hetenyi, G., Jr. (1969). *Horm. Metab. Res.* **1**, 179.

Fowler, W. M., Gardner, G. W., and Egstrom, G. H. (1965). *J. Appl. Physiol.* **20**, 1038.

Frenkl, R., Csalay, L., and Csákváry, G. (1969). *Acta Physiol.* **36**, 365.

Garren, L. D., Richardson, A. P., Jr., and Crocco, M. R. (1967). *J. Biol. Chem.* **243**, 630.

Gillespie, C. A., and Edgerton, V. R. (1970). *Horm. Metab. Res.* **2**, 364.

Glick, S. M., Roth, J., Yalow, R. S., and Berson, S. A. (1964). *Diabetes* **13**, 355.

Goldberg, A. L. (1967a). *Amer. J. Physiol.* **213**, 1193.

Goldberg, A. L. (1967b). *Nature (London)* **216**, 1219.

Goldberg, A. L. (1968). *Endocrinology* **83**, 1071.

Goldberg, A. L. (1969a). *Amer. J. Physiol.* **200**, 655.

Goldberg, A. L. (1969b). *Amer. J. Physiol.* **200**, 667.

Goldberg, A. L., and Goodman, H. M. (1969a). *Amer. J. Physiol.* **216**, 1111.

Goldberg, A. L., and Goodman, H. M. (1969b). *Amer. J. Physiol.* **216**, 1116.

Goldspink, G. (1964). *J. Cell. Comp. Physiol.* **63**, 209.

Goldspink, G. (1968). *J. Cell. Sci.* **3**, 539.

Goldspink, G. (1969). *Life Sci.* **8**, 791.

Goldspink, G. (1970). *In* "The Physiology and Biochemistry of Muscle as a Food" (E. J. Briskey, R. G. Cassens, and B. B. Marsh, eds.), Vol II, pp. 521–536. Univ. of Wisconsin Press, Madison.

Goldstein, M. S., Mullic, V., Huddlestun, B., and Levine, R. (1953). *Amer. J. Physiol.* **173**, 212.

Gordon, E. E., Kowalski, K., and Fritts, M. (1967a). *Arch. Phys. Med. Rehabil.* **48**, 296.

Gordon, E. E., Kowalski, K., and Fritts, M. (1967b). *Arch. Phys. Med. Rehabil.* **48**, 577.

Goss, R. J. (1966). *Science* **153**, 1615.

Greenbaum, A. L., and Young, F. G. (1953). *J. Endocrinol.* **9**, 127.

Grimm, F., de la Torre, L., and La Porta, M., Jr. (1970). *Circ. Res.* **26**, 45.

Grove, D., Nair, K. G., and Zak, R. (1969a). *Circ. Res.* **25**, 463.

Grove, D., Zak, R., Nair, K. G., and Aschenbrenner, V. (1969b). *Circ. Res.* **25**, 473.

Gutmann, E., Vodica, Zd., and Zelena, J. (1955). *Physiol. Bohemoslov.* **4**, 200.

Gutmann, E., Hanikova, M., Hajeck, I., Klicpera, M., and Syrovy, I. (1966). *Physiol. Bohemoslov.* **15**, 508.

Hall-Craggs, E. C. B., and Lawrence, A. (1970). *Z. Zellforsch. Mikrosk. Anat.* **109**, 481.

Hamosh, M., Lesch, J. B., and Kaufman, S. (1967). *Science* **157**, 935.

Hatt, P. Y., Ledoux, C., Bonvalet, J. P., and Guillemat, H. (1965). *Arch. Mal. Coeur Vaiss.* **58**, 1703.

Hazelwood, R. L., and Galaznik, J. G. (1970). *Can. J. Physiol. Pharmacol.* **48**, 85.

Helander, E. A. S. (1961). *Biochem. J.* **78**, 478.

Heywood, S. M., Dowben, R. M., and Rich, A. (1967). *Proc. Nat. Acad. Sci. U. S.* **57**, 1002.

Hoffer, B. J., Siggins, G. R., and Bloom, F. E. (1969). *Science* **166**, 1418.

Holloszy, J. O. (1967). *J. Biol. Chem.* **242**, 2278.

Holmes, R., and Rasch, P. J. (1958). *Amer. J. Physiol.* **195**, 50.

Hombrook, K. R., Burch, H. B., and Lowry, O. H. (1966). *Mol. Pharmacol.* **2**, 106.

Hunter, W. M., and Greenwood, F. C. (1964). *Brit. Med. J.* **1**, 804.

Hunter, W. M., Gonseka, C. C., and Passmore, R. (1965). *Science* **150**, 1051.

Ikai, M. and Fukunaga, T. (1970). *Int. Z. Angew Physiol. Einschl. Arbeitsphysiol.* **28**, 173.

Katz, S., Dhariwall, A., and McCann, S. M. (1967). *Endocrinology* **81**, 333.

Kerkut, G., Shapira, A., and Walker, R. J. (1967). *Comp. Biochem. Physiol.* **23**, 729.

Knigge, K. M. (1958). *Anat. Rec.* **130**, 326.

Kow, I. M., Wilkinson, P. N., and Chornock, F. W. (1967). *Science* **155**, 342.

Krulich, L., and McCann, S. M. (1966a). *Proc. Soc. Exp. Biol. Med.* **122**, 668.

Krulich, L., and McCann, S. M. (1966b). *Endocrinology* **78**, 759.

Laity, J. L. H. (1969). *Brit. J. Pharmacol.* **37**, 698.

Lentz, T. L. (1971). *Science* **171**, 187.

Levine, D. N., Burke, R. E., Tsairis, P., and Zajac, F. E. (1971). *Fed. Proc. Fed. Amer. Soc Exp. Brol.* **30**, 377 (abstr.).

Liebson, L. G., Vizek, K., and Hahn, P. (1968). *Physiol. Bohemoslov.* **17**, 505.

Lipman, R. L., Schnure, J. J., Bradley, E. M., and Lecocq, F. R. (1970). *J. Lab. Clin. Med.* **76**, 221.

Liu, F. T. Y., and Lin, H. S. (1970). *Proc. Soc. Exp. Biol. Med.* **133**, 1354.

Lucco, J. V., and Eyzaguirre, C. (1955) *J. Neurophysiol.* **18**, 65.

Machiya, T., Takagi, H., Sakuri, T., and Hosoya, N. (1969). *Endocrinol. Jap.* **16**, 473.

MacKay, B., and Harrop, T. J. (1969). *Acta Anat.* **72**, 38.

Maier, A. (1972). Doctoral Dissertation, University of California, Los Angeles.

Mansour, A. M., and Nass, M. M. K. (1970). *Nature (London)* **228**, 666.

Martin, R. E., and Wool, I. G. (1968). *Proc. Nat. Acad. Sci. U.S.* **60**, 569.

Maynard, J. A., and Tipton, C. M. (1971). *Int. Z. Agnew. Physiol. Arbeit.* **30**, 1.

Morkin E. (1970). *Science* **167**, 1499.

Nelson, C., Simpson, D. R., and Edgerton, V. R. (1971). Unpublished data.

Olson, C. B., and Swett, C. P. (1969). *Arch. Neurol. (Chicago)* **20**, 263.

O'Malley, B. W. (1969). *Trans. N. Y. Acad. Sci.* [2] **31**, 478.

O'Shea, J. P., and Winkler, W. (1970). *Nutr. Rep. Int.* **2**, 351.

Ou, L. C., and Tenney, S. M. (1970). *Resp. Physiol.* **8**, 151.

Ozand, P., and Narahara, H. T. (1964). *J. Biol. Chem.* **239**, 3146.

Perry, S. V. (1970). *In* "The Physiology and Biochemistry of Muscle as a Food" (E. J. Brisky, R. G. Cassens, and B. B. Narsh, eds.), Vol II, pp. 539–553. Univ. of Wisconsin Press, Madison.

Peterson, R. P. (1966). *Science* **153**, 1413.

Pruett, E. D. R. (1970). *J. Appl. Physiol.* **29**, 155.

Quaade, F. (1969). *Acta Med. Scand.* **186**, 77.

Rasio, E., Malaisse, W., Franckson, J. R. M., and Conard, V. (1966). *Arch. Int. Pharmacodyn. Ther.* **160**, 485.

Reitsma, W. (1970). *Acta Morphol. Neer.-Scand.* **7**, 229.

Richter, G. W., and Kellner, A. (1963). *J. Cell Biol.* **18**, 195.

Ring, G. C., Bosch, M., and Chu-Shek Lo (1970). *Biol. Med.* **133**, 1162.

Robbins, N., Karpati, G., and Engel, W. K. (1969). *Arch. Neurol. (Chicago)* **20**, 318.

Robert, E. D., and Oester, Y. T. (1970). *Arch. Neurol. (Chicago)* **22**, 57.

Roch-Norlund, A. E., Bergstrom, J., Castenfors, H., and Hultman, E. (1970). *Acta Med. Scand.* **188**, 445.

Rose, R. M. (1969). *Psychosom. Med.* **31**, 405.

Roth, J., Glick, S. M., Yalow, R. S., and Berson, S. A. (1963a). *Metab., Clin. Exp.* **12**, 577.

Roth, J., Glick, S. M., Yalow, R. S., and Berson, S. A. (1936b). *Science* **140**, 987.

Rowe, R. W. D., and Goldspink, G. (1968). *Anat. Rec.* **161**, 69.

Salmon, S., and Vrbova, G. (1969). *J. Physiol. (London)* **201**, 535.

Schiaffino, S., and Hanzlíková, V. (1970). *Experientia* **26**, 152.

Schreiber, S., Oratz, M., and Rothschild, M. A. (1966). *Amer. J. Physiol.* **211**, 314.

Schreiber, S., Oratz, M., and Rothschild, M. A. (1967). *Amer. J. Physiol.* **213**, 1552.

Schreiber, S., Oratz, M., Evans, C. D., Gueyikian, E., and Rothschild, M. A. (1970). *Amer. J. Physiol.* **219**, 481.

Schuh, F. T., and Albuquerque, E. X. (1971). *Fed. Pro., Fed. Amer. Soc. Exp. Biol.* **30**, 557 (abstr.).

Schwarz, F., ter Haar, D. J., van Riet, H. G., and Thijssen, J. H. H. (1969). *Metabo., Clin. Exp.* **18**, 1013.

Shahab, L., and Wollenberger, A. (1970). *J. Mol. Cell. Cardiol.* **1**, 143.

Sobel, B. E., and Kaufman, S. (1969). *Physiologist* **12**, 360.

Sola, O. M., and Martin, A. W. (1953). *Amer. J. Physiol.* **172**, 324.

Sperling, M. A., Kenny, F. M., and Drash, A. L. (1970). *J. Pediat.* **77**, 462.

Steel, R. G. D., and Torri, J. H. (1960). "Principles and Procedures of Statistics." McGraw-Hill, New York.

Stein, O., and Cross, J. (1959). *Endocrinology* **65**, 707.

Steinbach, M. (1968). *Sportarzt. Sportmed.* **19**, 485.

Steiner, D. F., Randa, V., and Williams, R. H. (1961). *J. Biol. Chem.* **236**, 299.

Stewart, D. M. (1968). *Amer. J. Physiol.* **214**, 1139.

Stringfellow, C., and Brachfeld, N. (1970). *J. Mol. Cell. Biol.* **1**, 221.

Stull, A. G., and Clarke, D. H. (1970). *Res. Quart.* **41**, 189.

Susheela, A. K. (1964). *Experientia* **16**, 391.

Sutton, J., Young, J. D., Lazarus, L., Hickie, J. B., and Maksvytis, J. (1968). *Lancet* **2**, 1304.

Szabo, A. J., Mahler, R. J., and Szabo, O. (1969). *Horm. Metab. Res.* **1**, 156.

Tomanek, R. J., and Woo, Y. K. (1970). *J. Gerontol.* **25**, 23.

Tomita, K. (1966). *Jap. Heart J.* **7**, 566.

Viru, A., and Äkke, H. (1969). *Acta Endocrinol. (Copenhagen)* **62**, 385.

Vranic, M., and Wrenshall, G. A. (1969). *Endocrinology* **85**, 165.

Walsh, G., DeVivo, D., and Olson, W. (1971). *Arch. Neurol. (Chicago)* **24**, 83.

Wright, P. H. (1968). *Amer. J. Physiol.* **214**, 1031.

CHAPTER

2

Physical Activity and the Growth and Development of Bone and Joint Structures

Robert L. Larson

> That which is used develops and that which is not used wastes away.
>
> HIPPOCRATES

I. Introduction

There can be little question that physical activity and exercise promote improved physical fitness. The physical deterioration of younger people and a general decline in the fitness of American people was the concern of the political administration of the 1960's. Yale University reported that in 1947 49% of the freshmen failed a physical fitness test; whereas, in 1960 the failure rate of freshmen rose to 66%. These figures, confirmed by other studies, provided the impetus for the formation of The President's Council on Physical Fitness by President Kennedy in 1963.*

The result was an expanded program of physical fitness and athletic programs for our school children. Both the number of health and physical education teachers and the number of pupils participating in physical education rose appreciably during the years 1961 to 1969. The benefit of this program was revealed in an improvement in the number of children who met physical fitness standards—8 of 10 in 1965 compared with 6 of 10 in 1961.

The increased activity of our youth has quite understandably caused some to question the effects of increased stress on the immature skeleton. Particularly challenged has been the area of competitive athletics. Are there deleterious effects from too much activity too soon? Are there increased injury potentials in the immature athlete? Will growth and development be affected by exercise and physical activity? Summarily, these questions reduce themselves to the simple inquiry: Are the benefits worth the risks?

II. Growth and Development of the Skeletal Structures

A. EMBRYOLOGICAL AND PHYSIOLOGICAL DEVELOPMENT OF THE SKELETON

Long bones are pre-formed of cartilage in the embryo. Ossification of these bones begins as primary ossification centers which develop in the mid-

---

* President Eisenhower formed the President's Council on Youth Fitness in 1956. This became the President's Council on Physical Fitness in 1963. It was expanded into the President's Council on Physical Fitness and Sports by President Johnson in 1968.

portion of the bone. Usually at birth, ossification has progressed to include the entire shaft or diaphysis. At the ends of the long bones cartilaginous epiphyseal centers develop. Secondary ossification centers appear in the epiphyseal areas. Between the secondary ossification center and the diaphysis of the bone is a growth plate (epiphyseal plate). It is from this area that bone growth occurs. When full growth has been obtained, the epiphysis fuses to the metaphysis (the area of flare of the long bones between the epiphysis and diaphysis), and the growth plate is obliterated (Fig. 1).

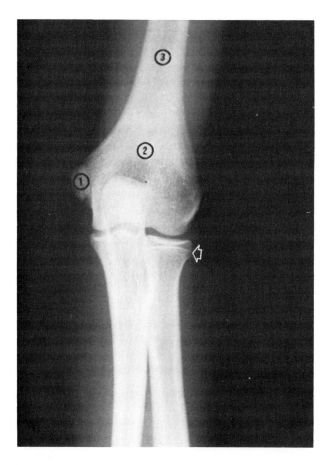

FIG. 1. Roentgenogram of a normal elbow of an adolescent: (1) medial epicondylar epiphysis of the humerus, (2) metaphysis of the humerus, and (3) diaphysis or shaft of the humerus. Arrow points to the growth plate (epiphyseal line) of the proximal radial epiphysis.

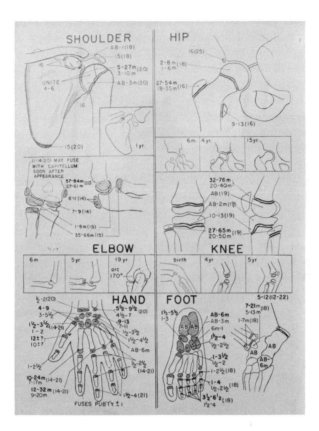

FIG. 2. Range of time of appearance of centers of ossification, tenth to ninetieth percentile; "m" indicates month; otherwise, years. Where two sets of figures are given for one center the upper, heavier figure indicates male and the lower, lighter figure indicates female. AB means visible at birth. Approximate age of fusion in parentheses. Reproduced with permission of Dr. B. R. Girdnay and The Williams and Wilkins Company.

There is a variation in the time of appearance of the secondary ossification centers in the epiphysis as well as the time of the closure of the growth plate in different bones (Fig. 2). Variations also occur relative to sex as well as individual variations. For example, the secondary ossification center of the distal femoral epiphysis is present at birth. The proximal femoral secondary ossification center is usually not present until between the first and sixth months in females and the second to eighth month in males. The ossification center of the medial condyle of the humerus does not appear until somewhere between the seventh and ninth year. Closure of the distal femoral epiphysis is usually complete by the age of 19 years, whereas the medial condyle of the humerus has generally fused to the metaphysis of the humerus by the age of 14 years. Because of these variations, exogenous effects related to growth are difficult to evaluate.

The growth plate is made up of growing cartilage cells which receive their nourishment from blood vessels that enter the plate from the epiphyseal side. Growth disturbances can occur when the vascular supply to these cells has been lost or the cells have been damaged. Affections of bone growth are essentially those of the growth plate.

B. Normal Growth and Development Patterns

Growth is a developmental increase in the total mass and size of the body. Development may be defined as the maturation and differentation of tissues and organs which are necessary for the formation and completion of the whole individual.

Two growth spurts normally occur. The first occurs in both males and females in the age range of 5½–7 years. This is called "The mid growth spurt." The "adolescent growth spurt" begins in females around the eleventh to thirteenth years and in males during the thirteenth to fifteenth year.

The female during her adolescent growth spurt grows faster and larger before slowing down, except for pelvic width which continues to increase because of the action of the female growth hormone. The male, because of the delay of onset of the adolescent growth spurt, has an extended growing time, allowing the legs to become longer. In males, the sex hormones alter the growth pattern by increasing the shoulder widths.

Many factors affect skeletal growth during the prenatal and postnatal periods. Such factors as genetic; maternal nutrition and disease; hormonal, nutritional, or dietary deficiencies; infections; malformations; and vascular insufficiency diseases; as well as environmental, sociological, and climatic influence skeletal maturation.

Finally, in assessing growth, the wide individual variation, as well as variation in relation to somatotypes, should be considered. The concept of chronologic age vs. physiological or developmental age is helpful in reducing the variation.

The physiological age is based on the time of appearance of sexual characteristics, menarche, and the skeletal and dental age of the child. Skeletal age of an individual is determined by a comparison of roentgenograms of the wrists and hands with standard age grouping covering the growth period. Such standards may be found in the work of Todd (1937) or Greulich and Pyle (1950).

C. EFFECTS OF PHYSICAL ACTIVITY ON BONE GROWTH

Professor M. F. Ivanitsky (1962), Director of the Central Pedagogogical Institute of Physical Culture of the U.S.S.R., has found through his studies that long participation in athletics effects a change in diameter of bones, in their internal structure, and to a lesser extent in their length. He gives the following examples:

1. The femur of long time soccer players is frequently larger in diameter than that of the nonathlete.
2. A study of a group of young athletes ages 11–13, observed individually over 3 years, revealed the radius of the tennis players was larger than that of the swimmers or gymnasists. In addition, the tennis players showed strongly asymmetrical enlargement of the radius, chiefly the right.
3. Enlargement of the marrow cavity of the tibia in runners active for more than 5 years.
4. Among women of the same height, the external conjugata of the pelvis was smaller in those girls who started gymnastics before the age of 14, when there were still cartilages in the pelvis, than in those women who started the sport later in life.

These studies, interesting as they are, are difficult to evaluate since it is generally agreed that there is an individual difference in morphological responses to exercise. The effects, in essence, are merely a reflection of Wolff's law which states that bones will adapt to suit the stresses and strains placed upon them. Such adaptation is not confined to developing bone. Special needs of adult bone, as well as stresses placed upon healing fractures, are met by development and readjustment of osseous struts.

Stresses and strain act on bones as they do on soft tissues. A mild degree of trauma may act as a stimulant, whereas a severe degree may result in necrosis. Between the two are various degrees of inflammation. Stresses and strains within the tolerance of bone will result in new supportive trabeculae. A severe or prolonged strain may result in inflammation and resultant osteoporosis. An example in an athletic environment is a fatigue fracture. The inflammatory process which results from the excessive or repetitive stresses to the bone results in osteoporosis and resultant fracture. Protection and rest result in healing and compensatory new bone formation.

III. Epiphyseal Development and Growth

A. GROWTH PLATE

To understand affections of and injury to the growth plate, a more detailed description of the cellular anatomy is indicated. The growth plate is divided into four cellular zones (Fig. 3). The first lies immediately adjacent to the epiphysis and is called the zone of resting cells. These cells act as a reservoir for future growing cartilage cells and are nourished by blood vessels through the epiphysis. The second layer is the zone of proliferating cells. The cells increase in size as an extracellular cartilaginous matrix develops. In the third zone—zone of hypertrophied cells or zone of provisional calcification—the cells begin to arrange themselves in vertical columns as hypertrophy and degeneration begin along with calcification of the cartilaginous matrix. The zone of enchondral ossification is the fourth layer and shows the outgrowth of capillaries from the metaphysis and the replacement of dying cartilage cells by trabeculae of bone.

The actual growing area of the growth plate encompasses zones I and II —the resting reservoir of cells and the proliferating cells. Multiplication and

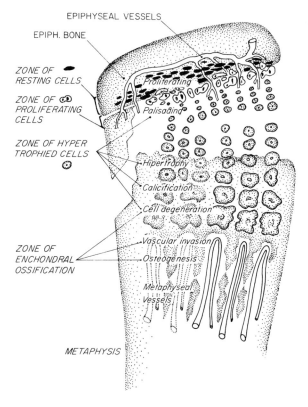

FIG. 3. The cellular anatomy of the growth plate (see text). This drawing shows the usual area of separation with an "epiphyseal fracture" Type I. Note the small triangular fragment of the metaphysis which is often displaced with the epiphysis.

growth of these cells is the mechanism of expansion within the growth plate. This results in a steady increase of distance between the epiphysis and the metaphysis of the bone. Growth rate depends on the age of the individual and the inherent potential for growth of each plate. Factors previously mentioned—nutritional and hormonal, general health, and disease of the body as a whole, as well as injury to the growing cells—will influence the ultimate growth of the growth plate.

As bone increases in length, the periosteal covering of bone contributes to increase in width of metaphysis and diaphysis. Widening of the epiphysis also occurs as the bone enlarges in all directions. Widening of the bone is called "appositional growth."

Those epiphyses which act as insertions or origins of muscles of growing bone contribute to bone shape but not length. These are called "traction epiphyses" or "apophyses." Examples of traction epiphyses are the tibial tuberosity at the proximal end of the tibia to which the distal end of the quadriceps muscle—the patellar tendon—attaches or the medial and lateral epicondyles of the humerus which act as origins for muscle groups going into the forearm.

"Pressure epiphyses" are located at the end of long bone. The normal forces to which these growth plates are subjected are the result of muscle tension and weight bearing; hence, the name "pressure epiphyses." The anatomic appearance and the response to physiological influence are the same in both pressure and traction epiphyses.

Hormonal influence is the basic controlling mechanism of growth. Though there are genetic limitations in skeletal stature, many factors such as maintenance of rate, growth spurts, and time of cessation of growth are primarily regulated by anterior pituitary function. A complex interplay of other hormones also contributes to the overall growth pattern. Physical activity, to the writer's knowledge, has no effect on these hormonal mechanisms.

The nourishment of the cells of the growth plate depend upon their vascular supply. Interference with vascular supply results in a failure or diminution of growth. Increased vascularity and hyperemia stimulate growth of these cells. Abnormality of cell production theoretically could result in production of abnormal matrix, calcifiability of matrix, changes in the orderliness of zone transformation, and, finally, changes in normal ossification patterns. Such changes may not be first recognizable but affect later growth potentials.

Much of the complex physiological and biochemical mechanism of growth and the growth plate remain a mystery. Mechanical forces to the growing skeleton are therefore difficult to evaluate, not only because of the complex interaction of the factors mentioned but also because of the wide individual variation of growth.

B. Osteochondroses

The term "osteochondrosis" is used to designate a derangement of the normal process of bone growth occurring at the various ossification centers at their period of greatest activity. They thus share the one common feature of involving an epiphysis that is undergoing ossification. Though the condition can occur in any epiphyseal area, its relative rarity in non-weight-bearing areas of the skeleton, as compared to the spine and lower limbs, suggests that transmission of body weight and minor stresses are important contributory causes.

The cause, though unknown, is associated with a vascular disturbance producing a necrosis of the epiphysis and fibrosis of the adjoining metaphyseal region. Relating the time of occurrence in the various epiphyses frequently involved with the general growth pattern indicates (Duthie, 1959) the following:

1. The condition develops shortly after the appearance of the ossification of the epiphyseal nucleus.
2. It occurs immediately before or during the midgrowth spurt or adolescent growth spurt which may aggavate it.
3. Because of earlier maturation, the condition generally clears up earlier in girls.

Though there are many theories of pathogenesis, trauma is generally the most favored. Duthie (1959) has suggested that the condition is initiated by a hormonal change, causing abnormal or extensive proliferation of osteogenic cells. This, in turn, causes an increased nutritional demand which may be unanswered as a result of trauma and interference with blood supply of that epiphysis.

The symptoms of osteochondrosis are generally gradual without the history of an acute precipitating cause. The symptoms, though mild, may be aggravated by vigorous activity. In a weight bearing joint, particularly the leg, a limp, sometimes without pain, will be the initial sign of the problem. Tenderness over the epiphysis, limitation of motion, and, at times, muscle spasm are other indicators of the condition. Occasionally there will be no symptoms and the condition discovered only when deformities develop.

The roentgenographic picture shows fragmentation of the epiphysis in the early stages (Fig. 4). At a later stage, dense calcified necrotic bone may be visulaized in the ossification center. As the process continues, loss of calcification and resorption of the necrotic bone occurs. This is followed by recalcification of the resorbed necrotic bone. This process may extend over many months to a few years, during which the ossification center may become

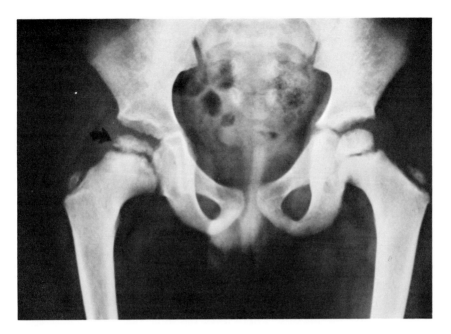

FIG. 4. Roentgenogram of the pelvis showing an osteochondritis of the proximal femoral epiphysis (Legg-Calvé-Perthes' disease). Note the difference of the involved epiphysis (arrow) and the normal epiphysis (opposite side).

compressed or deformed. Protection of the developing epiphysis while the process is active is felt to lessen the chances of ultimate deformity; however, even with adequate treatment, deformity may sometimes develop.

Unfortunately, the osteochondroses have come to be known by the original observer or concurrent observers of the condition.

1. Legg-Calvé-Perthes' Disease (Osteochondritis of the Proximal Femoral Epiphysis)

Legge-Calvé-Perthes' disease usually occurs near the midgrowth spurt between the ages of 4–7 years, although it can occur earlier or later. Atrophy of the proximal femoral epiphysis and deformity of the developing femoral head occurs as a result of the growth disturbance. In 10% of the individuals it is bilateral. It may take many months or years for the condition to subside and vascularity to reestablish itself.

During the period of abnormal epiphyseal development and ossification, the femoral head is felt to be in a particularly vulnerable situation with regard to strain and compressive forces. Treatment is directed toward protection of the developing femoral head from weight bearing, either by prolonged bed rest, various types of braces, or crutches until the condition resolves.

Obviously, any physical activity which would produce stresses or strains to the hip joint is contraindicated. Even with the most favorable conditions, deformity of the femoral head may result. If deformity of the femoral head does occur, an incongruity of joint surfaces is produced causing increased wear of the hip joint and ultimately degenerative joint disease. It should be mentioned that not all of the hip joints which are affected will result in incongruous joint surfaces. Approximately 50% of those having had this condition have no residuals.

2. Osgood-Schlatter's Disease (Osteochondritis of the Proximal Tibial Tubercle)

Osgood-Schlatter's disease is the most common form of osteochondritis seen. It involves the downward extension of the proximal tibial epiphysis— the tibial tuberosity. It is at this area that the extension of the quadriceps mechanism attaches as the patellar tendon (see Fig. 7). Its onset is near the adolescent growth spurt between the ages of 9 and 14 years. The abnormality does not involve the pressure epiphysis of the tibia; therefore, interference with bone length or involvement of joint surfaces is not a problem.

The onset is usually insidious with tenderness over the bone prominence of the tibial tubercle. The discomfort is aggravated by exercise or kneeling on the involved knee. Extension of the knee against resistance often aggravates the discomfort.

The condition should not be confused with traumatic epiphysitis or traumatic separation of the tibial tubercle. Both can be related to excessive strain to the tibial tubercle either by excessive repetition or single injury.

The roentgenographic picture is one of irregularity or fragmentation of the ossification center of the tibial tuberosity. Residuals of this condition can be seen radiographically as separated bony fragments overlying the tibial tubercle or an enlargement of this prominence (Fig. 5).

Protective padding and the avoidance of vigorous or repetitive activity is recommended during the active phase of the condition.

3. Scheuermann's Disease (Vertebral Epiphysitis)

Scheuermann's disease involves the secondary epiphyseal centers of the vertebral body, occurring near the adolescent growth spurt between the ages of 12 and 17. It tends to occur earlier in girls, and the osteochondritic process terminates at an earlier age in girls. As with most of the osteochondroses, it is more frequently seen in males. The onset is often detected by an in-

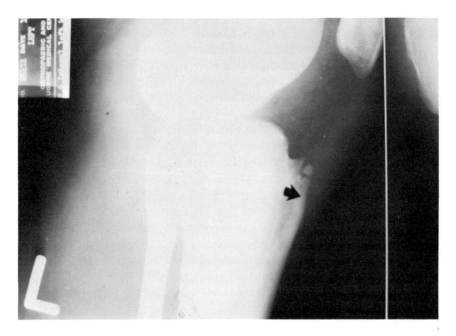

FIG. 5. Roentgenogram of a knee showing the residuals of osteochondritis of proximal tibial tubercle (Osgood-Schlatter's disease). Note the enlarged bone prominence (arrow) and the separated bone fragment.

creasing roundness of the back, associated with aching in the back. Pain or discomfort as the predominant feature occurs infrequently.

The radiographic appearance, as in other osteochondroses, shows an irregularity and fragmentation of the upper and lower epiphyses around the vertebral body. The body of the vertebra or vertebrae (since it can involve more than one) often becomes wedge-shaped as normal growth is impeded.

A residual permanency of the wedge-shaped vertebra may result. If the process involves more than one vertebra, particularly in the thoracic spine, such wedging may produce a permanent kyphosis.

4. Activity and the Osteochondritic Process

An osteochondritic process can develop in any epiphyseal area. The three discussed above are the more common. When symptoms of such a condition occur, medical evaluation should be obtained. Restriction from vigorous exercises should be instituted during the active phases of the abnormal growing process. Reevaluation for residual problems may be necessary and is desirable before return to active physical participation.

The insidious onset of these conditions and their relationship to periods of active epiphyseal growth lead one to consider the possibility of injury to epiphyseal areas by too vigorous activity during maturation of the skeleton. Indeed, the word "disease" attached to these common abberations of growing bone has been questioned. "Micro-trauma" to the maturing skeleton has been raised as a reason for placing limitations on the physical activity of the young.

In assessing such restrictions, one must realize that every strain or stress is a trauma in miniature. These miniature stresses and strains are necessary to produce mild stimulation with proliferation of the primitive mesenchymal cells and the laying down of new bone. Without the stimulant effect, cells become quiescent, their circulation becomes excessive, and there results a relative hyperemia with bone resorption. As mentioned earlier, stresses and strains to the developing bones within tolerance are necessary to develop supporting trabeculae where they will be of most benefit. Microtrauma to bone and joint structures is such an indefinite and immeasurable quantity that its use in a deleterious sense is to be ignored.

This is not to say that developing skeletal structures need not be protected. During the active growth spurts in children, thought should be given in developing activity and athletic programs which do not place excessive or repetitive stress on those epiphyseal areas most subject to osteochondrosis. Long distance running, high jumping, trampoline, or prolonged athletic con-

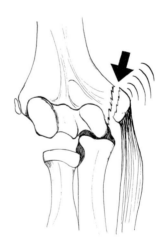

Fig. 6. Drawing of the elbow joint in an adolescent. The violent contraction of the flexor-pronator group of muscles to the forearm in the act of throwing causes a strain to the growth plate of the medial epicondylar epiphysis of the humerus (arrow). This is a traction epiphysis (compare with Fig. 1).

tests in the preschool and early primary age school children are examples of necessary restrictions of physical activity in this age group.

The question still remains as to what is excessive. This cannot be answered in relation to a certain age group. Each individual must be considered separately since, as mentioned, skeletal maturation varies widely in relation to chronological age. Muscular development is also widely variable, and the strength of muscle relates to the stresses that can be placed on the skeletal framework. If a group of the same chronological age is to participate in an exercise program, or engage in an athletic competition, the activity must be related to the fatigability of the least physically developed youngster or individual consideration must be given.

IV. Indirect Injuries to the Immature Skeleton

Any physical activity increases the danger of injury. Despite the most careful pre-activity physical examination, screening, and supervision, injuries will occur.

A. TRAUMATIC EPIPHYSITIS

The osteochondroses are strongly suspected of being related in some manner to trauma. Situations occur, however, where the onset of the discomfort and roentgenographic changes can definitely be related to either a single traumatic incident or to repetitive stress to a single epiphysis. When such is the case, "traumatic epiphysitis" is the preferable term to designate the condition.

Possibly the most notorious of the conditions is "Little Leaguer's elbow." This condition is initiated by repetitive strains to the medial epicondylar epiphysis of the humerus (Fig. 6). This traction epiphysis is the site of origin of the flexor-pronator group of muscles in the forearm. The youngster with this epiphysis still open, which includes the 8–14 year olds who play Little League baseball, exert considerable tension on this common tendon by the vigorous contraction of the flexor-pronator group during the relatively violent muscular action of throwing a baseball with maximum effort.

Little League studies have shown that 12-year-old boys can throw a baseball up to 70 mph. This sudden pull on the epiphysis may result in its separation or with repetitive stress set up an inflammatory response properly termed "traumatic epiphysitis of the medial epicondylar epiphysis."

Adams (1965), in a study of 162 boys in the age group of 9–14 years of age, showed by roentgenograms a discernible degree of epiphysitis in all of the 80 pitchers. Only a small percentage of the nonpitching baseball players and control group of nonplayers showed such roentgenographic changes.

The susceptibility of the elbow of youngsters of the age group of 8–14 years has led to the recommendation that pitchers of this age group be allowed to pitch only two innings per game and that curve ball throwing below age 14 years be eliminated. Official Little League baseball does have restrictions on the amount of pitching the youngster can do. These limitations are restriction to six innings that a regular Little League pitcher can pitch, a required 3-day rest after pitching four or more innings, and a 1-day rest after pitching less than four innings. Some means of discouraging prolonged practice sessions and an awareness of the possible harmful effects by the coach, the parent, and the youngster is necessary.

Traumatic epiphysitis of the tibial tubercle of the knee can also occur. This is closely similar to Osgood-Schlatter's disease and indeed any tenderness and sign of epiphysitis around the tibial tubercle is designated by many physicians as Osgood-Schlatter's. When, however, the onset of the epiphysitis can be related to a definite inciting cause, as running or jumping, or a bump or strain, and is aggravated by the vigorous activity, traumatic epiphysitis of the tibial tubercle explains the condition. No longer is the condition left in the basket of the speculative and unknown causes of osteochondroses.

The elbow and the knee are the two areas where a traumatic epiphysitis related to physical activity most often occur. The shoulder has on occasion developed similar discomfort related to vigorous throwing, and indeed the phrase "Little Leaguer's shoulder" has been coined. The incidence is much less frequent than the elbow. The condition involves the proximal humeral epiphysis. This is not a traction epiphysis; however, the strong pulling forces on the head of the humerus away from the glenoid at the completion of the throw subjects the proximal humeral epiphysis to repetitive traction strains. The radiographic picture is much less definite. Usually the only suggestive roentgenographic changes are a slight widening of the edge of the growth plate and slight demineralization without actual necrosis. These findings are consistent with a local inflammatory reaction from repeated stress. The condition usually rapidly subsides with rest and avoidance of the aggravating cause.

The treatment for traumatic epiphysitis, as for the osteochondroses, is generally rest to allow the inflammatory process to subside and healing to occur. Since these involve traction epiphyses and not the pressure epiphyses which control bone growth, length of bone will not be affected. Residuals do occur, however, consisting of permanent hypertrophy of the bony attachment of the tendon, residual loose bodies of bone overlying the tendon attachments, or a chronic inflammatory tendinitis at the tendon attachment.

Consideration is necessary to the complaints of youngsters, particularly when they complain of joint areas or areas of muscle attachment to epiphyses.

The risk of permanent damage is ever present. The residual handicap, though it may not be disabling, may affect the child's performance in related vigorous activity in later years.

B. TRAUMATIC SEPARATION OF THE EPIPHYSES

One degree higher in the scale of trauma to traction epiphyses is traumatic separation by the violent contraction of muscle.

The most common site for this to occur is at the tibial tubercle of the knee. The patellar tendon pulls on this tubercle with the mechanical advantage of a winch as the tendon, patella, and quadriceps tendon are pulled around the femoral condyle as the knee flexes (Fig. 7). The usual history is onset of sudden and severe pain at the tubercle site after landing from a jump. An overload of the isometrically contracting quadriceps with the sudden stopping or deceleration causes the epiphyseal plate to separate. The severity of the injury varies from a partial or complete avulsion of the tibial tubercle to a separation that involves the proximal tibial epiphysis of which the tibial tubercle is a distal extension.

History, examination, and radiographic study readily differentiate this condition from osteochondritis or traumatic epiphysitis. Common injury sit-

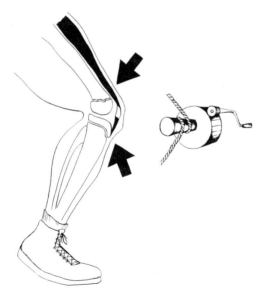

FIG. 7. Diagrammatic illustration of the mechanical force applied to proximal tibial tubercle by the contraction of the quadriceps muscles through the quadriceps tendon, patella, and patellar tendon.

uations are landing stiff-legged following a lay-in or rebound in basketball or an awkward or off-balance landing following a pole vault, high jump, or broad jump in a 12–15-year-old boy. Disability is immediate. Pain and tenderness over the tibial tubercle is easily identified. Roentgenograms show the avulsion varying from a slight separation to detachment of the entire epiphysis.

Treatment varies from rest in the mild cases with minimal separation to manual reduction of the displacement followed by cast immobilization. Rarely is it necessary to replace surgically the fragment prior to immobilization.

In the immature athlete who resumes activity too soon before healing has occurred, the bony fragments remain separated and are painful because of the repetitive tug of the extensor mechanism of the knee. Since the fragments may have lost some of their blood supply, avascular necrosis may occur. Reactive inflammation from the extensor tug may also result in enlargement of bone. As one can see, the symptoms and residuals of inadequate treatment are essentially the same as those found in osteochondritis and traumatic epiphysitis. In an individual with long-standing symptoms referable to the tibial tubercle, it may be impossible, particularly with an inadequate history, to say with certainty which condition produced the residual symptoms.

Analogous to the tibial tubercle of the knee is the separation or detachment of the medial epicondylar epiphysis of the elbow. Though this can occur as a result of a vigorous throwing motion, it is more often seen after a more violent stress such as a fall with a sudden valgus strain to the elbow. Occasionally, the violence of the injury will cause the epicondylar fragment to be pulled into the elbow joint (Fig. 8).

Treatment again varies from immobilization to allow healing to surgical reattachment of the displacement if it is severe enough. When the fragment has been pulled into the elbow joint, manipulative attempts to free it and manually replace it in its former bed are usually unsuccessful. Surgical reduction then becomes mandatory.

The residuals are the same as those found around the tibial tubercle. Since the muscular tension is less, the symptoms are proportionately less severe. Often the slight deformity that persists does not produce enough disability to warrant treatment. With the vigorous activity of throwing, minimal residuals may produce an irritative tendinitis at the tendon attachment of the flexor-pronator group. In the athlete, surgical relief is sometimes necessary to remove the source of chronic irritation to allow effective sports participation.

Other sites where an abnormally strong muscular contraction may cause separation of a traction epiphysis are the attachment of the sartorius muscle to the anterior–superior iliac spine, the rectus femoris muscle to the anterior–inferior iliac spine, the psoas muscle to the lesser trochanter of the femur, and the attachment of the adductor magnus and hamstrings to the ischial tuberosity. These epiphyseal avulsions result from the vigorous muscular contractions associated with jumping and sprinting. Closed reduction generally gives satisfactory results except with marked displacement of the apophysis of the ischial tuberosity where an open reduction and fixations may be necessary. Unreduced fractures of the latter area result in fibrous union, uncomfortable with sitting and painful with stress. Bony enlargement

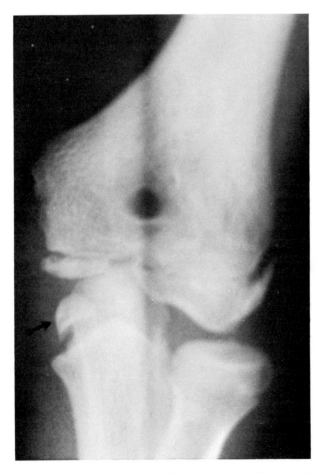

Fig. 8. Roentgenogram of an elbow in a youngster. The medial epicondylar epiphysis of the humerus has been pulled off and is entrapped in the elbow joint (small arrow). Compare with Figs. 1 and 6.

may occur with a resultant ischial mass. Bone enlargement may occur at the other sites mentioned, but generally these are asymptomatic.

V. Direct Injuries to the Immature Skeleton

A. FRACTURES OF THE LONG BONES

Fractures of the long bones cause no diagnostic problems. The injury is usually apparent, the disability readily accepted, and treatment immediately instituted. Fractures in children are different from fractures in adults. Because of lack of complete ossification, at certain ages minor twists and tumbles will cause fractures of long bones that do not occur in adults. Multiple fractures without adequate trauma in a child suggest a deficiency disease such as osteogenesis imperfecta. Isolated spontaneous fractures in children may result from benign bone cysts. Greenstick fractures, in which the fracture does not break through both cortices, occur in younger children because of their more pliable bone structure.

It is the growth factor which differentiates children's and adults' fractures. Further growth usually helps to correct deformities of long bones. Fractures involving the epiphysis may produce deformity as a result of interference with growth. These injuries are discussed in the next section.

Some stimulation of longitudinal growth of long bone will often result because of physiological response of increased blood supply produced at the fracture site to hasten healing. This is particularly important in the lower extremity where an overlap of the fractured ends is desirable to compensate for the increased longitudinal growth. Side-to-side apposition produces, in children, a strong, rapid union. Children's fractures generally heal more quickly than similar adult fractures.

During the period of immobilization required to allow the fracture to heal, physical activity is restricted to what can safely be engaged in without interfering with or prolonging the healing process. After immobilization has been discontinued, a program of rehabilitative exercises to restore joint motion and regain muscle strength is required before full and vigorous participation is allowed. The protection of strong muscles to the immature skeleton is necessary to provide maximum safety during the child's physical pursuits.

B. FRACTURES OF THE EPIPHYSES

In this section injuries to the pressure epiphyses will be discussed. Since these epiphyses control the longitudinal growth of bone, the complication of growth disturbance may be present after injury. Such a complication can usually be predicted and in some cases prevented. The assumption that any injury to an epiphysis will result in a growth disturbance is erroneous.

The microscopic anatomy and the zonal formation of the growth plate has been previously discussed in this chapter under Section III. The term "fracture of the epiphysis" is, in most cases, a separation of the growth plate nearly always occurring in the zone of hypertrophied cells (also called zone of provisional calcification) (Fig. 3). This is the weakest part of the epiphyseal plate. Separation through this zone does not involve the reservoir of future cartilage cells, nor does it interfere with the blood supply to these cells. Nature has provided a weak link, a safety valve, at an area where separation can do the least amount of harm.

1. Types of Epiphyseal Fractures

Epiphyseal fractures are most simply classified into three types. Type I fractures are those in which the cartilaginous growth plate separates, as mentioned above. Often there is a small triangular fragment of bone from the metaphysis carried along with the displaced epiphysis (Fig. 3). Type II fractures occur across the bony epiphysis and/or extend into the cartilaginous growth plates. Type III epiphyseal fractures extend across the epiphyseal plate and into the diaphysis. Types II and III fractures usually result from crushing injuries to the growth plate as the mechanism of force (Fig. 9). The crushing of the resting cartilage cells may cause cellular damage. Deformity may result from interference with growth and premature closure of the growth plate at the point of crush injury.

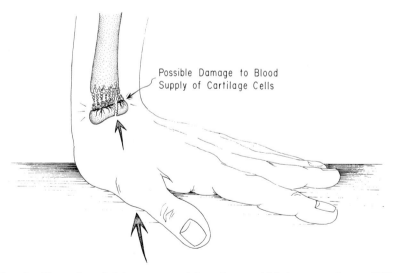

Possible Damage to Blood
Supply of Cartilage Cells

FIG. 9. Type of crush injury to an epiphyseal area which increases the possibility of growth disturbance. The illustration is of a Type II epiphyseal fracture.

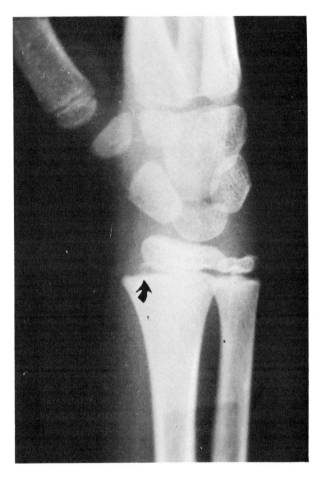

FIG. 10. Roentgenogram of a displaced distal radial epiphysis of the wrist. This is a Type I epiphyseal fracture. Separation has occurred through an area which rarely causes a growth disturbance. Compare with Fig. 3.

Fortunately, the great majority of epiphyseal fractures are Type I fractures (Fig. 10). This type of fractures generally do not produce any growth disturbance. Aitken (1965) stated that 98% of epiphyseal fractures are Type I. In a personal analysis of athletically caused epiphyseal injuries, 82% were Type I and 18% were Types II and III fractures. In approximately 10% of epiphyseal fractures there is significant disturbance of growth.

2. Prognosis for Growth Disturbance

Several factors determine the probability of growth disturbance after epiphyseal injury. The type of fracture, as mentioned above, is important prognostically. The age of the child at the time of injury influences the amount of deformity, should it occur. In an older child, where complete ossification and obliteration of the growth plate is about to occur normally, premature ossification of a section of the growth plate will not produce significant deformity. In the younger child with many years of growth left at the injured epiphysis, deformity is almost a certainty if cartilage cell damage has occurred.

The epiphysis injured has prognostic importance. Certain epiphyses, such as the proximal femoral epiphysis, are entirely intra-articular and therefore devoid of periosteal covering. Their blood supply must traverse the rim of the growth plate to reach the epiphysis. Should displacement occur, these epiphyseal vessels are more vulnerable to injury. Interference with the blood supply to the developing cartilage cells produces an avascular necrosis and developmental disturbance to the growth plate.

Whether the injury to the epiphysis is an open or closed one influences prognosis. Though open injuries to the epiphyses are rare, the factor of contamination and possible infection greatly increases the chances of damage to the cartilage cells by the infectious process.

3. Vulnerability of Certain Epiphyses to Injury

As some traction epiphyses are more liable to injury resulting from muscle action of vigorous sports activities, likewise certain pressure epiphyses are more vulnerable to injury. This vulnerability is because of ligamentous attachments, limited mobility of joints, or their exposed position. Ligamentous and fibrous capsular attachments around the joint are felt to be 2–5 times stronger than the weakest section of the epiphyseal plate. Though sprains of ligamentous structures around joints in children can occur, one must be ever mindful of the possibility of epiphyseal injury. Tears of the ligaments in children whose cartilaginous growth plates are still open are rare.

The distal tibial epiphysis is an example of one of the more vulnerable epiphyses to direct injury. This results from the limited lateral motion of the ankle and the strong ligamentous and capsular attachments supporting this joint. A force great enough to shear, distract, or compress this joint will cause injury to the growth plate.

The phalanges of young athletes are frequently injured. This is because of the younger athlete's lack of dexterity and the exposed position of these joints in physical activity. A common injury to the phalanx is a "mallet" or

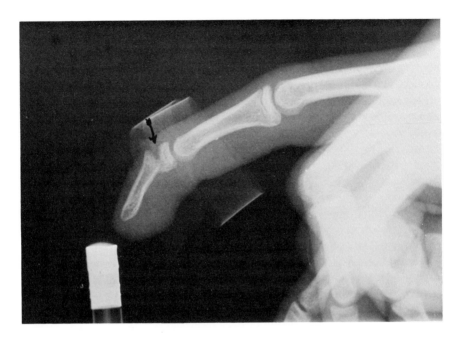

Fɪɢ. 11. Roentgenogram of a finger showing a "baseball" or "mallet finger" type of injury. Such an injury is usually caused by a blow on the end of the finger. The bone has separated through the growth plate (arrow). This is a Type I epiphyseal fracture.

"baseball finger." The distal phalanx of a finger has the flexor tendon attached to the metaphysis while the extensor tendon attaches to the epiphysis. A blow on the end of the finger causes the metaphysis to be suddenly flexed while the epiphysis remains extended with the resultant flexed deformity of the distal phalanx (Fig. 11). In an adult this type of injury could produce a rupture of the extensor tendon, but in the child it is the weaker cartilaginous growth plate that separates.

Another common site of epiphyseal injury in children is the distal radial epiphysis. The injury is usually a Type I fracture. Prognosis is good (see Fig. 10).

The knee joint shows the vulnerability of one epiphysis to injury compared with an adjacent epiphysis. The distal femoral epiphysis is the proximal attachment for the collateral ligaments of the knee. Distally these ligaments attach to the tibia distal to the proximal tibial epiphysis. A forceful valgus or varus strain to the knee will result in force transmitted to the distal femoral growth plate through this ligamentous attachment.

TABLE I

ATHLETIC INJURIES[a]

(15 YEARS AND UNDER)

Activity	Total	Epiphyseal	%
Football	188	22	11.7
Physical education classes	82	6	7.3
Track	78	9	11.5
Skiing	71	8	11.2
Basketball	70	5	7.1
Baseball	48	7	14.5
Miscellaneous	136	29	
	673	86	

[a] These were taken from a total series of 2745 athletic injuries of all ages.

The shoulder and hip, because of their wide range of motion, are least likely to sustain epiphyseal separation. Because of this wide latitude of motion, Types II and III fractures rarely occur.

C. INCIDENCE OF EPIPHYSEAL INJURIES

In a study of 2745 athletic injuries, 673 occurred in children 15 years of age and under. Eighty-six of the injuries occurring in the 15 and under age group were epiphyseal injuries, 12.7%. The frequency of epiphyseal injury in each sport is given in Table I.*

In the case of epiphyseal injuries, the likelihood of such an injury producing a growth disturbance is between 10 and 20%. A study of athletically caused epiphyseal injury revealed 18% of those followed for over 1 year and re-x-rayed, had some radiographic change.

In another study of 31 epiphyseal injuries requiring hospitalization, 23% resulted from athletically related activities. The remainder were from falling, usually from a height, and vehicular accidents.

D. REQUIREMENTS FOR RETURN TO PHYSICAL ACTIVITY AFTER EPIPHYSEAL INJURY

A proper time of immobilization to allow healing after epiphyseal fracture is necessary. Following healing, an assessment of joint function and muscle strength should be made before return to full physical activity. In the majority of epiphyseal fractures, once healing and rehabilitation have been

* These athletic injuries were those seen in the writer's orthopedic practice and would therefore be somewhat more severe than those seen by the general practitioner. They do not include those injuries not seen by an M.D.

obtained, no restrictions are necessary. In the more serious epiphyseal injuries, which actually involve joint surfaces or where the possibility of growth disturbance exists, greater caution should be exercised. An incongruous joint may be present, and a closer look for excessive wear changes in these joints should be made. Types II and III epiphyseal fractures demand follow-up for at least 1 year with yearly rechecks to ascertain if deformity is developing and roentgenographic evaluation of joint irregularity. In these types, where residuals are more common, abstinence from contact or collision sports or excessive physical exercise should be demanded for at least 1 year. This is necessary to protect the epiphysis and give it every opportunity to proceed with normal development. If growth irregularity does occur, activity restrictions may be required.

VI. Safeguards to the Health of the Younger Child in Physical Activity Programs

The Committee on Injury in Sports of the American Medical Association and The National Federation of State High School Athletic Associations (1959) issued a joint statement entitled "Safeguarding the Health of the High School Athlete." The check list they provided could be applied to any child who participates in a physical activity program. With some modifications, and taking into consideration the immature skeleton and its growth potentials, the following safeguards would be appropriate.

A. Adequate Medical Care

Before the youngster participates in a vigorous physical activity program, especially one developed by a school system or organization, a thorough history and physical examination should be given. This is given to assess the general health and fitness of the participant. It is important to determine if any conditions exist which would be precipitated or aggravated by physical activity. Conditions such as growth disturbances, previous injuries, osteochondroses, as well as other medical problems need to be evaluated before participation in an organized physical activity program. The insistence on a proper history and physical examination before competitive athletics is begun has long been recognized. The trend is away from the short, mass screening examinations done in the locker room or gymnasium. This does not provide an adequate past medical history, nor does it provide the completeness that is necessary for the neophyte athlete.

Certain physical and medical conditions disqualify an individual from some activities. The physician who makes the decision as to fitness for physical activity participation must evaluate these conditions and relate them to the degree of activity required in a given program. A condition which is disqualifying for a contact sport may not disqualify the child for another physical activity such as swimming.

Some body types are more prone to injury during the growth spurts. The tall, gangly, underdeveloped, uncoordinated adolescent is more prone to injury than his compactly build, coordinated counterpart. Often physical examination will reveal ligamentous laxity around joints or poor muscle tone. An individual of this body build should be discouraged from body contact sports until his muscle development and coordination catch up to his longitudinal growth. The obese, Fröhlich somatype is also a susceptible candidate for epiphyseal injury. He, too, should be restricted from physical contact sports until epiphyseal closure has occurred.

Since physical activity increases the chance of an injury, programs should have proper first aid by trained personnel immediately at hand and provisions for proper medical care by a physician readily available. An increasing number of contact sports at the high school level and above require a physician in attendance at such contests. This is not practical for the many athletic contests, such as Little League, in which growing youngsters are participants. A thoughtful plan on how to deal with an injury and, if serious, the patient's transport to a medical facility is an important aspect in developing a physical activity program.

After an injury the decision as to when return to participation can begin should be made by the physician. Too often overzealous physical education instructors, or coaches—and sometimes parents—attempt to make this judgment without sound medical advice. In the growing child, as indicated in previous sections of this chapter, the decision often depends on factors not readily apparent.

B. PROPER CONDITIONING

Competitive athletics requires proper conditioning to increase muscle strength, decrease fatigue, and lessen chances of injury. This means that the demands imposed upon the child should be gradually increased and that due attention be given to the correct mechanical use of the body in the performance of the skill characteristic of the sport. Within each age group there are wide individual differences in levels of growth and development and also in cardiovascular and muscular fitness. Insofar as possible, physical activity programs should be individualized. Both individual and team sports require grouping as to stature, coordination, and skill during the growing years.

Prior to competitive contest, an adequate period of warm-up is most beneficial before all-out effort is required. The contests of younger children often neglect this important period.

Youth often minimizes or ignores signs of fatigue. It is the responsibility of the supervisor or coach to provide adequate rest periods or adequate substitutions in a game situation to prevent excess fatigue. Fatigue decreases the capabilities of muscles to withstand stress and lessens the attentiveness of the player. "The unexpected poses far more risk to the athlete than the expected. The athlete who is unaware of, or does not identify, an upcoming hazard cannot use the defense of anticipation."*

C. Careful Coaching or Supervision

Skillful performance decreases the incidence of injury. Proper teaching of the skills of the activity and the safety aspects are important attributes of adult supervisors of the physical activity of children. If injury situations do develop or repeat themselves, analysis of their cause and methods of prevention is a function of a competent supervisor or coach.

In competitive athletics, tactics which increase the hazard to the player or his opponent should be discouraged. This is particularly true in children who are beginning athletes. To teach a potentially dangerous type of maneuver increases the chance of injury and may instill in the youngster a disregard for safe and sportsmanship-like participation.

Activity periods should be planned with regard to the age of the children, their endurance, and their fatigue. Teaching of the basic skills and techniques and the promotion of sportsmanship and fitness is the foundation on which youth physical activity and athletic programs are built. Many times the supervisors or coaches are "volunteers" without the proper training necessary to provide a safe and rewarding experience for youngsters. The goal of a competent coach goes beyond winning.

D. Appropriate Equipment and Safe Facilities

Children who participate in vigorous activity are entitled to the best in protective equipment. Proper fitting and adjustment are necessary for individual protective equipment used in contact sports. The used, hand-me-down equipment of the older echelon is not sufficient no matter what level of competition is being outfitted.

Gymnastics equipment and playground facilities must meet approved standards of safety and must be properly maintained and serviced. Trained supervisors should be present when students are using these facilities.

The play areas, whether they be the gymnasium or playground fields, should be adequately maintained to provide safety. Uneven surfaces, holes in the ground, protruding objects all increase the hazard.

* Committee on the Medical Aspects of Sports of the American Medical Association.

E. GOOD OFFICIATING AND RULE REVISIONS TO PROTECT THE YOUNGER ATHLETE

In athletic contests involving youngsters, good officiating provides not only enjoyment but also protection against injury. Often rule revisions or special rules are necessary to protect the growing child. The limitations on Little League pitchers, previously mentioned, are an example. A continual search for and analysis of injury situations are necessary so that revisions can be made if required.

As has been said many times, "Children are not miniature adults." They have a growing skeleton, a neophyte's experience, and an undeveloped knowledge of injury situations. For children to participate, no matter what the physical activity, demands trained, qualified, and understanding leadership. In the years ahead this leadership may determine the development and the progress of the youngster toward a healthful, physically fit life.

References

Acheson, R. M. (1957). *Clin. Orthop.* **10**, 19–39.

Adams, J. E. (1965). *Calif. Med.* **102**, 127–132.

Adams, J. E. (1968). *Clin. Orthop.* **58**, 129–140.

Aitken, A. P. (1965). *Clin. Orthop.* **41**, 19–23.

Blount, W. P. (1955). "Fractures in Children." Williams & Wilkins, Baltimore, Maryland.

Clarke, H. H. (1966). *Proc. Nati. Conf. Med. Aspects Sports, 1966,* pp. 49–57.

Duthie, R. B. (1959). *Clin. Orthop.* **14**, 7–18.

Greulich, W. W., and Pyle, S. I. (1950). "Radiographic Atlas of Skeletal Development of Hand and Wrist." Stanford Univ. Press, Stanford, California.

Ivanitsky, M. F. (1962). Quoted in *Med. Trib.* **3**, 34.

Larson, R. L. (1968). *Med. Times* **96**, 679–688.

Larson, R. L., and McMahan, R. O. (1966). *J. Amer. Med. Ass.* **196**, 607–612.

Mercer, W. (1959). "Orthopaedic Surgery," 5th ed. Williams & Wilkins, Baltimore, Maryland.

Morton, D. J., and Fuller, D. D. (1952). "Human Locomotion and Body Form." Williams & Wilkins, Baltimore, Maryland.

President's Council on Physical Fitness. (1965). "4 Years for Fitness, 1961–1965."

President's Council on Physical Fitness and Sports. (1969). "Physical Fitness Facts."

Siffert, R. S. (1966). *J. Bone Joint Surg., Amer. Vol.* **48**, 546–561.

Slocum, D. B., and Larson, R. L. (1964). *Amer. J. Orthop.* **6**, 248–259.

Todd, T. W. (1937). "Atlas of Skeletal Maturation," Part I. Mosby, St. Louis, Missouri.

Tupman, G. S. (1962). *J. Bone Joint Surg., Brit. Vol.* **44**, 42–67.

Weiss, P. (1939). "Principles of Development." Holt, New York.

Growth in Muscular Strength and Power

Erling Asmussen

I. Definitions

Muscular strength is an expression used to describe a person's ability to produce force by means of his muscles. Force is used here in its physical sense, as defined by the equation $f = m \times a$ (force equals mass times acceleration) and can be measured in physical units of force. For most physiological purposes the metric unit kilopond (kp) is appropriate. It is defined as the normal pull of gravity on the mass one kilogram (kg); i.e., it corresponds to the weight of 1 kg. (The British or U.S. pound-force can likewise be used as an expression for force: 1 lbf = 0.454 kp.) In the body, muscular force can be exerted under various conditions. If no—or practically no—movement takes place and the muscles maintain a constant length, the force is called "static" or "isometric" force. If movement takes place during the production of muscular force, the muscles can either contract and shorten or they can be lengthened by some other force, which they then attempt to resist. In both cases one can talk of "dynamic" (as opposed to static) force. In the first case the word "concentric" is used to describe the force developed during shortening, in the second case the force is called "excentric," produced while the muscles are being stretched.

The expression "muscular power" is colloquially used synonymous with "explosive force," the ability to produce maximum force within a very short time. Power in physics means rate of work and is measured, e.g., in watts or in kilopondmeter per minute. It also has the dimension of force $\times$ velocity, and it is in this sense that it is used when describing a short maximal burst of muscle activity as in a jump or throw. Since power is also used in expressions as maximum aerobic (or anaerobic) power and is measured in milliliter O_2 per minute or in kilocalories per minute, the word "muscular power" should only be used when it is made quite clear what is meant.

II. Physiological Considerations

In muscles isolated from the body it is possible by artificial stimulation, directly on the muscle fibers or indirectly via the motor nerves, to obtain maximum contractions either as twitches or as tetani. In this way the maximum force and the maximum power that a muscle can produce under controlled conditions can be recorded and measured. The information obtained in this way is fundamental for the understanding of muscular strength and power, but the data obtained on isolated, artificially stimulated muscles cannot uncritically be transferred to muscles *in situ*. The reasons for this are several, the most important probably being that in the body the muscles are stimulated by nervous impulses coming from the central nervous system, and the functional capacities of this system must, therefore, also be considered. Further, in the body there exists an extensive interplay between periphery and center in the form of various feedback mechanisms, both nervous and humoral. These feedback mechanisms may inhibit or facilitate the output of motor impulses and thus influence the mobilizable muscular force and power. Their extremely important role in coordinating force, in timing and recruitment of individual muscles in movements is well known.

The *"milieu interne"* in which the muscles are placed may also well influence their mechanical function: Temperature, osmolality, ions, hormones, etc., in the extracellular fluid compartments, and the amounts and availability of chemical energy in the muscle fibers may vary and be different from what they are in an experiment with an isolated muscle or muscle fiber.

Muscular strength and power is the integrated expression of what the neuromuscular system can do under the existing conditions. When considered in relation to growth of an individual it becomes clear that since the rate of anatomical growth, of maturation, and of changes in the internal milieu may vary independently, their combined influence on strength cannot be expected to correlate in a simple way with chronological age.

A. MUSCLE CONTRACTIONS

As mentioned before, the information gained from experiments on isolated muscles or muscle fibers is fundamental for the understanding of strength and power. A short description of the mechanical response of muscle fibers to artificial stimulation may be appropriate. An effective stimulus to a muscle fiber will produce an all-or-none contraction in the form of a twitch. If recorded under isometric conditions the outcome will be force (tension), and the whole process of developing force (time to peak) will last 25–100 msec in different human muscle fibers (Buchthal and Schmalbruch, 1970). If the stimulation is repeated rhythmically, with increasing frequency, the muscle fiber will perform a series of twitches that eventually will fuse into a sustained tetanus. The frequencies at which this happens vary between 50 and 125 stimuli/sec in the cat (Buller, 1970). The tetanic tension will be 3–5 times greater than the peak tension in the twitch. Since the contraction times for cat and human muscles vary within nearly the same limits, it seems justified to assume that the fusion frequency for man also will be 50–125 stimuli/sec. The highest possible tension in human muscle fibers should then be attained at these frequencies. The question now is: Can the motor neuron fire at this rate under voluntary contractions? To answer this question one might analyze electromyograms from voluntarily contracting muscles. It has, however, proved to be very difficult to identify action potentials from individual muscle fibers or motor units during maximal contractions because of interference from neighboring units. In partially denervated human muscles (polio sequelae) the frequency was found to vary between 35 and 90 per second at maximum effort (Seyffarth, 1940), but in most cases maximum frequencies were below 50 per second. By partially pressure blocking the nerve to m.adductor pollicis, Bigland and Lippold (1954) found that the maximum frequencies were 30–45 per second. Thus it still seems doubtful if tetanizing frequencies do ever occur even in maximal, voluntary contractions. If not, then it is ordinarily not possible to produce the highest tension of which the muscle should be capable, but there must be a reserve of force that will only be called forth under extraordinary conditions.

Ikai and Steinhaus (1961) showed that shouts or pistol shots could reinforce voluntary innervation and produce more force, and Ikai *et al.* (1967) showed that electric stimulation through the ulnar nerve could evoke contractions in the adductor pollicis of greater strength than could maximum voluntary exertions [although Merton (1954) in a similar experiment found no difference]. Finally, it may be recalled that several observations have shown that humans in real emergencies may exhibit "superhuman" strength.

What it is that normally limits the exertion of strength and power is not known. There are several levels in the central nervous system where an inhibition may set in. The inhibition may result from a negative feedback from peripheral receptors or it may be a persistent inhibition from centers in the brain, e.g., cortex and reticular formation. A decline in inhibition, through growth and maturation, or maybe as a result of systematic training will thus be able to influence strength and power independently of the growth of the muscles themselves.

B. FAST AND SLOW FIBERS

Another fact that has to do with strength is the existence of two or three populations of muscle fibers. It was mentioned earlier that the contraction times—and hence the minimum frequencies for tetanic contractions—varied widely in humans, between 25 and 100 msec. The "slow" and the "fast" muscle fibers are distributed in two populations and seem to be identical with muscle fibers that differ with respect to histochemical composition and probably also to nervous connections. Such differences are well known from several animals, where "white, fast" muscles can be distinguished from "red, slow" muscles. In humans, three types of muscle fibers may be recognized by their affinity to certain dyes. They are usually called type A, B, and C. Type A is relatively poor in mitochondria, rich in glycolytic enzymes, and probably corresponds to the fast, white muscles. Type C is rich in mitochondria and in respiratory enzymes and corresponds to the red, slow muscles. Type B is an intermediate form. These fibers are practically all present at the same time in human muscles, but their relative number varies. Table I (from Buchthal and Schmalbruch, 1970) shows their relative number in some human muscles.

From animal experiments it has further been found that the A fibers are organized in large motor units, i.e., that a large number of fibers are connected to and innervated by the same, relatively large motor neuron in the

TABLE I

RELATIVE NUMBER OF FAST, WHITE (A), INTERMEDIATE (B), AND SLOW, RED (C)
MUSCLE FIBERS IN SOME HUMAN MUSCLES[a]

Muscle	A (%)	B (%)	C (%)
Biceps br.	43	32	25
Triceps br.	50	48	2
Tibialis ant.	19	35	46
Gastrocnemius	10	8	82
Soleus	4	7	89

[a] Data from Buchthal and Schmalbruch (1970).

ventral horns. Type C fibers are combined in small motor units comprising a small number of fibers and innervated from relatively smaller motor neurons in the medulla. (The even smaller gamma-neurons that innervate the intrafusal muscle fibers in the muscle spindles are not considered.) Henneman, *et al.* (1965; Henneman and Olson, 1965) deduced from experiments with reflex innervation of the "motor pool" in the ventral horns that the larger motor neurons had a higher threshold than the smaller ones.

If these observations can be transferred to human muscles *in situ* it will give the following picture of submaximal and maximal contractions. Weak contractions are elicited by weak voluntary stimulation of the motor pool in the ventral horns. The smaller motor neurons with the lowest threshold will respond first and will activate small units of slow, mitochondria-rich C fibers. If the firing frequency surpasses about 50 impulses/sec these fibers will go into tetanic contractions. Stronger and maximum contractions will need stronger innervation of the motor pool, strong enough eventually to surpass the threshold of the larger motor neurons, thus activating the larger motor units consisting of fast, mitochondria-poor A muscle fibers. This will add large steps to the development of force because of the large motor units, but in order to get maximum force the firing frequency probably must exceed about 100 impulses/sec. Is this possible by normal voluntary innervation?

The foregoing considerations have been made in order to make it clear that "strength" is not simply a question of muscles but rather a complicated phenomenon involving both muscles and nervous system. Growth may influence various parts of the neuromuscular system differently, asynchronously, and strength in one situation may therefore be different from strength in another situation because the cooperation of these various parts may be different in the two situations.

C. "Power" and "Strength"

"Power" and "strength" are not independent variables. From experiments on single muscles and muscle fibers it is well known that the velocity with which force can be produced in a contraction depends on the force in a characteristic way: When force is great, velocity becomes low and vice versa (Hill, 1922). This was first believed simply to result from viscosity in the muscles, but Fenn and Marsh (1935) pointed out that the force-velocity relationship was not rectilinear as it should be if only viscosity were involved, but rather it is hyperbolic. On the basis of these and other observations the force-velocity curve was constructed, and Hill (1938) worked out an equation that describes it mathematically.

It is not immediately obvious that because an isolated muscle's or muscle fiber's contractions against different loads can be described by a hyperbolic force-velocity curve, this should also hold for the much more complicated contraction of a whole muscle *in situ*. First, even though the exertion from the point of view of the subject may be considered maximum, it is quite possible that the feedback from the periphery is different in slow and fast contractions, thus modifying the efferent motor outflow. Further, the fact that a large number of motor units are contracting at different frequencies, some maybe in tetani but others in series of single twitches, makes it look as pure luck that the force-velocity curve of human voluntary contractions resemble those from isolated muscles in every respect. Figure 1 shows two force-velocity curves. Similar curves have been found in humans by, e.g., Wilkie (1949) and Kaneko (1970). It appears that as speed of shortening increas-

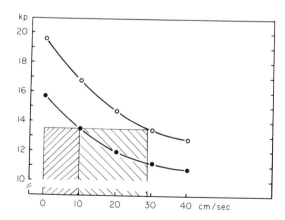

FIG. 1. Force-velocity curves (elbow flexors) before (below) and after (above) 8 weeks of strength training. Average values from 6 girls and 6 boys. The shaded areas represent power (force × velocity) with the load 13.5 kp (unpublished data).

es less and less force can be developed. At a definite velocity it would be zero.

Maximum power is the maximum rate at which work can be produced by the active muscles. For each instant of a movement it will be the product of the maximum force and the corresponding velocity at that instant, because, as mentioned before, force × velocity has the dimension kilopond × meter/time, i.e., power or rate of work. In Fig. 1 this product is shown for a certain force (load) as the shaded areas under the force-velocity curves. Training has increased the maximum force to a higher value, and consequently the initial curve has shifted upward. It can easily be seen that with the same load maximum power has increased considerably (cf. the added area). In practical situations (e.g., in shot putting and jumping) it must, however, be borne in mind that movements are restricted anatomically. A very high power, which for a fixed load of necessity also means a high velocity (cf. Fig. 1), will give a short work period. What actually counts in a throw, a jump, or other "explosive" events is the velocity v of the implement, or the body, at the takeoff; and this is determined, e.g., by the formula: $f \times d = \frac{1}{2} m \times v^2$, in which f is average muscular force developed through the distance d and m is the mass of implement or body. Since d and m are anatomically or technically defined and may be considered constants, it follows that $v \sim f^{0.5}$; i.e., the height or length of a throw or jump will depend on muscular strength.

D. "WIND-UP"

The ability to mobilize maximum force immediately at the onset of a movement is a question of neuromuscular control. It is probably an ability that has to be learned. There are inhibitions to be overcome, and as all other motor skills it presupposes a certain degree of maturation of the central nervous system. But even with this degree fully developed, maximum force lags a little behind in the beginning of a contraction. This is probably a result of the elastic elements in the muscles and tendons. It can be overcome by a "wind-up" movement, i.e., a countermovement, preferably fast, that must be braked by the same muscles that are going to perform the actual movement. By the braking, which takes the form of an excentric muscular contraction, the elastic elements in the muscles will be pre-stretched, and no force will be lost during the first part of the movement. This fact is illustrated by Fig. 2, from a laboratory experiment on the arm-shoulder muscles in a pulling movement.

To summarize, muscular strength and power are physiological functions that depend on several factors: the size of the muscles, their type and *milieu interne;* their connections with the central nervous system, both by efferents

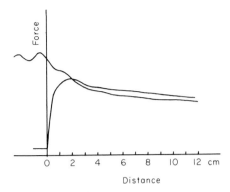

FIG. 2. Maximum force exerted during a horizontal pull on a handle that moves with a velocity of 9 cm/sec. Lower curve: start from relaxed condition. Upper curve: a countermovement (wind-up) precedes the pull (left of vertical), and as a result force × distance = work becomes greater (upublished data).

and by afferents; and the different levels of organization of the motor centers in the brain, maturation and learning. These different factors may follow different patterns during growth, resulting in complex situations that often are difficult to analyze. The following is an attempt at an analysis of some such situations during the period of growth.

III. Isometric Muscular Strength and Growth

As in other investigations of growth, a longitudinal study would have been preferable. Only few such studies are to be found in the literature (e.g., Jones, 1949), and in most cases one will have to make do with cross-sectional studies. However, for general purposes this seems to be justifiable. In unpublished longitudinal studies of children, who had had polio some few years earlier and who subsequently were followed over a period of 5 years, it was found that their strength, with few exceptions, followed the same pattern as that set by a cross-sectional study of normal children (Fig. 3). As material for the following discussion, data collected by the author and his co-workers (Asmussen and Heebøll-Nielsen, 1955, 1956; Asmussen *et al.,* 1959) are mainly used in cross-sectional studies of Danish school children. The very young (below about 7 years of age) are not included for lack of data.

The analysis will take the form of a hypothetical prediction of strength from certain assumptions that will be mentioned and justified below. These predictions will be compared to actual measurements of strength or manifes-

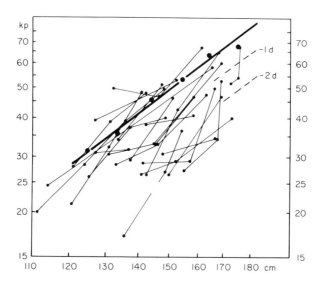

Fig. 3. A longitudinal study of the growth of strength in the back muscles in boys with polio sequelae. Heavy line: normal standards. Thinner lines connect values from same individuals over a period of up to 5 years (unpublished data).

tations of strength, and it will then be attempted to explain the eventual deviations from theory.

A. Preliminary Assumptions

It is assumed that children from the age of about 7 and upward, as well as adults, are geometrically similar. The only changes taking place in growth would then result from changes in size, and these can be predicted if only one parameter is known. The body height h is chosen as this parameter, not least because it is so easy to measure. It will be understood, therefore, that all linear dimensions in the body will be proportional to and vary in proportion to variations in body height. This goes for lengths of extremities, muscles and their lever arms, etc. Further, it follows that all areas in the body varies as h^2, e.g., body surfaces, but also cross-sectional areas such as the muscles' cross-sectional areas. Finally, volumes and weights will vary as h^3, the latter provided body composition is constant.

The justification for this first assumption is that studies of body proportions within the age brackets 7–16 years show that they only change slightly, at least in boys (see Fig. 4). Further, the derived proportionality between height h and weight h^3 in several studies have come fairly close to that predicted [e.g., for boys, weight $\sim h^{2.68}$; for girls, weight $\sim h^{2.88}$ (Asmussen and Heebøll-Nielsen, 1956)].

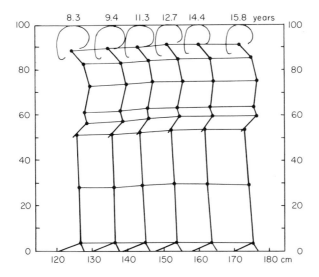

FIG. 4. Body proportions in about 200 boys, arranged in height groups at 10 cm intervals. The points are from below: malleolus lat., epicond. fem. lat., trochanter maj., spina ill. post. sup., deepest point in lordosis lumb., highest point in kyphosis dors, vert. prominens, and meatus acust. ext. Data from Heebøll-Nielsen (1958).

A second, preliminary assumption is that maximum isometric muscle force is proportional to the cross-sectional area of the muscle and hence to h^2. Ikai and Fukunaga (1968) measured isometric strength of the elbow flexors and cross-sectional areas of the same muscles by ultrasonic photog-

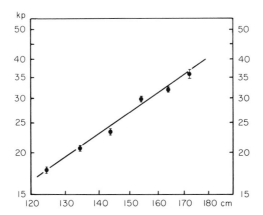

FIG. 5. Muscular strength in a pull downward with right hand vertically over shoulder, elbow flexed 90°. Girls. Logarithmic coordinates. Ordinate: kilopond abscissa: body height. Data from Asmussen et al. (1959).

TABLE II

TABLE II

ACTUAL VALUES OF THE EXPONENT b IN THE PROPORTIONALITY (STRENGTH $\sim h^b$)[a]

Condition	b
Arm pull, downward	2.18
Arm push, downward	2.33
Handgrip	2.80
Back Muscles	2.36
Abdominal muscles	2.23
Hip flexion	2.75
One leg extension	2.73
Both legs extension	3.01

[a] Data from Asmussen *et al.* (1959) based on 300 girls.

raphy and found the maximum force to be 6.3 ± 0.81 kp/cm², independent of age (12–20 years), and nearly the same in males and females.

For a working hypothesis it thus seems justifiable to accept the presumptions that little and big children and adults are geometrically similar and that muscle force varies as height squared.

B. MUSCULAR STRENGTH AND SIZE

Muscular strength was measured as the best results of voluntary isometric (static) contractions applied to a strain gage dynamometer (Asmussen *et al.*, 1959). Standard positions and good fixation were secured. The subjects included 300 boys and 300 girls, ages 6–16 years. They were grouped according to sex and in height classes with 10 cm intervals. The average values were plotted on double logarithmic scales.

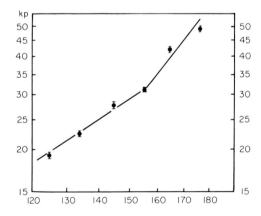

FIG. 6.   As in Fig. 5, but measured on boys. Data from Asmussen *et al.* (1959).

Figure 5 shows one example from girls performing a pull downward on a handle vertically over the shoulder, elbow flexed 90°. The line has been fitted to the points by eye. The slope of the line can be calculated to be 2.18; thus, strength in pulling would be proportional to $h^{2.18}$, i.e., fairly close to the predicted $h^{2.0}$. Table II presents corresponding values for other muscle groups in the upper and lower extremities as well as in the trunk. It is obvious that in most other muscle groups strength increases with height at a greater rate than corresponding to h^2, and in some, e.g., finger flexors and leg muscles, even closer to h^3. Apparently some other factor besides increase in size must influence growth in strength in these girls. What this other factor might be will be discussed later (cf. Section III, D).

C. MUSCULAR STRENGTH AND SEX

All the data collected on girls could—after transcription to logarithms—be fitted to straight lines as seen in Fig. 5. If the data from the boys were treated in the same way, most would result as shown in Fig. 6 (pull downward, arm). It is clear that the data are best fitted to a broken line. The lower part of the line has practically the same slope as that of the girls, although it parallel-shifted a little upward, but the upper part rises at a much steeper rate. The break-off point coincides with body height about 155 cm, and from a height–age curve of these Danish boys it can be read that this corresponds to about 13 years of age. Thus, there seems to be two points on which girls and boys differ with respect to growth in strength. First, boys tend to be stronger than girls of the same height at all ages. This difference is most apparent in the upper extremities and the trunk but seems to be completely absent for the lower heights in the lower extremities. Second, in boys a new factor. that influences muscular strength very much, seems to set in at about age 13 and to persist into adulthood. Its influence is very conspicuous in the upper extremities, less so in trunk and lower extremities, although still discernible.

It seems natural to identify this factor with sexual maturation. Puberty in boys is associated with an increase in production of male sex hormones which influence muscularity. Female adolescence is not accompanied by any noticeable increase in production of androgens, nor by any increase in muscularity. The extra gain in strength in the boys is probably more the result of a quantitative change than of a qualitative change, because, as Ikai and Fukunaga (1968) found, strength per square centimeter is nearly the same for males and females. In adults, the average difference in strength between men and women is about 35% of the men's strength. It will be understood that part of this difference must result from differences in size between the two sexes. But even with this difference eliminated, women will still possess only about 80% of the strength of their male counterparts.

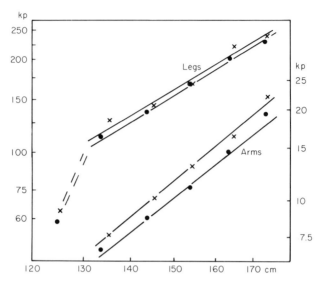

FIG. 7. Muscular strength in hip-knee extensors (right leg) and in arm-shoulder muscles (pull, right arm) in relation to body height. Logarithmic coordinates: × older subgroups and (●) younger subgroups. Average age difference between subgroups about 1.5 year. Data from Asmussen and Heebøll-Nielsen (1956).

D. Muscular Strength and Age per se

It was pointed out earlier that muscle strength increased more rapidly than muscle size during the growth period (Table II). In order to analyze the possible effect of age per se on strength, the boys from each height group were divided into two age groups, an older and a younger. The average difference in years turned out to be approximately 1.5 years. Strength measurements, e.g., in arms and legs, showed (Fig. 7) that at corresponding heights the older subgroups were stronger than the younger. The difference seemed to be a percentage difference since the lines relating log strength to log height turned out to run nearly parallel. Calculated per year the difference in strength amounted to 5–10%.

The reason for this difference may only be speculated on. A plausible explanation may be that in growing older a certain maturation of the centers that control motor activity takes place. Thus the "skill" of performing a maximal isometric contraction improves with chronological age independently of the growth in size. It was further an almost general observation that the data from the very youngest groups investigated tended to fall below the line that was characteristic for the strength–height relationship, although usually much less than seen in Fig. 7 for the leg muscles. As will be seen later, this influence of a possible central nervous maturation was also clearly observable in functional tests involving more complex motor coordination than simple isometric strength testing.

E. "HANDEDNESS" and STRENGTH

Generally speaking, there is a very high degree of correlation ($r = .8$) between the strength of the muscles in the right and the left side of the body, much higher than between different muscle groups in the body, where the correlation between, e.g., handgrip and leg muscle strength may be quite low ($r = .4$).

In about 50 boys, in the height brackets 130–140 cm (age about 10 years) it was found that 55–65% were stronger in their right side extremities but 50–55% were stronger in their left side trunk and hip musculature (Asmussen and Molbech, 1958). The stronger left side trunk muscles are possibly a compensatory consequence of the more extensive use of the stronger right side extremities. The opposite was the case with those stronger in their left side extremities.

When the stronger side muscles, independently of right or left, were compared to the weaker side, the difference was about 5% of the stronger side in the arms but up to 11% in the legs. Girls showed a comparable difference (Heebøll-Nielsen, 1964). When these differences are compared to the differences in young male adults, it is found that the difference between stronger and weaker side becomes greater in the adults, especially in the upper extremity. This increasing asymmetry is doubtless the result of adaptation to everyday usage of certain tools. It is noteworthy that in this investigation there were 20–40% that showed higher isometric strength in their left than in their right side arm and hand muscles, although the normal frequency of functional left-handedness is only 4–8%.

F. MUSCULAR STRENGTH IN HANDICAPPED CHILDREN

Muscular strength in handicapped children will be only briefly dealt with. With physical handicaps, as for instance polio sequelae, strength is naturally decreased in all the muscles that have been seriously attacked. As mentioned earlier, and as shown in Fig. 3, the development with growth of the remaining strength seems to follow the same pattern as in normal children. Serious deformities will curb the possibilities of the existing muscles to exert their normal functions, but no systematic data are available. In mentally handicapped children strength is slightly decreased in relation to height. As an expression for the general mental status I.Q. is the easiest and most commonly used parameter. In an investigation on Danish boys (Asmussen and Heebøll-Nielsen, 1956) we found no correlation between I.Q. and strength in the normal range of I.Q. (90 and upward). Boys with lower I.Q. (70–90) showed lower strength than normals at comparable heights. It must be emphasized that all the boys came from the same geographical and social environment as the standard group.

To summarize the above, it has been found that isometric strength increases with growth at a rate that in some muscle groups come close to the predicted value, i.e., to being proportional to height squared. The deviations found were as follows: In boys, from about age 13, a new factor sets in, which presumably is sexual maturity with changed hormonal levels. In girls, no such break in development of strength is seen. It was further found that the difference in sex, even before puberty, would give the boys a certain lead, at least in arms and trunk. Age per se had an incremental effect on strength, possibly because of maturation of the central nervous system. Low mental ability usually was accompanied by low strength.

Strength is hardly ever used in the way it is in an isometric muscle test. It is used in some function that besides strength demands neuromuscular coordination. The following will be an attempt at analyzing strength under functional conditions.

IV. Functional Muscular Strength

Strength applied to a definite task demands skill. Therefore, as growing older usually also means aquiring skills and ability to learn skills, it cannot be expected that strength used under such conditions would follow the simple dimensional laws. From available data we shall follow the development of strength as applied to two tasks: a vertical, standing high jump and sprint running.

A. THE VERTICAL JUMP

The height a of a vertical jump depends on the velocity v the body can reach at takeoff: $a = v^2/2g$, in which g is the acceleration of gravity. To obtain this velocity the muscles must produce work on the body and implant a certain amount of kinetic energy in it. The work of the muscles will be the average force f they exert on the ground, multiplied by the distance d over which it works, usually from a semisquatting position to toe standing. We get

$$f \times d = \tfrac{1}{2} mv^2$$

in which m is the mass of the body. It follows that $v^2 = 2f \times d/m$, and inserting this in the formula for the vertical jump, one gets $a = f \times d/w$ as $m \times g$ is equal to the weight w of the body.

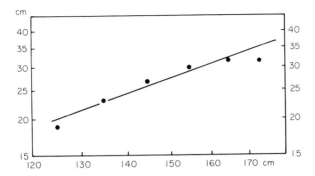

FIG. 8. Height of a standing, vertical jump in relation to body height. Girls. Logarithmic coordinates. Data from Asmussen and Heebøll-Nielsen (1956).

From a dimensional point of view, a, the height of a vertical jump, should be independent of dimensions because $f \simeq h^2$; $d \simeq h$; $w \simeq h^3$; and, consequently, $a \simeq h^3/h^3 = 1$. Actual measurements on 200 girls gave the results shown in Fig. 8, i.e., an increase in jumping height that approximately was proportional to $h^{1.71}$. For boys in the lower height group, the data were identical to those of the girls, but they showed the usual break upward at height about 155 cm (age about 13 years), also in this case best explainable by the advance of sexual maturity. Considering only the girls, in whom this complicating factor is absent, we can rearrange the proportion: $a \simeq f \times d/w$ to give $f \simeq a \times w/d$. By introducing the proportionalities: $a \simeq h^{1.71}$ and $w \simeq h^{2.88}$, both found experimentally, we get $f \simeq h^{1.71} \times f^{2.88-1} = h^{3.59}$. That means that strength of the leg muscles as applied to a vertical jump increases with growth at a much faster rate than could be expected from the increase in size alone $(f \simeq h^2)$, and also faster than the noted increase in isometric force $(f \simeq h^{3.01})$. The reason for this must probably be sought in an increase with the growth in neuromuscular coordination, in skill, and thus in muscle power.

It will be noticed in Fig. 8 that the value for the lowest, i.e., youngest, group lies below the regression line. This shows that skill is less well developed in these 7–8-year-old girls. Also, the two tallest groups tend to show a leveling off, which here in the girls may signify a relatively greater increase in body weight.

B. ACCELERATION

Accelerating the body up to maximum speed in a sprint is another example in which functional strength must play an important part: In a series of "explosive" contractions demanding a high degree of coordination and skill,

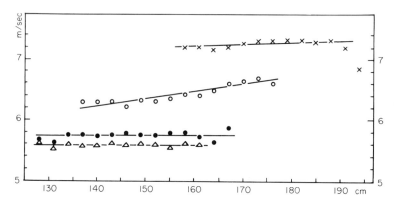

Fig. 9. Speed in running, plotted against body height, for boys of different ages: (×) 18 years, (○) 14 years, (●) 12 years, and (△) 11 years. Asmussen (1957); data from Bach (1955).

the steady state of running is approached. The acceleration during the first 4 m of a sprint was determined in 200 boys and as usual plotted against body height in a log-log system. It was found from the regression that acceleration was proportional to $h^{1.86}$. From the equation $f = m \times a,$ and introducing the found proportionalities mass $\simeq h^{2.68}$ and $a \simeq h^{1.86}$, one gets $f \simeq h^{4.54}$. Again, functional muscle strength (f) increases with growth at a much higher rate than expected. The very fast increase in functional strength for these boys include all the changes resulting from increasing size, age per se, and sexual maturing. Corresponding data for girls are not available.

C. Horizontal Running

In the steady state of horizontal running maximum speed should be independent of height. Roughly, it may be assumed that maximum speed is pro-

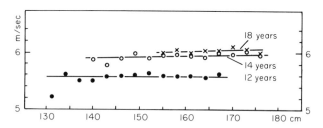

Fig. 10. Same as in Fig. 9, for girls: (×) 18 years, (○) 14 years, and (●) 12 years.

portional to the maximum work of the muscles and inversely proportional to the body weight. This gives the relation: maximum speed $\simeq h^2 \times h/h^3 = 1$. If a large enough number of children of the same sex and age are compared one finds that their height may vary considerably. Such a large number was found in a report published by Bach (1955) containing the results from school competitions in Bavaria. Figures 9 and 10 show the results obtained by grouping together and converting times for sprints to velocities (Asmussen, 1957). Figure 9 shows that (1) with constant age, maximum speed in running is independent of height, as predicted; (2) age per se adds to performance capacity, as exemplified by the difference between the 11- and 12-year-old boys; (3) at 14 years of age the taller boys ran faster than the shorter boys of the same age, probably because of the concurrent impact of sexual maturing and the adolescent "growth spurt"; and (4) at age 18, speed again is independent of height, but at a much higher level for the same heights as before sexual maturity.

Figure 10 shows corresponding results from girls. It can be seen that (1) at the same prepuberty age girls run a little slower than boys; (2) there is the same effect of age per se (neuromuscular maturation?) as in the boys; and (3) the enhancing effect of sexual maturity is lacking. Figures 9 and 10 thus demonstrate the three significant factors that influence functional strength besides anatomical size: age, sex, and sexual maturity, the latter factor only in boys.

Summarizing, the three examples have shown that functional strength increases with growth at a faster rate than would be expected from a purely dimensional growth. The rate of increase is larger than that found in the same muscle groups for isometric strength. It is probably a question of an increasing ability to use the available muscular strength through better coordination. This increase is probably more pronounced in the more difficult to learn tasks (cf. Table III).

TABLE III

LEG MUSCLES (STRENGTH $\sim h^b$) [a]

Test	b
Horizontal running, age eliminated	2.00
Horizontal running, age included	2.08
Isometric strength	3.01
Vertical jump	3.59
Start to sprint	4.54

[a] The exponent b in the proportionality strength $\sim h^b$ derived from measurements of various functions. All data except those in lowest line are from girls.

V. Growth, and Training of Muscular Strength

As discussed in the foregoing, muscular strength increased throughout the growing period, more or less in accordance with general physiological laws. It is, however, possible to accelerate this growth of strength, at least over certain periods, by systematic training. Correspondingly, a decrease in the rate of strength gain must be expected when systematic training is discontinued or in periods of illness. The "trainability" of muscle strength has been investigated by several authors. Rohmert (1968) studied the increase in isometric strength in several muscle groups in 8-year-old boys and girls who were given series of standard training programs consisting of one daily, maximal isometric contraction of either 1 sec duration (training I) or of 6 sec duration (training II). In all except 10% of the training series, the strength increased in the children more with training II than with training I. In parallel studies on adults, there were 42% who did not show any strength increase. The training was continued until a steady state of strength had been achieved *("Grenzkraft")*, different, of course, for the two programs. Rohmert found that when the initial strength was expressed as a percentage of this final strength the children had started at a relatively lower level than the adults. The rate of increase with time (up to 10 weeks), however, followed identical curves in children and adults when the percentage of final strength *(% Grenzkraft)* was used as ordinate. These results may be interpreted as showing that children are naturally adapted to a lower degree of strength utilization in daily life and therefore may be easier to train than adults. This interpretation is supported in experiments by Ikai (1967) who found that another parameter, muscular endurance, as measured by the number of contractions of one-third maximal strength performed at 1 sec intervals, could be increased considerably more in children, especially in the age brackets 12–15 years, than in adults. This poses a very important question: Is there a certain age—e.g., the years of the adolescent growth spurt —when children are especially susceptible to physical training? Endurance may depend to a higher degree on vascularization and blood flow than on muscular strength. Ikai and co-workers (Kagaya and Ikai, 1970) have shown that maximum blood flow increases very much by training of 13-year-old children, about four times more than in adults. Endurance for maximal contractions as those studied by Ikai and co-workers is especially a feature of the slow, mitochondria-rich muscle fibers. But if these have a period during the growing age, when they are especially sensitive to systematic training, it seems justifiable to assume that also the fast, mitochondria-poor fibers on which maximum force depends will be especially susceptible to the stimuli of training during the same period of life.

This emphasizes the very great importance of physical activity and systematic training in childhood and adolescence.

References

Asmussen, E. (1957). *Tidskr. Legemsøvelser* No. 1.
Asmussen, E., and Heebøll-Nielsen, K. (1955). *J. Appl. Physiol.* **7**, 593.
Asmussen, E., and Heebøll-Nielsen, K. (1956). *J. Physiol.* **8**, 371.
Asmussen, E., and Molbech, S. M. (1958). *Comm. Dan. Nat. Ass. Infant. Paral.* No. 2.
Asmussen, E., Heebøll-Nielsen, K., and Molbech, S. (1959). *Comm. Dan Nat. Ass. Infant. Paral.* No. 5, suppl.
Bach, F. (1955). *In* "Ergebnisse von Massenuntersuchungen über sportliche Leistungsfähigkeit." Limpert-Verlag, Frankfurt am Main.
Bigland, B., and Lippold, O. C. J. (1954). *J. Physiol. (London)* **125**, 322.
Buchthal, F., and Schmalbruch, H. (1970). *Acta Physiol. Scand.* **79**, 435.
Buller, A. J. (1970). *Endeavour* **29**, 107.
Fenn, W. O., and Marsh, B. S. (1935). *J. Physiol. (London)* **85**, 277.
Heebøll-Nielsen, K. (1958). *Tidskr. Legemsøvelser* No. 2.
Heebøll-Nielsen, K. (1964). *Comm. Dan Nat. Ass. Infant. Paral.* No. 18.
Henneman, E., and Olson, C. B. (1965). *J. Neurophysiol.* **28**, 581.
Henneman, E., Somjen, G., and Carpenter, D. D. (1965). *J. Neurophysiol.* **28**, 560.
Hill, A. V. (1922). *J. Physiol. (London)* **56**, 19.
Hill, A. V. (1938). *Proc. Roy. Soc. Ser., B.* **126**, 136.
Ikai, M. (1967). *ICHPER, 10th Int. Congr., Vancouver, Canada* pp. 29–35.
Ikai, M., and Fukunaga, T. (1968). *Int. Z. Angew. Physiol. Einschl. Arbeitsphysiol.* **26**, 26.
Ikai, M., and Steinhaus, A. H. (1961). *J. Appl. Physiol.* **16**, 157.
Ikai, M., Yabe, K., and Ischii, K. (1967). *Sportartzt Sportmed.* **5**, 197.
Jones, H. E. (1949). "Motor Performance and Growth," pp. 1–181. Univ. of California Press, Berkeley.
Kagaya, A., and Ikai, M. (1970). *Res. J. Physiol. Ed.* **14**, 127.
Kaneko, M. (1970). *Res. J. Physiol. Ed.* **14**, 141.
Merton, P. A. (1954). *J. Physiol. (London)* **123**, 553.
Rohmert, W. (1968). *Int. Z. Angew. Physiol. Einschl. Arbeitsphysiol.* **26**, 363.
Seyffarth, H. (1940). *Skr. Nor. Videnskaps Akad. Oslo, I: Mat.-Naturv. Kla.* No. 4, pp. 1–63.
Wilkie, D. R. (1949). *J. Physiol. (London)* **110**, 249.

CHAPTER

4

Factors Affecting the Working Capacity of Children and Adolescents

Forrest H. Adams

During the past two decades a great amount of information has been accumulated on the effect of exercise on the body in health and in disease. Such information has potential usefulness to physicians, physical educators, physical therapists, epidemiologists, physiologists, and other members of the health profession. For a more detailed discussion on the effects of exercise, the reader is referred to three recent symposia (Adams et al., 1963; Shephard, 1967; Chapman, 1967) and to two excellent monographs (Carlsten and Grimby, 1966; Falls, 1968).

The purpose of this chapter is to summarize what is known about the effects of physical activity on the well being of the growing child. No attempt will be made to relate these findings to physical fitness, since the term "physical fitness" defies exact definition or measurement. Thus, the scores of the many different tests available to measure fitness are not closely correlated (Fowler and Gardner, 1963; Cumming and Keynes, 1967).

I. Physiology of Exercise in Normal Individuals

In order to evaluate exercise properly, it is important to understand the sequence of events that take place in normal individuals during exercise.

A. GENERAL EFFECTS OF EXERCISE

During resting conditions the work of the heart is minimal and the metabolism of the body is near minimal, particularly after prolonged rest under basal conditions. With increasing physical activity the metabolism of the body increases; this requires an increased utilization of oxygen. The increased demands by the body for oxygen are mainly met by an increase in cardiac output (Bevegard et al., 1960; Holmgren et al., 1960a; Åstrand et al., 1964). In the untrained individual, the increase in cardiac output is almost entirely brought about by an increase in heart rate with no change in stroke volume, whereas in the trained individual there is also a small increase in stroke volume.

The regulation of the cardiac output during exercise is complex. It is likely that both neural and humoral factors are important. Thus, its regulation has both rapid and slow components.

B. PREPARATION FOR EXERCISE

Anticipation of and preparation for exercise produces tachycardia and an increase in myocardial contractility. This is very likely the result of inhibition of vagal centers and a generalized discharge of the sympathetic nervous system. The cholinergic sympathetic vasodilator system is promptly activated and produces dilation of the larger resistance vessels in the muscles. Si-

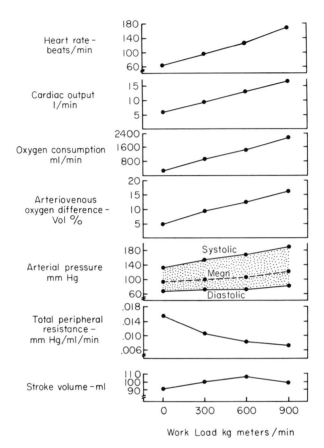

FIG. 1. Effect of increasing work loads on several cardiovascular parameters in 13 healthy men. Data from Grimby and Nilsson (1963).

multaneously, the sympathetic vasoconstrictor fibers produce an increase in vascular resistance in the skin, kidneys, and splanchnic regions, thus diverting blood away from these areas.

C. MODERATE EXERCISE

The cardiovascular events initiated in anticipation of exercise continue to unfold as the individual performs increasing amounts of work. As shown in Fig. 1, the heart rate, cardiac output, oxygen consumption, and arteriovenous oxygen difference all increase in a linear fashion in response to increasing work loads (Grimby and Nilsson, 1963). The arterial blood pressure rises gradually and at the same time there is a drop in the calculated total peripheral vascular resistance. As stated before, the stroke volume changes very little in response to increasing work loads in the average untrained individual. Thus, the increase in cardiac output observed with exercise is accomplished principally by an increase in heart rate.

D. SEVERE EXERCISE

In severe exercise a number of the cardiovascular variables such as heart rate, oxygen consumption, cardiac output, and arteriovenous oxygen difference reach a maximum level and then plateau (Mitchell *et al.*, 1958). If the exercise is continued to exhaustion the situation is commonly referred to as "the working capacity" of that individual (Åstrand, 1952).

As the heart attains its maximal output the demand of the active muscles for oxygen exceeds the supply of oxygen so that anaerobic energy metabolism increases (Hill *et al.*, 1924). This results in a buildup of lactic acid in the bloodstream. Throughout severe exercise ventilation and the arterial oxygen concentration remain normal, indicating that the heart's ability to pump blood is the limiting factor (Linderholm, 1960).

II. The Working Capacity and Its Measurements

Investigators in the field of exercise physiology have tried for many years to develop some satisfactory method for evaluating physical performance. One obvious approach is to consider the working capacity of the individual.

A. MAXIMAL WORKING CAPACITY

The highest work load performed during exercise prior to physical exhaustion has been considered the maximal working capacity. Such a definition of physical performance, however, creates several problems, both potential and real. For instance, it is known that some individuals will stop a work performance test prior to reaching their working capacity. This might be for any one of a number of reasons: disinclination to work harder, discomfort, and poor instructions by the tester. Likewise, some physicians have been unwilling to push a candidate to exhaustion because of the possibility of some cardiac catastrophe, such as cardiac arrhythmia, cardiac failure, and cardiac arrest. These concerns, although real, do not need to eliminate the usefulness of the working capacity test (Bruce *et al.*, 1963; Goldberg *et al.*, 1969).

B. SUBMAXIMAL WORKING CAPACITY

In light of the criticisms of the maximal working capacity tests, Sjöstrand in 1947 reported on the potential usefulness of the submaximal working capacity test. His test requires that the individual perform three consecutive work loads of mild, moderate, and severe degree lasting 6 min each such as to produce in the last period a final heart rate of approximately 170 beats/min. Sjöstrand (1947) has shown that the work load necessary to produce a heart rate of 170 beats/min is associated with near maximal values for cardiac output and oxygen consumption. For most individuals, including children, the submaximal working capacity test does not produce exhaustion or discomfort and does not seem to carry the risk of a cardiac catastrophe.

Values of the submaximal working capacity have been reported for normal Swedish and California children (Åstrand, 1952; Bengtsson, 1956a; Adams *et al.,* 1961a,b). The working capacity was found to increase with age, height, weight, surface area, heart volume, and degree of physical training in the Swedish school children (Adams *et al.,* 1961b). The boys had greater working capacities than girls for the same age, size, and heart volume. Generally, there were no differences in the working capacity between country and city Swedish children and between Swedish and California children of comparable size.

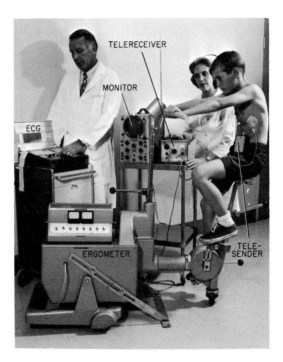

FIG. 2. Equipment and arrangement for evaluating cardiovascular response to increasing work loads on the bicycle ergometer.

TABLE 1

Exercise Work Load According to Surface Area[a]

Duration (min)	Surface area		
	Under 1 m²	1 to 1.2 m²	Over 1.2 m²
2	100	100	100
2	300	400	500
2	500	600	800
2	+100 increments	+100 increments	+200 increments

[a] In kilogram meters/min.

C. Maximal Endurance

Recently, the working capacity test has been modified to combine some features of both the maximal and the submaximal tests (Goldbert *et al.,* 1966a). The individual is asked to perform at increasing work loads on the bicycle ergometer (Fig. 2), each of 2 min duration, until exhaustion.

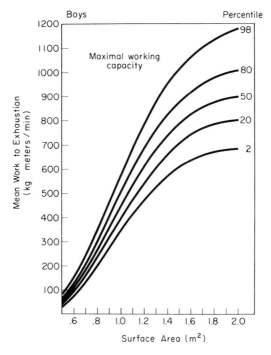

Fig. 3. Relationship of mean work (to exhaustion) to body surface area in 113 normal boys ages 6–17 years.

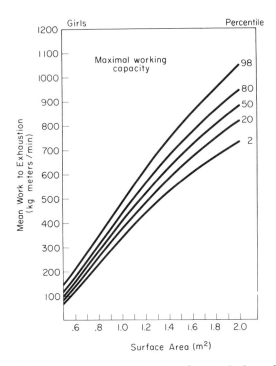

Fɪɢ. 4. Relationship of mean work (to exhaustion) to body surface area in 124 normal girls ages 6–17 years.

The electrocardiogram is recorded during the terminal part of each work load as well as during the recovery period at 30, 60, and 120 sec. The child is constantly encouraged to produce maximal effort. The test is terminated prematurely if the child develops chest pain or electrocardiogram ST depression greater than 2.5 mm over the control value. The total amount of work performed during the entire test is then divided by the total length of time taken to perform this test. This value is called the "mean work to exhaustion." The work load schedule for each individual is predetermined, based upon body size as shown in Table I.

The mean work to exhaustion in the maximal endurance test has been determined for 113 normal boys and 124 normal girls, ages 6–17 years. When these values were plotted against the surface area for each individual tested, a good correlation existed: .88 for boys and .93 for girls. The relationship, including percentiles, of mean work to exhaustion to surface area for both boys and girls is shown in Figs. 3 and 4.

III. Other Methods of Evaluating Exercise Performance

The working capacity test, although used by many individuals, is not totally accepted by all investigators as the best method for evaluating physical performance. It has the advantage that it can be more readily used in large population surveys since it does not require extensive equipment. Many physiologists, however, prefer either the determination of maximal oxygen consumption or maximal cardiac output.

A. MAXIMAL OXYGEN CONSUMPTION

To many individuals the best single measure of overall fitness with severe exercise is aerobic power or the maximal oxygen consumption (Åstrand, 1956). When the test is properly performed it gives a good measure of the function of the heart, lungs, peripheral vascular system, and cellular metabolism. Its main drawback is that it is a very different test to perform outside the research laboratory. The equipment necessary to perform the test is extensive, and there are chances for errors in measuring the oxygen uptake.

Although the oxygen consumption during mild to moderate exercise is linearly related to work load, heart rate, and cardiac output, it is known that during severe exercise the cardiac output levels off before the maximal oxygen consumption is reached (Bates, 1967).

Maximal oxygen consumption is also closely correlated with heart volume, total hemoglobin, lean body weight, and active tissue (Taylor et al., 1963). Active tissue has been defined as lean body mass minus the bone mass. Studies have shown that the maximal oxygen consumption per unit of active tissue in athletes is higher than that in sedentary subjects (Buskirk and Taylor, 1957). This suggests that the oxygen transport and muscular systems of the athlete are qualitatively and quantitatively superior to those of the untrained individual. The higher oxygen consumption of the trained individual is not simply the result of an increase in relative muscle mass but also seems to be related to greater capacity of the heart and lungs to deliver oxygen and of the muscles to utilize oxygen. Good predications of maximal oxygen consumption can also be made with ±12% accuracy from various submaximal, bicycle, treadmill, and step tests (Åstrand and Ryhming, 1954).

B. MAXIMAL CARDIAC OUTPUT

Maximal cardiac output, like maximal oxygen uptake, is a test preferred by many physiologists (Åstrand et al., 1964). Like the latter, it is a difficult test to perform, requiring intubation of the body, and probably is not practical outside of the research laboratory. From a theoretical point of view, many assumptions must be made and met if the conclusions are to be valid. For instance, use of the Fick principle assumes a steady state, which may or may not be present under the conditions of the test. It is now known that a steady state is reached faster at mild rather than at severe work loads (Grimby and Nilsson, 1963). At mild to moderate work loads the cardiac output rises mainly during the first 2 min of exercise, but there is a further small increase between the second and fifth minutes. At severe work loads, the cardiac output does not reach a plateau until between the fifth and seventh minutes.

Other new methods now available for determining cardiac output include indicator dilution techniques and Doppler flow determinations.

C. Pulse Recovery Index

Although no longer commonly used as a method of evaluating exercise performance, the pulse recovery index has been used in longitudinal studies of normal children ranging in age from 7.5 to 20 years (Sexton, 1963). In this test the heart rate is determined at rest, during 4 min of exercise, and then during a 5-min recovery period. A complicated scoring procedure is used from which normal standards have been developed for both boys and girls (Sexton, 1963).

From a theoretical point of view the test has limited usefulness. Only one work load is employed, and this of necessity must produce varying degrees of severity of exercise for children of the same size. Furthermore, the calculation of the index makes use of the resting pulse which is known to vary considerably depending upon many factors, including emotional status.

D. Electrocardiographic Changes during Exercise

The effect of exercise on the electrocardiogram in the adult has been extensively studied. The most common test employed is that described by Masters and Oppenheimer (1923). For the child and adolescent, this test has little application since its major use has been to detect coronary disease which rarely occurs during childhood. Some studies on electrocardiographic changes during exercise have been reported in children (Bengtsson, 1956b; Goldberg, 1969), but much more needs to be learned about this age group. At the present time the test is poorly standardized, including lead placement, and the results are difficult to interpret. Interestingly, the amplitude of the T-wave in lead CR6 always increases in normal children during exercise and reaches is peak during maximal exercise or in the first minutes thereafter. Alterations in this normal response pattern are usually seen in children with aortic stenosis and aortic insufficiency (Goldberg, 1969).

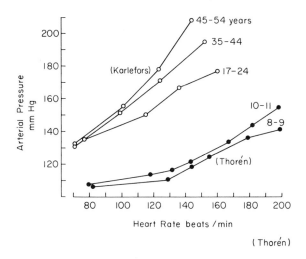

Fig. 5. Relationship of systolic arterial blood pressure to heart rate during exercise according to age.

E. Blood Pressure Response during Exercise

As already stated, in adults the arterial blood pressure rises gradually and the total peripheral vascular resistance decreases with increasing work loads (Grimby and Nilsson, 1963). In children the same phenomenon occurs, as shown in Fig. 5, but to a lesser degree, and there are definite age differences (Thorén, 1968). Such information is particularly helpful in evaluating the blood pressure response in patients suspected of having cardiovascular disease such as coarctation of the aorta.

IV. Factors Affecting the Working Capacity

A. Body Size

Of all the normal variables that affect the working capacity, the most important relate to body size (Robinson, 1938; Åstrand, 1952; Adams *et al.,* 1961a) such as weight, height, surface area, lean body weight, and active tissue (Buskirk and Taylor, 1957). As can be seen from Figs. 3 and 4, the mean increase is approximately eightfold in the working capacity in children ages 6–12 years. For some, the increase is considerably greater, and for others it is less. Since children of the same age vary considerably in size, particularly during adolescence, it is important to consider body size and not age in evaluating the working capacity.

B. Sex

Except for the very young child, differences in the working capacity are known to exist between males and females (Åstrand, 1952; Adams *et al.,* 1961a,b). This is also illustrated in Figs. 3 and 4.

The sex differences in working capacity are also associated with differences in maximal oxygen consumption, maximal cardiac output, and pulse recovery index, as might be expected because of their close relationship to each other. This is in spite of the fact that the maximal heart rate response is the same for both sexes (Goldberg *et al.,* 1966a). The explanation for these differences is currently unknown.

C. Heart Volume, Stroke Volume, and Total Hemoglobin

In normal individuals heart volume, stroke volume, and total circulatory hemoglobin all correlate directly with the working capacity (Kjellberg *et al.,* 1949). Since all three of these parameters increase with growth during childhood, such correlations seem reasonable. On the other hand, it is also known that all three of these variables increase somewhat with training (Holmgren *et al.,* 1960b).

D. Environment

Although no specific studies have been performed in children, it is known that environmental conditions can affect the working capacity in adults. Factors believed to be of importance are temperature, humidity, and oxygen tension of inspired gas (Carlsten and Grimby, 1966). Thus, it is apparent that environmental factors need to be considered in evaluating the working capacity when comparisons are to be made with other individuals or groups of individuals.

E. Training

The athlete or the well-trained individual has a much greater working capacity than the untrained individual (Bevegård *et al.,* 1963). For instance, very well-trained girl swimmers 12–16 years of age have aerobic capacities up to 3.8 liters/min (Åstrand *et al.,* 1963) as compared to 2.6 liters/min in average girls of the same age (Åstrand, 1952).

As stated earlier, this greater ability to perform work is associated with a large heart volume, stroke volume, total hemoglobin, and a qualitatively better oxygen transport and utilization system (Buskirk and Taylor, 1957).

The effect of training can be studied by sequentially following the various circulatory parameters. It is known that fairly short training periods of 1 hr three times a week will cause the heart rate to become lower both at rest and during exercise (Holmgren *et al.,* 1960b). If the training is continued for 3 months there is a parallel increase in heart volume, stroke volume, blood volume, and total hemoglobin.

In children, the effect of training is transitory. If the training is discontinued, the work performance returns to the original level (Sexton, 1963). In one study there was no effect of training on the working capacity (Cumming *et al.,* 1969).

F. DISEASE

A number of diseases can reduce the working capacity of an individual, the most notable being cardiovascular disease, pulmonary disease, neuromuscular disease, and infectious disease.

1. Cardiovascular Disease

In children the most common form of heart disease in the developed countries is congenital. In those who survive the infancy period, many have normal or near normal exercise performance (Duffie and Adams, 1963; Kramer and Lurie, 1964). Thus, for many children with congenital heart disease restricted physical activity is not warranted (Adams and Moss, 1969).

Marked reduction in physical working capacity has been observed in all forms of cyanotic congenital heart disease and in those noncyanotic individuals with large left-to-right shunts or severe valvular disease (Duffie and Adams, 1963; Goldberg *et al.,* 1966a, 1969). It would appear that the maximal endurance test separates the child with heart disease from his normal peers better than does the submaximal working capacity test (Goldberg *et al.,* 1966a). The favorable effect of surgical correction of the congenital heart lesion on the working capacity has also been demonstrated (Goldberg *et al.,* 1967).

2. Pulmonary Disease

Although the lung is seldom the limiting factor in providing oxygen to the muscles during exercises, occasionally patients are seen where this is the case. Children with cystic fibrosis or severe asthma have a reduced physical working capacity.

3. Neuromuscular Disease

Since coordination is important in the performance of physical activity, neuromuscular disorders producing incoordination also cause a reduction of working capacity. Although not strictly a disease of the neuromuscular system, vasoregulatory asthenia is associated with a marked reduction in physical working capacity (Holmgren *et al.*, 1957a,b). Such patients complain of vague symptoms referable to the heart, including fatigue, without any signs of heart disease. It is believed that during exercise these patients have a faulty distribution of peripheral blood flow to the skin rather than to the working muscles. Physical training seems to improve the capacity in many with this condition.

4. Infectious Disease

The effect of infectious diseases on the working capacity of individuals is variable. It is important to recognize, however, that such infections are capable of producing a reduction in performance (Bengtsson, 1956c).

V. Techniques and Equipment Used in Studying Working Capacity

A. Natural Conditions

For some questions the best answer can be obtained by a direct approach to the problem. This may be difficult, but not necessarily so. For instance, it is possible to determine the heart rate and oxygen consumption under various natural conditions of exercise such as running, swimming, and heavy work. The heart rate and electrocardiogram can be telemetered to a central station where it can be recorded (Goldberg *et al.*, 1966b).

Using this type of arrangement, the response of 21 healthy children 11–16 years of age to various forms of exercise was compared to the maximal working capacity as performed on the bicycle ergometer (Goldberg *et al.*, 1966b). The results are shown in Tables II and III and a typical indi-

TABLE II

RESPONSE TO EXERCISE IN ELEVEN NORMAL BOYS

Exercise	Heart rate	Range	% MWC[a]	Range
Half-mile walk	120	115–130	25	15–32
Stairs	138	130–150	43	26–57
Baseball	143	130–160	48	32–77
Calisthenics	140	135–155	54	47–68
170 HR (WC)[a]	170	—	73	56–85
400 yard run	190	178–200	94	90–100
Basketball	198	180–210	99	92–100

[a] Abbreviations: WC, submaximal working capacity; MWC, maximal working capacity; and HR, heart rate.

TABLE III

RESPONSE TO EXERCISE IN TEN NORMAL GIRLS

Exercise	Heart rate	Range	% MWC[a]	Range
Half-mile walk	127	105–160	27	21–43
Volleyball	129	115–150	28	14–72
Stairs	139	120–155	40	25–51
Calisthenics	163	145–180	64	50–81
170 HR (WC)[a]	170	—	70	62–83
Folk dance	192	180–200	92	90–100
200 yard run	195	180–210	95	89–100

[a] Abbreviations: WC, submaximal working capacity; MWC, maximal working capacity; and HR, heart rate.

vidual response is shown in Fig. 6. With appropriate instrumentation, oxygen consumption can also be determined while swimming (Åstrand et al., 1963) and performing other forms of exercise (Åstrand and Saltin, 1961).

B. SIMULATED CONDITIONS

Three techniques have received wide usage in evaluating the response to exercise: the step test, the treadmill, and bicycle ergometer.

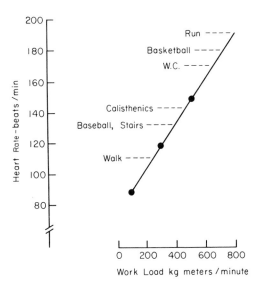

FIG. 6. Relationship of heart rate to known work loads, working capacity (WC), and certain common forms of exercise in a normal boy.

1. Step Test

The step test is the cheapest to use but also has several disadvantages. It does not provide exhaustive work performance for the trained individual, and it does not permit easy calibration of the amount of work performed.

2. Treadmill

The treadmill is widely used in many research laboratories. It has the advantage that it is easy to increase the work performance to exhaustion by either increasing the angle or the speed of the treadmill, or both. It has the disadvantages of being large and heavy for mobile studies, and it is also difficult to quantitate the work load.

3. Bicycle Ergometer

The bicycle ergometer is probably the most widely used instrument for evaluating exercise performance. It is not excessively expensive. It is compact and therefore easily movable. The work load is easily calibrated.

It has the disadvantage that it generally cannot be used by children under 6 years of age. Furthermore, there are some individuals who find it difficult for mechanical reasons to exercise on the bicycle. It has also been shown that the maximal heart rate, oxygen consumption, arteriovenous oxygen difference, cardiac output, and stroke volume for the same individul is generally higher on the treadmill than on the bicycle ergometer (Hermansen *et al.,* 1970). This may be because larger groups of muscles are required to exercise on the treadmill than on the bicycle.

VI. Summary

The physiological factors affecting the physical performance of children have been reviewed. The most important of these are body size, sex, and the presence or absence of disease. Various work performance tests have been discussed including their advantages and disadvantages. The maximal endurance test is recommended as an excellent technique for evaluating the working capacity of both normal children and those with disease.

Training appears to have little effect on the working capacity of normal children. The favorable effect of training is more evident in the individual who is already an athlete. On the other hand, there is no physiological evidence to indicate that severe training has any harmful effect on the body.

References

Adams, F. H., and Moss, A. J. (1969). *Amer. J. Cardiol.* **24,** 605.
Adams, F. H., Linde, L. M., and Miyake, H. (1961a). *Pediatrics* **28, 55.**

Adams, F. H., Bengtsson, E., Berven, H., and Wegelius, C. (1961b). *Pediatrics* **28**, 243.

Adams, F. H., Linde, L. M., Hall, V. E., and Fowler, W. M. (1963). *Pediatrics* **32**, 653.

Åstrand, P.-O. (1952). "Experimental Studies of Physical Working Capacity in Relation to Sex and Age." Munksgaard, Copenhagen.

Åstrand, P.-O. (1956). *Physiol. Rev.* **36**, 307.

Åstrand, P.-O., and Ryhming, I. (1954). *J. Appl. Physiol.* **7**, 218.

Åstrand, P.-O., and Saltin, B. (1961). *J. Appl. Physiol.* **16**, 977.

Åstrand, P.-O., Engström, L., Eriksson, B., Karlberg, P., Nylander, I., Saltin, B., and Thorén, C. (1963). *Acta Paediat., (Stockholm) Suppl.* **147**.

Åstrand, P.-O., Cuddy, T. E., Saltin, B., and Stenberg, J. (1964). *J. Appl. Physiol.* **19**, 268.

Bates, D. V. (1967). *Can. Med. Ass. J.* **96**, 704.

Bengtsson, E. (1956a). *Acta Med. Scand.* **154**, 91.

Bengtsson, E. (1956b). *Acta Med. Scand.* **154**, 225.

Bengtsson, E. (1956c). *Acta Med. Scand.* **154**, 359.

Bevegård, S., Holmgren, A., and Jonsson, B. (1960). *Acta Physiol. Scand.* **49**, 279.

Bevegård, S., Holmgren, A., and Jonsson, B. (1963). *Acta Physiol. Scand.* **57**, 26.

Bruce, R. A., Blackman, J. R., Jones, J. W., and Strait, G. (1963). *Pediatrics* **32**, 742.

Buskirk, E., and Taylor, H. L. (1957). *J. Appl. Physiol.* **11**, 72.

Carlsten, A., and Grimby, G. (1966). "The Circulatory Response to Muscular Exercise in Man." Thomas, Springfield, Illinois.

Chapman, C. B. (1967). *Circ. Res.* **20**, Suppl. 1, 1.

Cumming, G. R., and Keynes, R. (1967). *Can. Med. Ass. J.* **96**, 1262.

Cumming, G. R., Goulding, D., and Baggley, G. (1969). *Cana. Med. Ass. J.* **101**, 69.

Duffie, E. R., and Adams, F. H. (1963). *Pediatrics* **32**, 757.

Falls, H. B. (1968). "Exercise Physiology." Academic Press, New York.

Fowler, W. M., and Gardner, M. S. (1963). *Pediatrics* **32**, 778.

Goldberg, S. J. (1968). *In* "Heart Disease in Infants, Children, and Adolescents." (A. J. Moss and F. H. Adams, eds.), Chap. 12, Williams & Wilkins, Baltimore, Md.

Goldberg, S. J. (1969). *Isr. J. Med. Sci.* **5**, 589.

Goldberg, S. J., Weiss, R., and Adams, F. H. (1966a). *J. Pediat.* **69**, 46.

Goldberg, S. J., Weiss, R., Kaplan, E., and Adams, F. H. (1966b). *J. Pediat.* **69**, 56.

Goldberg, S. J., Adams, F. H., and Hurwitz, R. A. (1967). *J. Pediat.* **71**, 192.

Goldberg, S. J., Mendes, F., and Hurwitz, R. A. (1969). *Amer. J. Cardiol.* **23**, 349.

Grimby, G., and Nilsson, N. J. (1963). *Scand. J. Clin. Lab. Invest.* **15**, Suppl. 69, 44.

Hermansen, L., Ekblom, B., and Saltin, B. (1970). *J. Appl. Physiol.* **29**, 82.

Hill, A. V., Long, C. N. H., and Lupton, H. (1924). *Proc. Roy. Soc. Ser. B* **97**, 84.

Holmgren, A., Jonsson, B., Levander, M., Linderholm, H., Sjöstrand, T., and Ström, G. (1957a). *Acta Med. Scand.* **158**, 413.

Holmgren, A., Jonsson, B., Levander, M., Linderholm, H., Mossfeldt, F., Sjöstrand, T., and Ström, G. (1957b). *Acta Med. Scand.* **158**, 437.

Holmgren, A., Jonsson, B., and Sjöstrand, T. (1960a). *Acta Physiol. Scand.* **49**, 343.

Holmgren, A., Mossfeldt, F., Sjöstrand, T., and Ström, G. (1960b). *Acta Physiol. Scand.* **50**, 72.

Kjellberg, S. R., Rudhe, U., and Sjöstrand, T. (1949). *Acta Physiol. Scand.* **19**, 152.

Kramer, J., and Lurie, P. (1964). *Amer. J. Dis. Child.* **108**, 283.

Linderholm, H. (1960). *In* "Clinical Cardiopulmonary Physiology," p. 648. Grune & Stratton, New York.

Masters, A. M., and Oppenheimer, E. T. (1923). *Amer. J. Med. Sci.* **177**, 223.

Mitchell, J. H., Sproule, B. J., and Chapman, C. B. (1958). *J. Clin. Invest.* **37**, 538.

Robinson, S. (1938). "Arbeitsphysiologie" **10**, 251.

Sexton, A. W. (1963). *Pediatrics* **32**, 730.

Shephard, R. J. (1967). *Can. Med. Ass. J.* **96**, 695.

Sjöstrand, T. (1947). *Acta Med. Scand.* **128**, Suppl. 196, 687.

Taylor, H. L., Wang, Y., Rowell, L., and Blomquist, G. (1963). *Pediatrics* **32**, 703.

Thorén, C. (1968). *Z. Aerztl. Fortbild.* **62**, 938.

CHAPTER

5

Body Composition and Exercise during Growth and Development

Jana Parízková

Body composition, that is the proportion of lean, fat-free body mass and depot fat, is one of the most important morphological features characterizing a human organism (Behnke *et al.*, 1953; Keys and Brozek, 1953; Novak, 1963). The relative proportions of these components, while different for males and females through much of the life-span, are dynamically dependent on developmental level and thus are of interest to those concerned with human growth and development. Furthermore, the significant interac-

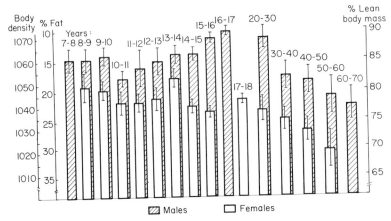

FIG. 1. The development of body composition (percent of lean body mass and depot fat, body density) during growth and aging in males and females.

tion between body composition and energy turnover is, among other things, closely related to the functional capacities of the organism; hence, body composition (Keys and Brozek, 1953; Parízková, 1959) is of considerable consequence in physical fitness evaluation of children and adults (Parízková, 1966; 1968d).

I. Age Changes in Body Composition

The proportion of lean body mass in man of mean, normal weight (Fig. 1) is relatively the highest and the proportion of fat lowest at approximately

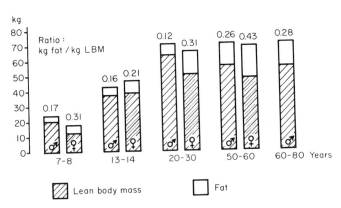

FIG. 2. Changes in absolute amount of lean body mass and depot fat and their ratio (kg fat/kg lean body mass) in males and females of different ages.

20 years of age (Parízková, 1959, 1961a, 1963, 1968d). The ratio of these components (Fig. 2) has an important impact on the economy of work performance, and it corresponds well to age changes in functional aerobic capacity (Åstrand 1967). Thus, the aerobic working capacity in man is high at the younger ages but decreases with the decline in proportion of lean body mass that comes with advancing age.

Results similar to the above have come from research on experimental animals. Observations on young growing rats have invariably shown that these animals have a lower proportion of fat tissue (Fig. 3) which, in addition to being more cellular in nature, contains more deoxyribonucleic acids (DNA) (Tenorová and Hruza, 1962) and is therefore more metabolically active (Altschuler et al., 1962; Parízková, 1966) than in older animals. Thus, the young animal is characterized by an increased ability to release free fatty acids from adipose tissue, both spontaneously and after addition of

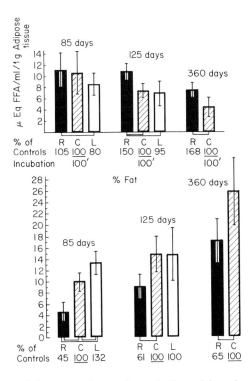

FIG. 3. Changes of fat proportion and metabolic activity of adipose tissue [expressed as free fatty acid (FFA) released *in vitro* after 2 of epinephrine per 1 gm of adipose tissue] in male rats of different ages and different physical activity. Abbreviations: R, running; C, controls; and L, limited physical activity.

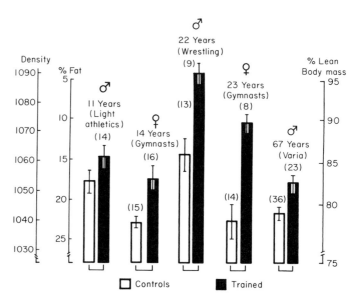

FIG. 4. Differences between trained and untrained control subjects of different ages in body density, fat, and lean body mass proportions.

epinephrine (Parízková and Stanková, 1967; Parízková, 1969). This means, among other things, that at this stage of life there is a more rapid availibility of fat metabolites as a fuel for muscle work (Fritz, 1961). Correspondingly, there is in the young growing organism an increased activity of an enzyme which facilitates the utilization of fat metabolites in the heart and skeletal muscles (lipoprotein lipase). This is characteristically found in young growing organisms (Parízková et al., 1966b; Parízková and Koutecký, 1968).

II. The Impact of Exercise on Body Composition

Systematic physical activity and athletic training can change body composition in a characteristic way (Parízková, 1959a, 1963). Under these conditions the proportion as well as the absolute amount of lean body mass increases significantly at the expense of fat (Parízková and Poupa, 1963a,b; Parízková, 1965). This applies to growing children, as well as adults, and the aged (Fig. 4). For example, trained adolescent boys have a higher lean body mass and also slightly increased creatinine excretion than untrained youth (Novak, 1963, 1966).

Cross-sectional comparisons, however, do not bring into clear focus the changes in body composition and physique which are attributed to intensive muscular exercise. This is more clearly evident when repeated observations are made on the same children over a period of several years. During the growing years this interrelationship may be noted, for children of better physical development and those with advanced skills seem to be more interested in and engage more in sports and systematic exercise than those who are less skilled and less advanced physically.

III. Dynamic Changes in Body Composition and Body Build in Boys with Different Physical Activity Regimens

A group of boys ($n = 96$) subdivided into four subgroups according to physical activity and followed longitudinally from 11 to 15 years showed no

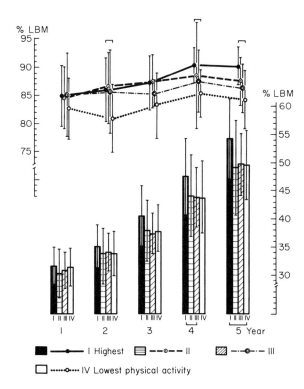

FIG. 5.   Changes in relative (%) and absolute (kg) amounts of lean body mass in groups of boys with different physical activity (I-IV) from 11 to 15 years (1961–1965).

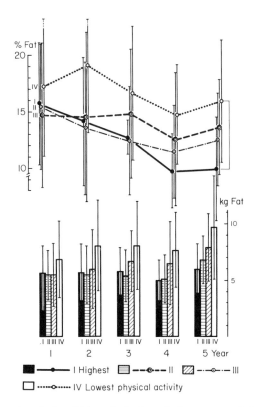

FIG. 6.   Changes in relative (%) and absolute (kg) amounts of depot fat in groups of boys with different physical activity (I–IV) from 11 to 15 years (1961–1965).

significant differences in growth in height and weight. There was, however, a slight tendency for the height of the boys with the most intensive exercise to be greater. Comparison of the results on numerous anthropometric measurements did not reveal differences in chest circumference; sitting height; biacromial, bicristal, and bitrochanteric breadths; and length and circumferential measurements of the extremities during the entire period from 11 to 15 years (Parízková, 1968a). Robusticity of the skeleton, measured as breadth of the wrist and femoral condyles, and skeletal age (which in this instance was identical with chronological age) also did not differ according to the regimen of physical activity (Parízková, 1968a, 1968d; Parízková and Sprynarová, 1968).

Both the percentage of lean body mass and its absolute amount [assessed by densitometry (Keys and Brozek, 1953; Parízková, 1959)] were the

TABLE I

CHANGES IN THE PROPORTION OF LEAN BODY MASS IN GROUPS OF BOYS WITH
DIFFERENT PHYSICAL ACTIVITY[a]

	1961	1962	1963	1964	1965	1966	1967	1968
			% Lean body mass					
	Group with highest physical activity							
x	86.2	86.1	88.6	90.6	91.6	91.5	90.8	92.4
SD	5.6	7.2	5.6	3.8	2.9	4.1	2.9	7.9
	Group with lowest physical activity							
x	83.2	81.9	83.2	85.8	85.1	85.0	88.1	89.4
SD	4.7	6.6	5.6	7.2	4.6	5.5	5.3	6.2

[a] Followed longitudinally from 11 to 18 years.

same in all groups in the first year (Fig. 5). In the second fourth, and fifth years lean body mass was significantly higher in group I (the group with the most intensive training) than in the other three groups. Likewise the proportion of body fat (Fig. 6) in the first year was not significantly different among the four groups. In the last year there was a significantly lower percentage of fat in group I in comparison to that in the other groups. The absolute amount of fat in group I did not change from the first to the fifth year. In the group with lowest intensity of exercise (group IV), the absolute amount of fat was significantly higher in the last year than in the first. A smaller number of these boys ($n = 41$) was followed for 8 years (Parízková and Sprynarová, 1970). Similar differences between groups with highest and lowest intensity of physical activity were still apparent at 18 years of age (Table I), the greatest differences being at the age of 14–15 years when the interest in exercise was most intensive.

Changes in body composition and weight clarify the lack of differences between groups when only girth measurements of the extremities are used. Increased development of the muscle was compensated by loss of subcutaneous fat which was also confirmed by skinfold measurements.

Relationships of selected anthropometric measurements (calculated as relative dimensions characterizing body proportionality and build) vary with the regimen of physical activity, following a pattern similar to that of body composition.

Relative breadth of the pelvis

$$\frac{\text{bicristal breadth} \times 100}{\text{height}} \quad \text{and} \quad \frac{\text{bicristal breadth} \times 100}{\text{biacromial breadth}}$$

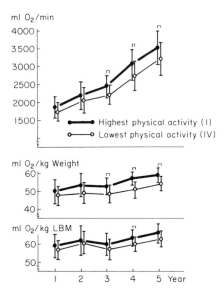

FIG. 7. Changes in absolute and relative (per kg total and lean body weight) values of maximal oxygen uptake in groups of boys with different physical activity (I and IV) from 11 to 15 years (1961–1965).

while similar for the four groups in the first year, were then significantly lower in group I (15.0 and 70.0) in the last year than in group IV (16.0 and 75.5). This indicated a relationship between physical activity, body fat, and body build—the more fat, the relatively broader pelvis and vice versa. Boys with the higher level of physical activity showed a greater increase in lean body mass as well as characteristic changes in body build, i.e., relatively narrower pelvis (Parízková, 1968a). Evidently physical activity and exercise during adolescence is related not only to soft tissue development but also to changes in the proportionality of the skeleton.

IV. Body Composition Changes as Related to Aerobic Capacity in Boys

Measurements of maximal oxygen uptake during graded work loads on a horizontal treadmill (Sprynarová, 1966) have shown that the morphological development under reference parallels the improvements in aerobic capacity and better overall physical fitness (Parízková and Sprynarová 1968, 1970; Sprynarová and Parízková, 1968). Figure 7 shows the maximal oxygen uptake in groups with the greatest and the least physical activity. In the first year when group I had just started training no differences were noted in the maximum oxygen uptake of this group and group IV (Fig. 7). Differences did appear in the third year and persisted until the last year. Maximal oxygen uptake as related to body weight (Fig. 7) was also significantly higher in group I after 2 years, but only at age 14 (i.e., after 3 years of training) was the maximal oxygen uptake per kilogram of lean body mass significantly the greatest in group I (Parízková and Sprynarová, 1968). At this age the physical activity was most intensive in all groups and then declined spontaneously.

Increased functional aerobic capacity as related to lean body mass can be brought about by improved efficiency of the cardiovascular and respiratory systems as has been shown by Buskirk and Taylor in adults (1957). In the study referred to above, oxygen pulse values were likewise found to be highest in group I during the last year of the study. Interestingly, ventilation and heart frequency were not influenced by physical activity (Sprynarová, 1966; Sprynarová and Parízková, 1968).

Body composition and maximal oxygen uptake were found to be the most revealing characteristics of the groups throughout the experiment, together with low pulse frequency during standard work load. Oxygen consumption, ventilation, and carbon dioxide output during standard bicycle ergometer work (60 and 85 W) did not clearly differentiate the groups (Ulbrich, 1966).

The morphological and functional adaptations under reference may become identifiable, however, only after a certain period of latency. With regard to relationships between total body weight and lean body mass (LBM), on the one hand, and aerobic power on the other, maximal oxygen uptake is most closely related to total body weight at the ages of 11 and 12, but at ages 13–15 correlation coefficients are higher with lean body weight (Table II) as in the case of adults (Parízková and Sprynarová, 1970).

V. Body Composition and Step Test Results

Body composition in growing boys has been shown to be related to cardiac responses in exercise. In a study involving Tunisian boys of ages 11 and 12 years, fat correlated negatively with heart rate recovery immediately following exercise and likewise negatively with the step test index. The higher the proportion of fat, the poorer the results in terms of physical fitness as characterized by the step test (Table III). The reverse applied for those with a high percentage of lean body mass (Merhautová and Parízková, 1972).

TABLE II

CORRELATION COEFFICIENTS OF THE RELATIONSHIPS BETWEEN HEIGHT, WEIGHT, AND LBM AND MAXIMAL OXYGEN CONSUMPTION AT REST, STANDARD, AND MAXIMAL WORK LOADS[a]

	Oxygen consumption								
	Work load								
	Maximal			Standard			Rest		
Years	Height (cm)	Weight (kg)	LBM (kg)	Height (cm)	Weight (kg)	LBM (kg)	Height (cm)	Weight (kg)	LBM (kg)
11	0.571	0.661	0.599	0.459	0.722	0.680	0.330	0.376	0.409
12	0.531	0.692	0.680	0.508	0.765	0.610	0.477	0.453	0.468
13	0.663	0.731	0.779	0.555	0.862	0.806	0.504	0.629	0.599
14	0.755	0.804	0.868	0.687	0.861	0.812	0.581	0.615	0.628
15	0.718	0.821	0.850	0.527	0.790	0.721	0.375	0.505	0.555

[a] In boys followed longitudinally from 11 to 15 years ($n = 90$).

TABLE III

CORRELATION COEFFICIENTS OF THE RELATIONSHIP BETWEEN PROPORTION OF DEPOT
FAT AND LEAN BODY MASS AND THE RESULTS OF STEP TEST

Age (years)		Years	
		$(n = 29)$ 11 r	$(n = 100)$ 12 r
% Fat:	Increase in pulse frequency during recovery period	0.547	0.262
	Step-test-index	−0.465	−0.379
% LBM:	Increase in pulse frequency during recovery period	−0.517	−0.256
	Step-test-index	0.502	0.286

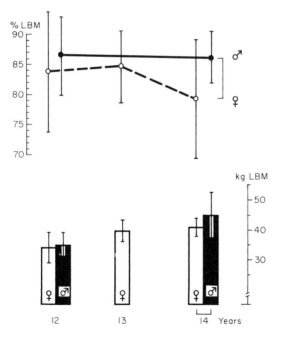

FIG. 8. Changes in relative (%) and absolute (kg) amounts of lean body mass in groups of girls and boys trained in swimming during 3 years (12–14 years of age).

VI. Effects of Physical Conditioning on Body Composition of Girls

Longitudinal observations of body size and body composition were made over a 3-year period by the writer on a group of twelve 12-year-old boys and a similar number of 12-year-old girls who had just begun training for competitive swimming. Mean height of the boys and girls did not differ significantly in the first year of study (girls: 151.2 cm, SD ± 6.3; boys: 150.8 cm, SD ± 6.6) nor were the differences significant the last year at the age of 14 years (girls: 160.6 cm, SD ± 3.4; boys: 165.0 cm, SD ± 7.8). The same applied for body weight (girls: 40.9 kg, SD ± 6.0 and 51.6 kg, SD ± 4.9; boys: 40.3 kg, SD ± 5.6 and 52.2 kg, SD ± 8.2). The relative and absolute amounts of lean body mass did not differ at the age of 12 and 13 years according to sex (in girls it was markedly higher than in normal untrained girls of the same age, see Fig. 1). In the last year lean body mass was significantly less in girls than in boys (Fig. 8) and was the same as in untrained girls of the same age. Apparently the increase in body fat which normally occurs in girls with the development of clinical puberty was not prevented by systematic training in swimming. The (relative dimension)

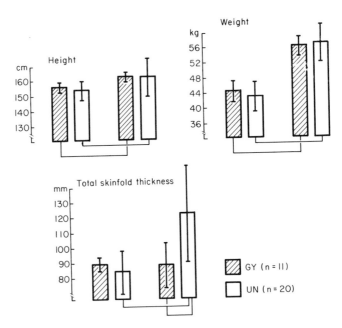

FIG. 9.   Changes of height, weight, and skinfold thickness in groups of trained (gymnasts) and untrained control girls after a 5-year period (12–17 years). Abbreviations: gy, gymnasts and un, untrained. Ten skinfolds were measured (Parízková, 1961b; see Fig. 73).

$$\frac{\text{bicristal breadth} \times 100}{\text{height}}$$

was significantly higher in girls than in boys both at 12 (girls, 15.8; boys, 15.3) and 14 years of age (girls, 16.5; boys, 15.8). On the other hand, the (relative dimension)

$$\frac{\text{bicristal breadth} \times 100}{\text{biacromial breadth}}$$

differentiated the sexes only at the age of 14 years (girls, 75.3; boys, 72.2).

In a similar study carried on over a 5-year period, annual measurements of height, weight, and skinfold thickness were secured on 10 girl gymnasts and a group of 7 untrained girls. At the end of the 5-year period (12–17 years) both groups achieved the same mean height and weight. The gymnasts however had less subcutaneous fat at the end of the 5-year period (Fig. 9).

VII. The Reversibility of Body Composition

The adaptive processes in man are plastic and flexible and at times reversible. The question of how long the exercise-induced increase of lean

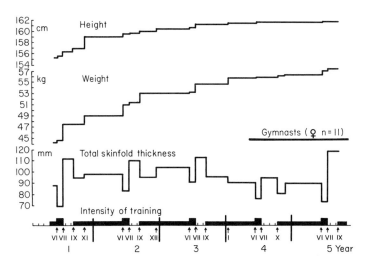

FIG. 10. Changes in height, weight, and skinfold thickness in girl gymnasts during periods of various intensity of training (summer camps with very intensive training, VI–VII; interruption of training, VII–IX; normal training during school year, IX–VI) followed 5 years longitudinally (12–17 years).

body mass at the expense of fat will persist without exercise is interesting to contemplate. This was studied using a group of 12-year-old girl gymnasts ($n = 11$) who trained 2–3 times a week for 1½–2 hr throughout the school year. The training was continued at even greater intensity during the summer vacations but was subsequently discontinued for a period of several weeks. Hence, it was possible to secure measurements on girls when they were under the influence of intensive physical activity as well as during periods of relative inactivity.

The height and the weight of the girls increased during the 5-year period as expected. Body fat estimated from the skinfold measurements (Parízková, 1961b) decreased with training and increased with inactivity (Fig. 10). Body weight did not prove to be a significant indicator of body composition (Parízková, 1963).

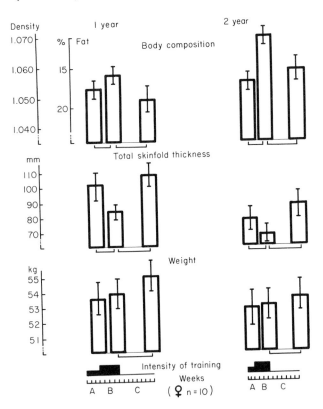

FIG. 11. Changes in body composition, skinfold thickness, and total body weight in a group of girl gymnasts during periods of various intensity of training (see Fig. 10) followed 2 years longitudinally.

The results obtained with skinfold measurements (Parízková, 1959; Parízková, 1963) were verified by body density measurements. A selected sample of 10 girls was studied by both methods (Fig. 11). The changes in body density paralleled the changes in skinfold thickness (Parízková, 1963; Parízková and Poupa, 1963). Although the girls gained in weight during the training, the weight increments involved increases in lean body mass with decreases in body fat. During the periods of inactivity there were notable increases in body fat. In some girls the increases in lean body mass accompanying intensive training were therefore greater than anticipated on account of body weight increments.

The above findings also indicate that discontinuation of intensive training is followed by a rise in body weight resulting from an increase in body fat. In healthy normal girls such a weight increase may reflect imbalance between caloric intake and output. Such imbalances also seem to explain the increased deposition of fat during periods of decreased physical activity in individuals

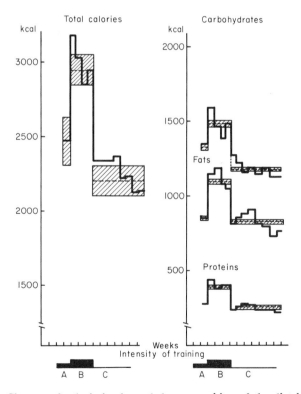

FIG. 12. Changes of caloric intake and the composition of the diet in a group of girl gymnasts during periods of various intensity of training (16 years).

whose bodies have become adapted to a high energy output. When such an individual maintains a high caloric intake while the exercise program is curtailed fat is invariably deposited.

VIII. Caloric Intake as Related to Body Composition during Training

Food intake estimated from questionnaires in the above-mentioned group of girl gymnasts in the period of varying intensity of training (Fig. 12) showed that during normal training (period A) the caloric intake amounted to about 2460 kcal/day, the ratio of proteins to carbohydrates to fat being 11:54:35. During the time of most intensive training when there was the maximum increase in lean body mass (period B) the intake of calories increased initially, then slightly decreased (Fig. 12), the ratio of proteins to carbohydrates to fat being 13:50:37. During the period of relative inactivity (period C) the calorie intake levels fell significantly to 2100–2300 kcal/day, a value somewhat lower than in period A. Between periods B and C the caloric intake values decreased significantly (by 25%). Protein intake in period C was reduced by 36%, that of fat by 25%, and that of carbohydrate by 20%. The ratio of proteins, carbohydrates, and fat was practically the same in periods A and C. During the period of most intensive training, period B, the girls consumed relatively more carbohydrates (significant rise of 2.7%) and less proteins (significant decrease of 2.1%), while the relative fat intake remained unchanged (Fig. 12).

It appears therefore that during periods of reduced activity the caloric intake accompanying intensive training does not persist and cannot be the main cause of fat increase. In fact, it decreased in our sample of girls to levels lower than that found prior to the intensive training (Parízková and Poupa, 1963a,b). Yet in light of such dietary changes body fat increased during the period of inactivity, which followed the period of intensive physical activity. This did not seem to be the result of a disturbance of the food intake regulating mechanisms. Adaptation to intensive muscular work in both man and animal no doubt evokes metabolic changes, among them an increase in utilization of fatty acids in the muscle (Drummond and Black, 1960; Fritz, 1961; Carlsson, 1967; Ekelund, 1969). Inactivity carries with it reduced utilization of fatty acids in muscles and slower release of these products from fat depots. This results in a greater accumulation of fat. The reverse apparently operates as a result of periods of increased physical activity.

IX. Metabolic Changes in Fat Tissue as a Result of Exercise

Adaptation to increased physical activity obviously has a profound impact on metabolic processes in the tissue which finally results in spectacular changes in depot fat and lean body mass proportions. Intensive muscular exercise and adaptation to it brings about a decrease in body fat to the values found at younger ontogenetical stages (Figs. 1 and 3; Parízková and Stanková, 1967: Parízková, 1968d). Moreover, selected metabolic characteristics change in the adipose tissue, which are in proportion to the intensity of the muscular activity. For example, the ability to release free fatty acids (FFA) from adipose tissue, both spontaneously and after administration of epinephrine, is increased in exercised animals (Fig. 3: Parízková and Stanková, 1964, 1967; Parízková, 1966, 1968d). Increased metabolic activity of fat tissue in exercised tissue runs parallel with its higher DNA content, i.e., higher cellularity (Parízková et al., 1966a) as well as decreased proportion of fat tissue (Fig. 3). This is strikingly similar to what is observed in earlier periods of life (Tenorová and Hruza, 1962). Increased metabolic activity of adipose tissue in a trained organism means a greater ability to pour FFA into the bloodstream as a fuel for muscle work as a result of epinephrine output which is liberated in higher amounts during muscle work. These findings correspond to the data of Issekutz et al. (1965) who found greater utilization of labeled FFA in trained dogs undergoing work than in untrained controls. In heart and skeletal muscles of animals adapted to daily exercise on a treadmill there was also greater activity of lipoprotein lipase (Parízková et al., 1966b; Parízková and Koutecky, 1968; Parízková, 1969).

Physical inactivity results in changes of an opposite character. Adipose tissue is metabolically less active, i.e., less FFA is released after the same stimulus; hence fat constitutes an ever-increasing part of the body. Inactivity has a relatively more profound effect on body composition during the period of growth and development when spontaneous activity tends to be high (Parízková, 1968c.d). Also, lipoprotein lipase in the muscles of animals which have been immobilized during early periods of development is significantly reduced (Parízková et al., 1966b; Parízková and Koutecky, 1968).

X. Body Composition and Aerobic Capacity in Obese Children

The incidence of child obesity is increasing in most industrialized countries. Abnormalities in endocrine activity determined by routine clinical and

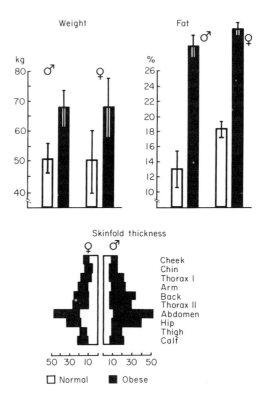

FIG. 13. Differences in weight, fat proportion, and skinfold thickness between normal and obese boys.

laboratory examinations are indeed rare. The main cause of obesity is a by-product of the way of life in most mechanized countries. Even when participating in some exercise or game an obese child proves less active than a normal one (Mayer, 1965).

For an accurate diagnosis of incipient childhood obesity the measurement of body composition is recommended (Parízková and Vamberová, 1967, 1969; Parízková, 1968c). With respect to somatic characteristics, the increased body weight in obese children is not only the result of enhanced deposition of fat [displaced in a manner reminiscent of a woman with enlarged fat pads over the hips and abdomen (Fig. 13) (Parízková, 1959,b 1970)] but also of lean body mass (Forbes, 1964; Parízková and Vamberová, 1967, 1969; Parízková, 1970) as determined by densitometry (Table IV). Bicristal breadth has been shown to be significantly greater in obese than in normal boys, the increase in pelvic breadth reflecting the

TABLE IV

ANTHROPOMETRIC DIMENSIONS AND BODY COMPOSITION IN NORMAL AND OBESE BOYS AND GIRLS (13–14 YEARS)

Condition	Height (cm)	Weight (kg)	Fat (%)	LBM (kg)	Length (cm)	Biacromial (cm)	Bicristal (cm)	Bitrochanteric (cm)
Boys								
Normal (n = 28)								
Mean	161.8	50.4	12.5	43.9	100.2	34.4	22.8	29.5
SD	6.3	6.7	5.9	5.1	4.2	1.6	1.8	2.4
Obese (n = 12)								
Mean	161.2	68.9	29.5	48.6	99.9	33.8	27.7	30.5
SD	2.1	6.9	3.2	5.0	3.1	1.4	1.0	1.0
Girls								
Normal (n = 25)								
Mean	156.9	50.4	18.1	40.7	—	34.6	26.8	—
SD	5.7	10.7	6.0	4.9	—	2.6	2.3	—
Obese (n = 9)								
Mean	157.5	68.9	31.9	46.7	95.8	33.0	27.4	31.5
SD	4.0	10.2	3.7	6.4	3.1	2.0	1.1	2.1

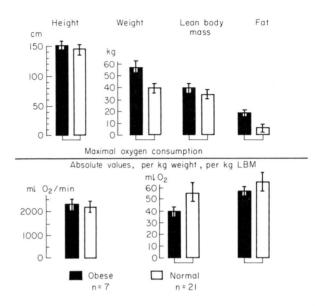

Fig. 14. Differences in height, weight, body composition, and absolute as well as relative (per kg of total and lean body weight) values of maximal oxygen consumption between normal and obese boys.

higher percentage of total body fat. [This finding cannot be explained by fat deposition over hips since the fat layers in boys measured as skinfolds (obese 29.1 ± 8.4, normal 7.0 ± 5.5 mm) do not correspond to the absolute differences in pelvic breadth.] This difference is not found between obese and normal girls (Parízková, 1970). Yet the onset of puberty was normal in all these children, and no other endocrine abnormality was found (Parízková *et al.,* 1961a,b; Parízková and Vamberová, 1967, 1969).

The absolute value of maximal oxygen consumption during graded work loads on a horizontal treadmill (Sprynarová and Parízková, 1965; Sprynarova, 1966) was found to be the same in a group of obese boys (*n* = 7, mean age = 11.7 years) in comparison with a group of normal boys of the same age (*n* = 21). Nevertheless, the maximal oxygen uptake in the obese boys was achieved in a shorter period of time on the treadmill at a lower speed than in the boys of normal weight (Parízková and Sprynarová, 1970). Maximal oxygen uptake as related to total weight and to lean body weight was significantly lower in the obese (Fig. 14). Excess fat obviously has an important negative impact on physical efficiency.

TABLE V

CHANGES IN WEIGHT, LEAN BODY MASS PROPORTION, OXYGEN CONSUMPTION, VENTILATION, AND VITAL CAPACITY[a]

	Boys ($n = 18$)				Girls ($n = 15$)			
	Before reduction		After reduction		Before reduction		After reduction	
	Mean	SE	Mean	SE	Mean	SE	Mean	SE
Weight in kg	65.75	3.26	58.52	2.82	70.33	2.85	63.13	2.35
% LBM	68.6	0.97	74.8	1.14	68.1	0.95	72.9	1.01
ml O_2/1 min	770	30	570	30	730	50	620	80
ml O_2/1 min/1 kg weight	11.6	0.2	9.7	0.4	10.2	2.0	9.7	2.0
ml O_2/1 min/1 kg LBM	17.2	0.3	12.9	0.06	15.4	1.0	13.3	0.2
Ventilation/1 min absolute in 1	15.74	0.59	13.26	0.60	15.48	0.65	15.0	0.77
Ventilation/1 kg LBM	0.34	0.01	0.29	0.00	0.34	0.02	0.32	0.01
Vital capacity	2942	105	3073	115	2640	124.8	2838	133.8

[a] In obese boys and girls (12.6 years) before and after reducing treatment in a summer camp using increased physical exercise.

XI. Changes in Body Composition and Aerobic Capacity in the Obese after Exercise Therapy

Since physical inactivity is believed to be one of the most important causes of childhood obesity, a complex therapeutic program consisting of exercise, directed diet of 1700 kcal/day, and an instructional program has been employed in the treatment of obese children in special summer camps (Parízková *et al.*, 1961a,b; Mayer, 1965; Parízková and Vamberová, 1967, 1969; Vamberová, 1958). In one such program (Parízková *et al.*, 1961a,b) children lost approximately 10% of their initial weight after a period of 7–8 weeks. Body composition and aerobic capacity were tested prior to and after the camp experience. The proportion of lean body mass increased significantly during this period (Table V) and body weight decreased by reduction of excess fat. Repeated measurements of oxygen uptake at a standard work load (one-sixth of the maximal work load calculated for the initial weight) (Janda, 1963) taken on a bicycle ergometer before and after weight reduction showed a significant decrease in oxygen up-

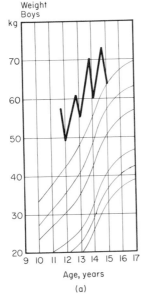

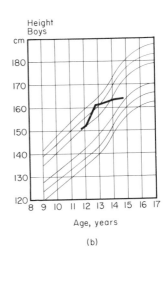

Fig. 15. Changes over a 4-year period in (a) weight and (g) height of obese boys (presented on the background of the growth grid of Tanner *et al.*, 1965, 1966) before and after treatment in special summer camps (reducing regimen by increased exercise) (↔).

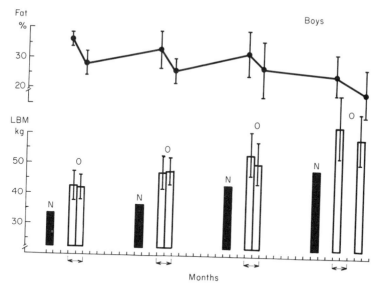

FIG. 16. Changes over a 4-year period in fat proportion (%) and lean body weight (kg) of obese boys before and after treatments in special summer camps. (↔).

take. The pulse frequency both during and after the standard exercise test was also significantly lower at the conclusion of the experimental period. These changes clearly showed a decrease in caloric output and diminished energy demands for the same work load. After reduction of excess fat, children performed the same work load with smaller caloric output than before the weight loss. The vital capacity and muscle force (Vanečková and Vamberová, 1958) of these children also increased after the fat reduction. Blood cholesterol level decreased significantly in the boys (Parízková *et al.*, 1963a) but not in the girls.

A 4-year longitudinal study was completed on a small group of boys (*n* = 4) who received special treatment in the above-mentioned summer camps. Body weight evaluated against the standards of Tanner *et al.* (1965, 1966) varied according to the dietary therapy and exercise program (Fig. 15). Similar changes occurred in the proportion of fat (Fig. 16). Lean body weight was significantly greater in the obese as compared with normal boys of the same age (Fig. 16). During the first 2 years, i.e., at the age of 11.8 and 12.8 years the absolute amount of lean body mass did not change significantly following the treatment. After the third and fourth treatment periods a reduction of lean body mass was observed, which follows the general leveling off of the growth curve (Fig. 15).

For the above group there was a decrease in oxygen uptake and pulse frequency under a standard work load (Fig. 17) following each of the four periods of diet and exercise therapy. Since the work load either was the same or was increased after the period of therapy, the task was performed with relatively lower energy expenditure. Also, the oxygen uptake relative to weight and to lean body mass decreased significantly under the standard work load during each of the experimental periods with the exception of the last year. A similar experiment with a small group of girls ($n = 4$) produced results which were nearly identical in respect to changes in body composition and economy of work.

Repeated measurements of maximal oxygen uptake taken before and after a weight reduction program showed unchanged absolute values in 4 boys who underwent treatment for the first time at 14.5 years of age. The loss in body weight chiefly resulted from a reduction of fat. The slight decrease in lean body weight was not significant (Fig. 18). Neither pulse frequency nor oxygen pulse changed materially. Measurements on an additional 7 obese boys showed a decrease in absolute values of maximal oxygen uptake after weight reduction, but in this case there was a slight but significant decrease

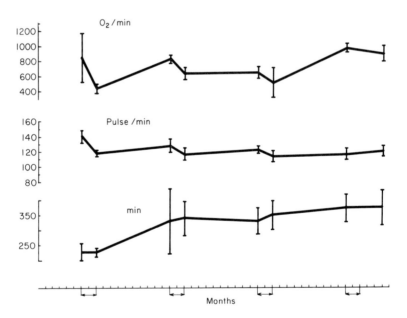

FIG. 17. Changes over a 4-year period in oxygen consumption and pulse rate during standard work load on a bicycle ergometer (kg/min) of obese boys before and after treatment in special summer camps ↔).

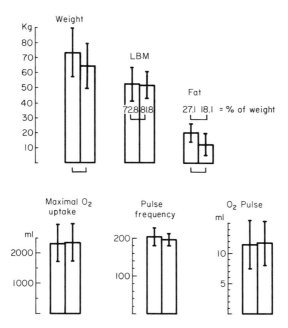

FIG. 18. Changes in weight, body composition, maximal oxygen uptake, pulse frequency, and oxygen pulse before and after reducing treatment in a group of obese boys.

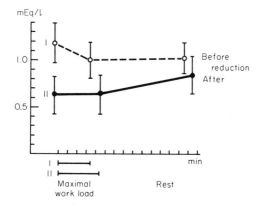

FIG. 19. Effect of weight reduction on resting FFA blood levels and changes in these levels during and after a maximum work load.

of lean body mass. In both cases, however, the maximal oxygen uptake after weight reduction was achieved after a longer run at higher speeds than before the weight loss.

The adaptation to increased work load in a summer camp resulting in a decrease of body fat brought about a reduction in FFA blood levels in the resting state and their different reaction to maximal work load on a treadmill (Fig. 19; Parízková et al., 1963b, 1965). Before weight reduction, initial rest values decreased significantly during maximal work load and did not change further after 10 min rest. After reduction of weight, FFA remained the same during work load test and increased significantly after rest manifesting an increased ability to release FFA from adipose tissue into the bloodstream as a fuel for muscle work (Fritz, 1961). Blood levels of esterified fatty acids and glucose did not react significantly to weight reduction (Parízková et al., 1963b, 1965).

XII. Summary

Exercise has a profound impact on body composition during growth as in all other periods of life. A physically active life develops lean body mass at the expense of fat both in boys and girls. These changes are dependent on the actual intensity and duration of the exercise. After interruption of training there is an increase in body weight, largely fat, which is not caused solely by a relative or absolute disproportionate increase in caloric intake but rather from metabolic adaptations in adipose as well as muscular tissues. During training there is an increased ability to utilize fat metabolites as a fuel for muscle work, and correspondingly increased metabolic turnover in the adipose tissue, which results in its decreased proportion. When energy output is suddenly decreased, fat is laid down.

Changes in soft tissues (lean body mass and depot fat) tend to be paralleled by changes in body build. Leaner subjects develop relatively narrower pelvis and vice versa. The broader pelvis is characteristic of the obese person, especially the heavy boy.

Adaptation to exercise resulting in increased lean body mass is characterized simultaneously by increased functional aerobic capacity (measured as maximal oxygen uptake on a treadmill) and overall better fitness. Leaner subjects work more economically both under standard and maximal work loads. This is manifested in a most marked way in obese children after weight reduction. The same work load is performed with smaller oxygen uptake, ventilation, and heart rate.

References

Altschuler, H., Lieberson, M., and Spitzer, J. J. (1962). *Experientia* **18**, 91.
Åstrand, P.-O. (1967). *Clin. Pediat.* **6**, 234.
Behnke, A. R., Osserman, E. F., and Welham, W. C. (1953). *Arch. Intern. Med.* **91**, 585.
Buskirk, E. R., and Taylor, H. L. (1957). *J. Appl. Physiol.* **11**, 72.
Carlsson, L. A. (1967). *Fed. Proc. Fed. Amer. Soc. Exp. Biol.* **26**, 1755.
Drummond, G. I., and Black, E. C. (1960). *Annu. Rev. Physiol.* **22**, 169.
Ekelund, L. A. (1969). *Annu. Rev. Physiol.* **31**, 85.
Forbes, G. F. (1964). *Pediatrics* **31**, 308.
Fritz, I. B. (1961). *Physiol. Rev.* **41**, 52.
Hampton, M. C., Huenemann, R. L., Shapiro, L. R., Mitchell, B. W., and Behnke, A. R. (1966). *Amer. J. Clin. Nutr.* **19**, 422.
Issekutz, B., Jr., Miller, H. I., Paul, P., and Rodahl, K. (1965). *J. Appl. Physiol.* **20**, 293.
Janda, F. (1963). *Rev. Czech. Med.* **9**, 82.
Keys, A., and Brozek, J. (1953). *Physiol. Rev.* **33**, 245.
Mayer, J. (1955). *Nutr. Abstr. Rev.* **25**, 597–871.
Mayer, J. (1965). *Med. Clin. N. Amer.* **49**, 421.
Merhautová, J., and Parízková, J. (1972). *J. Physiol. (Paris)* (in press).
Novak, L. (1963). *Ann. N.Y. Acad. Sci.* **110**, 545.
Novak, L. (1966). *J. Amer. Med. Ass.* **197**, 891.
Parízková, J. (1959a). *Cs. Fysiol.* **8**, 426–427.
Parízková, J. (1959b). *Physiol. Bohemoslov.* 7, 112–117.
Parízková, J. (1960). *Physiol. Bohemoslov.* **9**, 516–523.
Parízková, J. (1961a). *J. Appl. Physiol.* **16**, 173–174.
Parízková, J. (1961b). *Metab. Clin. Exp.* **10**, 794–802.
Parízková, J. (1963). *Anna. N.Y. Acad. Sci.* **110**, 661–674.
Parízková, J. (1965). *In* "Body Composition." Pergamon Press, Oxford, pp. 161–176.
Parízková, J. (1966). *Proc. Nutr. Soc.* **25**, 93–99.
Parízková, J. (1968a). *Hum. Biol.* **40**, 212–225.
Parízková, J. (1968b). *Curr. Anthropo.* **9**, 273–287.
Parízková, J. (1968c). *Borden's Rev. Nutr. Res.* **29**, p. 40–54.
Parízková, J. (1968d). Memoirs (Vol. I.), XIIth *Int. Congr. Pediat. 1968* pp. 32–35.
Parízková, J. (1969). "Biochemistry of Exercise," Vol. 3, p. 137. Karger, Basel.
Parízková, J. (1970). *Int. Congr. Anthropol. Ethnol. Sci. 8th,1968* Mon. Soc. Res. Child Devel. **35**, 28–32.
Parízková, J., and Koutecký, Z. (1968). *Physiol. Bohemoslov.* **17**, 177–189.
Parízková, J., and Poupa, O. (1963). *Brit. J. Nutr.* **17**, 341–343.
Parízková, J., and Sprynarová, S. (1967). *Proc. Int. Semin. Ergometry, 2nd,* Berlin, *1967* pp. 115–128.
Parízková, J., and Sprynarová, S. (1970). *Proc. Int. Congr. Nutr. 8th, 1969, Excerpta Med.,* p. 316–320.
Parízková, J., and Stanková, L. (1964). *Brit. J. Nutr.* **18**, 325–332.
Parízková, J., and Stanková, L. (1967). *Nutr. Dieta* **9**, 43–53.
Parízková, J., and Vamberová, M. (1967). *Develop. Med. Child Neurol.* **9**, 202–211.
Parízková, J., and Vamberová, M. (1969). *Antropologia* **7**, 25–29.

Parízková, J., Vanečková, M., and Vamberová, M. (1961). *Čs. Fysiol.* **10**, 273–274.

Parízková, J., Vanečková, M., and Vamberová, M. (1962). *Physiol. Bohemoslov.* **11**, 351–357.

Parízková, J., Vamberová, M., Opplt, J., and Vanečková, M. (1963a). *Proc. Nat. Congr. Czech. Physiol. Soc., 5th, 1961* pp. 66–69.

Parízková, J., Stanková, L., Sprynarová, S., and Vamberová, M., (1963b). *Čs. Fysiol.* **12**, 332.

Parízková, J., Stanková, L., Sprynarová, S., and Vamberová, M. (1965). *Nutr. Dieta* **7**, 21–27.

Parízková, J., Stanková, L., Fábry, P., Koutecký Z. (1966a). *Physiol. Bohemoslov.* **15**, 31–37.

Parízková, J., Koutecký, Z., and Stanková, L. (1966b). *Physiol. Bohemoslov.* **15**, 237–243.

Sprynarová, S. (1966). *2nd Int. Congr. Phys. Fitness Youth, Prague 1966* p. 374–378.

Sprynarová, S., and Parízková, J. (1965). *J. Appl. Physiol.* **20**, 934.

Sprynarová, S., and Parízková, J. (1968). *Teor. Prakt. Fiz. Kult.* **31**, 70–73.

Tanner, J. M., Whitehouse, R. G., and Takaishi, M. (1965). *Arch. Dis. Childhood* **41**, Part I, 454.

Tanner, J. M., Whitehouse, R. G., and Takaishi, M. (1966). *Arch. Dis. Childhood* **41**, Part II, 613.

Tenorová, M., and Hruza, Z. (1962). *Cs. Fysiol.* **11**, 485.

Ulbrich, J. (1966). *2nd Int. Congr. Phys. Fitness Youth, Prague 1966* p. 185–191.

Vamberová, M. (1958). *New Czech. Med.* **4**, 135.

Vanečková, M. and Vamberová, M. (1958). *Physiol. Bohemoslov.* **11**, 351.

CHAPTER

6

Growth, Physique, and Motor Performance

Robert M. Malina and G. Lawrence Rarick

I. Introduction

Man has had a long-standing interest in human morphology. For over 2000 years attempts have been made to develop a feasible system for classifying humans into constitutional types. The impetus for such efforts has come largely from the belief that constitutional factors play a significant role in man's susceptibility to disease and that one's constitutional makeup is related to temperament and behavior. In more recent times it has become evident that the individual's constitutional makeup should be considered in any assessment of his capacity to adapt to the stresses of life.

Athletic coaches have for some time contended that high quality performance in certain sports is in no small measure dependent upon body build. Studies by Cureton (1951), Tanner (1964a), and Carter (1970) lend credence to this point of view. Tanner, for example, concluded that the somatotype of the Olympic athlete differed markedly from that of the general population in that over half of the somatotypes present in the general population were not found among the Olympic athletes. Furthermore, there were well-defined morphological differences in athletes competing in different events. Such differences were substantial when comparing sprinters, middle distance runners, and long distance runners. Sprinters were relatively short and very muscular men, all their limb muscles being larger in relation to their bones than was characteristic of other runners. Four-hundred-meter men tended to to be large, long-legged, broad-shouldered, and fairly heavily muscled. On the other hand, long distance runners were small, narrow-shouldered, and relatively small muscled. High hurdlers were for the most part tall, long-legged sprinters, as heavily muscled as the sprinters, but without the sprinters short legs. Without exception Olympic high jumpers were tall, having the longest legs relative to the trunk of all athletes measured. The weight men, namely, discus, shot, javelin, and hammer throwers, differed generally from other athletes. They tended to be taller and heavier with large muscles in relation to their limb bones and longer arms in relation to their legs. Weight lifters had physiques similar to the throwers, when allowances were made for a much smaller body size. Malina *et al.* (1971) reported similar morphological trends for female track and field athletes.

Clearly, factors other than body build are of great significance in accounting for differences in performance levels in a particular event. Yet there is strong reason to believe that constitutional type is a factor of considerable importance in limiting the athlete's ultimate attainment. Of immediate concern to those interested in physical activity programs for children and youth is the question of the extent to which differences in physical performance during the growing years may be attributable to differences in body build. Of perhaps equal significance is the consideration of the effect which exercise regimens and sports participation may have in altering physique in the growing years. Before focusing attention on these two considerations, attention will be directed to the methods available for assessing physique, a matter of major significance in any evaluation or discussion of the relationship between physique and motor performance. Consideration will also be given to some of the innate factors which account for variations in physique in humans during the period of growth.

I. Physique and Body Composition

The study of physique or body form is a single dimension of a multifaceted area of study termed "human constitution." Constitution is a general and an almost all-encompassing term representing an attempt to combine the variability characteristic of an individual into a usable set of dimensions. Rees (1960) used constitution as referring ". . . to the sum total of morphological, psychological and physiological characters of an individual, all being mainly determined by heredity but influenced, in varying degrees, by environmental factors" (p. 344). Damon (1970) defined constitution as ". . . the sum of a person's innate and relatively fixed biological endowment" (p. 180). The foregoing would seem to imply that no two people (except perhaps monozygotic twins) have an identical constitution. However, persons demonstrating similar characteristics can be grouped into various constitutional categories or "types."

Constitution, therefore, is a variably used term. It implies a kind of biological preconditioning to structural, functional, and behavioral patterns. Physique, it should be obvious, is only a single aspect of the constitutional continuum, which has been classified by Rees (1960) into four broad categories: (1) *morphological,* including the study of physique, body size, body composition, dysplasia, etc.; (2) *physiological,* including functional characteristics of various bodily systems; (3) *psychological,* including personality, temperament, character, etc.; and (4) *immunological,* emphasizing relationships between the above constitutional correlates and susceptibility to disease, be it somatic, psychic, or psychosomatic. The various attributes of constitution are obviously interrelated and perhaps interdependent, reflecting common factors for structural, functional, and behavioral manifestations.

Physique refers to an individual's body form, or more specifically, the conformation of the entire body as opposed to emphasis on specific features (Tanner, 1953a). It is the single aspect of constitution that is, perhaps, most amenable for study because it can be readily observed (anthroposcopy) and measured (anthropometry). Physique has been related to a variety of behavioral, occupational, disease, and performance variables. For detailed discussions of physique as a variable in behavioral and disease conditions the reader is referred to Anastasi (1958), Damon (1970), Domey *et al.* (1964), Parnell (1958), Paterson (1930), Rees (1960), Sheldon *et al.* (1940, 1954), and Sheldon and Stevens (1942).

III. Assessment of Physique

Methods used to assess physique are varied, ranging from traditional anthropometry to sophisticated measurement of the body's composition. Nevertheless, results are no better than the methods used in obtaining them.

A. TRADITIONAL ANTHROPOMETRY

Anthropometry is a systematized technique for measuring man. Height and weight, independently or relative to each other, are perhaps the most commonly used anthropometric dimensions in physique assessment. Norms derived from height and weight are used to assess growth status and progress and nutritional status. These measurements are similarly used as indicators of physique on a linearity–laterality or a lean–stocky scale as well as in classifying children for various physical activities. Height and weight, however useful, are limited in that they vary with maturational status and in that no treatment of height and weight has resulted in a growth timetable from which maturity status can be accurately assessed. Further, ratios of height and weight do not differentiate the possible components of linearity and laterality. For example, is a stocky individual fat or is he muscular; is a linear individual long-legged or long trunked; or are these characteristics demonstrated in combination, a muscular trunk with lean, linear lower extremities?

Other commonly used anthropometric dimensions relate to body proportions. Limb lengths, sitting height, bicristal breadth, biacromial breadth, condylar breadths of the femur and humerus, and limb circumferences are accurate indicators of various body proportions and build. Many of these dimensions are used in attempts to refine anthroposcopic physique assessments (see below).

Although anthropometry has many limitations, particularly in terms of the training necessary for making valid assessments, selection and meaning of measurements, age- and sex-associated variation, it has been and will continue to be a useful tool, especially as the techniques become more automated and refined.

B. INDICES

Numerous indices relating various body measurements to each other have been used in physique assessment. Perhaps the most commonly used index is the ponderal index (height divided by the cube root of weight). It is frequently used independently as a general indicator of linearity or laterality, particularly in epidemiological studies (Damon, 1970; Florey, 1970; Seltzer, 1966). The ponderal index is likewise incorporated into the more elaborate body form rating systems of Sheldon (Sheldon *et al.*, 1954) and Parnell (1958).

More recently, however, the use and interpretation of the ponderal index has been questioned. Florey (1970) evaluated the ponderal index and two other weight–height ratios, weight ÷ height (W/H) and weight ÷ height² (W/H²) in a large series of adult men and women. Relating these indices to two criteria, independence of height and highly correlated with weight, Florey noted the ratios are simply indices of "corrected weight," i.e., weight corrected for height, and should not be used as indicators of adiposity and physique ". . . in the belief that they are valid measures of these qualities" (p. 102). In Florey's sample, W/H² was most likely the best ratio for males and W/H was probably the best for females, with the ponderal index least favored for both sexes. Huber (1969) reported similar results for the ponderal index in a sample of young adult males. The ponderal index increased with increasing stature, thus indicating the dependence of the ponderal index on height *(r* ponderal index and height = .32, .32, and .45 for three samples grouped by age). Tall men tend to be more linear, and weight does not necessarily increase according to the cube of linear dimensions (Huber, 1969).

The foregoing does not mean that the ponderal index should be eliminated. Height and weight data are easily and accurately collected, and there are no taboos against taking such measurements. Hence, such data should be utilized. Scales derived from the ponderal index and other weight–height indices perhaps need revision and better definition. Also, their applicability to the growing child needs reexamination, especially since the ponderal index has been used frequently as a criterion of physique in a variety of studies dealing with children, social stereotypes, and body image.

Other indices of body form are available but generally require other measurements in addition to height and weight. The Rees–Eysenck Body Build Index, incorporating stature and transverse chest diameter, has received some use in behavioral research (Rees, 1960; Domey *et al.*, 1964; Segraves, 1970). On the assumption that growth is channelized according to body build, Pryor's width weight tables (Pryor, 1940, 1966) relate body weight as an index of nutritional status to body build using chest and bicristal breadth measurements in addition to age, height, and weight.

Indices have limitations. Used alone, they are not completely adequate and accurate indicators of body form. They can, however, be used as rough guides in physique assessment.

TABLE I

Summary of Factor Analysis Classification of Physique in Terms of Orthogonal Subdivided Group Factors[a]

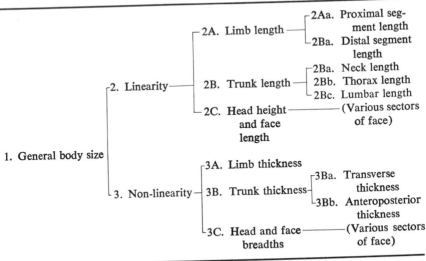

[a] Reprinted with permission. From Tanner (1964b).

C. Factor Analysis

Factor analysis is aimed at the minimum expression of variation in physique via statistical manipulation of a large series of measurements. It attempts to reduce a large number of measurements to a small number of factors accounting for the correlations or covariations among the variables studied (Tanner, 1953a, 1964b; Rees, 1960). Statistically identified factors are common (intercorrelated or oblique) or unique (independent or orthogonal). Factors, however, are statistics, and ". . . they cannot be equated *a priori* with any genetic or physiological mechanisms" (Tanner, 1953a, p. 762). Table I summarizes factor analysis classification of physique according to the subdivided group factors.

Factor analysis of physique has been applied to the growth years. McCloy (1940), for example, identified three factors in a series of children and adults. The first was associated with the development of subcutaneous fat, the second related to linearity, and the third related to cross-sectional dimensions, i.e., girths, depths, and breadths. A fourth factor was recognized but not clearly identified. It related to chest dimensions and shoulder breadth. Hammond (1953, 1957) reported factor analysis results for children through 18 years of age. He identified a general factor for size, a type

TABLE II

REPORTED INTERRELATIONSHIPS OF RADIOGRAPHIC MEASUREMENTS OF SUBCUTANEOUS FAT, MUSCLE, AND BONE BREADTHS IN EXTREMITIES[a]

Region	Age (years)	Sex	Correlations		
			Bone–muscle	Bone–fat	Muscle–fat
Arm	6.0–15.5	M	.256	.154	.145
		F	.226	.229	.202
Arm	Young adult	M	.29	−.04	−.01
		F	.17	.08	.16
Arm	Athletes	M	.23	.02	.02
Arm	Young adult	M	.36[b]	.03[b]	.26
Thigh	Young adult	M	.19	−.05	.18
		F	.00	.11	.08
Thigh	Athletes	M	.04	.12	−.06
Calf	0.5–5.0	M	.011	−.083	.050
		F	.160	.008	−.033
Calf[c]	1.0	M	.65	.19	.12
		F	.78	.63	.34
Calf[c]	3.0	M	.56	.47	.23
		F	.67	.60	.37
Calf[c]	6.0	M	.61	.34	.10
		F	.71	.51	.38
Calf	6.5	M	—	.24	.17
		F	—	.23	.02
Calf	7.5	M/F	−.17	.30	−.02
Calf	6.0–15.5	M	.095	.280	.120
		F	.040	.235	.180
Calf	Young adult	M	−.23	−.14	.04
		F	−.01	.07	−.02
Calf	Athletes	M	−.22	−.01	−.12
Calf	$\bar{X}$-37.8	M	.077	−.033	.224
Calf	$\bar{X}$-38.0	F	.053	.008	.160
Calf[d]	0.0–6.0	M	.60	.14	.13
		F	.14	−.09	.23
Calf[d]	0.0–18.0	M	.44	.04	.12
		F	.21	.07	.37
Calf[e]	Young adult	M	.09	.07	.08
		F	.13	.09	.09
Calf[e]	Athletes	M	.02	.04	.02

[a] Reprinted with permission, *Clinical Orthopaedics and Related Research, 1969.* From Malina (1969c, p. 30). See original for references on individual sources.
[b] Compact bone breadth used.
[c] Tissue areas used.
[d] Average correlations among summed tissue widths from several limb areas.
[e] Average within tissue correlations from arm, thigh, calf.

factor associated with length on one hand and girths and fat on the other, and a type factor related to trunk–limb development. Using anthropometric and photogrammetric dimensions in a factor analysis of physique of preadolescent boys 7 through 11 years, Barry and Cureton (1961) identified three factors related to physique: ponderosity (transverse dimensions and fat), lankiness (vertical dimensions and attenuated extremities), and leg–trunk development (disproportionate development of the trunk and legs). In addition, two tentative growth factors related to asymmetrical development of upper and lower extremities and androgyny were suggested.

D. BODY COMPOSITION

Considerable effort has been expended in recent years toward a more complete understanding of the body mass. Body composition studies attempt to partition the body mass into its elemental constituents or its primary tissues. Many approaches deal with the body's composition on a whole body basis, i.e., they identify the quantity of lean vs. fatty tissue in the total body. Such approaches, however, provide little information on the relative distribution or development of the primary tissue components responsible for considerable variation in body weight and form, namely, bone, muscle, and fat (see Malina, 1969c).

Measurement of bone, muscle, and fat offers the possibility of identifying the pattern of relationships, if any, between the three tissues. Tables II and III summarize tissue correlations from various sources. Independence of bone, muscle, and fat within the extremities is apparent during growth and adulthood in both sexes as well as in athletes and nonathletes (Table II). Moderate correlations between bone and muscle are apparent in two studies dealing with early ages; both studies, however, used somewhat different techniques from the others reviewed. Nevertheless, bone and muscle might be more highly related early in life than later, possibly reflecting a continuation of the generally uniform pattern of growth characteristic of bone and muscle prenatally and neonatally.

Correlations between corresponding tissues in different limbs or limb segments (Table III) indicate a moderate to high relationship between fat breadths and moderate to low relationships for both bone and muscle widths.

Using data derived from calf radiographs, Reynolds (1949) and Reynolds and Asakawa (1950) reported the largest amounts of fat and the highest fat–bone index* in endomorphs; smallest amounts of bone, muscle, and fat and a low fat–bone index in ectomorphs; and largest amounts of

---

* The fat–bone index is the ratio of fat breadth to bone breadth measurement from calf radiographs.

TABLE III

REPORTED INTERRELATIONSHIPS BETWEEN CORRESPONDING TISSUE BREADTHS IN THE ARM, THIGH, AND CALF[a]

Regions compared	Age (years)	Sex	Muscle	Bone	Fat
Arm–calf	6.0–15.5	M	.53	.62	.88[b]
		F	.42	.66	.82
Arm–calf	7.5	M	—	—	.64
		F	—	—	.64
Arm–calf	11.5	M	—	—	.69
		F	—	—	.62
Arm–calf	15.5	M	—	—	.60
		F	—	—	.61
Arm–calf	Young adult	M	.42	.49	.64
		F	.36	.38	.62
Arm–thigh	Young adult	M	.54	.46	.83
		F	.49	.29	.69
Thigh–calf	Young adult	M	.51	.50	.76
		F	.43	.43	.68
Arm–calf	Athletes	M	.42	.35	.50
Arm–thigh	Athletes	M	.30	.50	.60
Thigh–calf	Athletes	M	.45	.50	.55
Arm–calf[c]	Adult	M	—	—	.62–.56
Arm–calf[c]	Young adult	F	—	—	.39–.21
Arm–thigh[c]	Young adult	F	—	—	.42
Thigh–calf	Young adult	F	—	—	.61–.51

[a] Reprinted with permission, *Clinical Orthopaedics and Related Research, 1969.* From Malina (1969c, p. 31). See original for references on individual sources.

[b] Does not include 14.0- and 15.5-year-old males. Values for these two age groups were significantly different from the values for 6.0–13.0 years of age. The correlations were + .20 at 14.0 years and + .39 at 15.5 years suggesting differential adolescent fat loss in males.

[c] Arm fat measured at deltoid insertion.

muscle and bone and a low fat–bone index in mesomorphs. Adding anthropometric variables to radiographic measurements of bone, muscle, and fat, Tanner, Healey, and Whitehouse (unpublished, see Tanner, 1964b) carried out a factor analysis of data for young adult males and females. Five independent factors were postulated by the analysis: skeletal frame size, limb bone width, limb bone length, muscle width, and fat thickness.

E. BODY FORM AND BODY TYPES

The rating and classification of body form as a whole (as opposed to specific features) has a long history. Hippocrates offered a twofold classification, habitus phthisicus (linear) and habitus apoplecticus (lateral). Since then, there have been numerous classification efforts, all eventually describing physique in two or three types (for a more detailed discussion of the various taxonomies of physique, see Rees, 1960; Comas, 1957; Petersen, 1967; Domey *et al.*, 1964).

Sheldon's approach (Sheldon *et al.*, 1940, 1954) to physique assessment is perhaps the most widely used today, although several modifications of his method are presently available (see below). Sheldon's method is based upon the premise that there is continuous variation in physique based upon the contribution of varying components to the conformation of the entire body. These components are termed "endomorphy," "mesomorphy," and "ectomorphy." The method is basically photographic and anthroposcopic, coupled with the ponderal index. Each component is assessed individually from three rigidly standardized photographs. Rating is based upon a 7-point scale, with 1 representing the least expression and 7 the fullest expression of the specific component. Emphasis is placed upon the contribution of each component to total physique, which can be extreme or balanced, thus removing the necessity of forcing individuals into rigid categories. The ratings of each component comprise the individual's *somatotype,* which by definition does not change with age (Sheldon's method was developed upon 4000 college students), with nutritional state, nor with the state of physical or athletic training. Some data (see below) suggest somatotype alterations during growth, leading Tanner (1953a, p. 756) to comment that the ". . . somatotype is the appearance of the individual when growth ceases, that is about twenty, given adequate nutrition at that time." More recently, Petersen (1967) extended Sheldon's methods to children by using Sheldon's principles of classification and rating.

Parnell (1954, 1958) modified Sheldon's method by using a variety of anthropometric dimensions to derive a *phenotype, ". . .* the living body as it appears at a given moment. The phenotype has by definition no continuance in time" (Parnell, 1958, p. 4). Height, weight, three skinfolds, two limb circumferences, and two limb bone breadths are used to derive estimates of three components, fat (F), muscularity (M), and linearity (L) in place of Sheldon's three components.

Heath (1963) modified the somatotyping method by eliminating adjustments for age, by opening the component rating scales at both ends in order to accommodate a greater range of physique variation,* and by establishing a linear relationship between somatotype ratings and height–weight ratios. Heath and Carter (1966) compared Heath's (1963) and Parnell's (1954, 1958) approaches, the result being a somatotype

---

* Roberts and Bainbridge (1963), for example, had to extend Sheldon's ectomorphy scale up to 9 for a series of Nilotes.

method incorporating both anthropometric and anthroposcopic procedures, the Heath–Carter method (Heath and Carter, 1967).

Other variations of the somatotype approach have also been presented. Lindegard (1953) devised a system for describing body build based upon direct measurement. Four factors are recognized in the system—length, sturdiness, muscularity, and fat—the first three being interdependent. Ratings for each factor are evaluated relative to a graph of standard deviation units, providing an estimate of the conformation of the entire body. Damon *et al.* (1962) attempted to predict Sheldonian somatotype ratings from anthropometric dimensions in young adult Negro and white males. Separate regression equations were necessary for whites and Negroes. Munroe *et al.* (1969) presented regression equations for predicting each somatotype component in 12-year-old white boys from a series of anthropometric dimensions.

IV. Growth and Physique

A. GENETICS

One's body form is under genetic control. The relative significance of hereditary and environmental components in physique variation, however, is difficult to assess. Twin studies indicate higher heritability estimates for measurements of length (trunk or extremities) than for measurements of breadth, width, or girth (Tanner, 1953b; Osborne and De George, 1959; Vandenberg, 1962), Parent–child or family line comparisons also indicate the role of inheritance in size and build. Tall parents tend to have tall children, while short parents tend to have short children (Garn, 1962, 1966; Garn *et al.*, 1960; Malina *et al.*, 1970). Large parents (bony breadth of the chest as measured on anteroposterior chest radiographs) tend to have large-chested children, while small parents tend to have small-chested children (Garn, 1962, 1966; Garn *et al.*, 1960). Parnell (1958) reported comparisons of physiques of children and their parents *(n* = 45), noting a strong tendency for children (possibly up to 70–75%) to follow the "parental line principle," i.e., their physique ratings fell on or near a line connecting parental physiques on a three-dimensional somatochart. In approximately 22.5% of the families, however, children deviated in varying degrees from parental physiques. Davenport (1923, cited by Tanner, 1953b) reported that linear parents tended to have a greater percentage of linear offspring while medium and thickset parents had correspondingly more medium and thickset children. This is in agreement with numerous observations that indicate higher heritability estimates for linear compared to breadth and girth measurements. While a multifactorial process of inheritance may be inferred for the variables under consideration, linear parents do not always have linear children, nor do large parents always have large-chested children. Nevertheless, family line trends are apparent for physique.

In a comprehensive twin study, Osborne and De George (1959) found a relatively high degree of genetic control over rated somatotype, especially for the ectomorphic component. Also, measurements of arm girth (an index of relative muscularity), wrist breadth (bone), and skinfolds (to a lesser degree), indicated the function of a genetic component in these physique-related variables.

B. Physique Changes during Growth

Although the genetics of body form are complex and largely unknown, hereditary factors governing general growth of the body and its component parts manifest themselves throughout the years of growth. Two sets of genetically determined and independent factors, for example, are believed to affect an individual's size at any time during the growing years. One factor controls the final size attained, i.e., the genetic potential, and the other controls the rate of growth, i.e., the rate at which the ultimate size will be attained (Tanner, 1962). The genetic regulation of ultimate size and rate of growth is well documented. Adult height, for example, can be reliably predicted from 2 or 3 years of age. Monozygotic twin girls attain menarche within 2–3 months of each other (Tanner, 1953b, 1962).

According to Sheldon's concept of somatotype physique should remain constant over age. Petersen (1967), who applied Sheldon's principles to children, is of the opinion that changes in size, proportions, and body composition during childhood and adolescence do not fundamentally alter the child's body type. Petersen however, recognized that there are somatotyping problems during the adolescent years and suggested ". . . it will occasionally be wise to postpone somatotyping for the time being" (p. 29). In this regard, Hammond (1953, 1957) noted a reasonable degree of constancy of individual body type values over a 3-year period. Consistency of type was lowest in infancy and increased during the school ages. In children 5 through 18 years of age, for example, the correlations between the first and second measurements ranged from + .65 to + .92 for males and females. The correlations, however, tended to be lower during puberty. In Hammond's pachysome (girth and fat)–leptosome (length) type factor, 11% of the males up to 13 years of age changed their type, while 18% of the males 14 years of age and older changed their type over the 3-year observation period. Among girls, 8% up to 9 years of age changed their type, 22% who reached 10–13 years of age changed their type, and only 12% 14 years of age and older changed their type. Hammond interpreted these observations in terms of differential timing of the adolescent spurt and the effects of the spurt on the physical type.

Parnell (1958) assessed the physiques of 72 children, both males and females, at 7 years of age and again at 11 years. Linearity (ponderal index) had the highest constancy ($r = .66$ for males and $.69$ for females), followed by fat ($r = .68$ for males and $.60$ for females), and muscularity ($r = .48$ for males and $.62$ for females).

Constancy of somatotype in the adolescent years was examined by Hunt and Barton (1959), who correlated somatotype ratings of the same boys at a prepubertal (11.5 years) and subadult (16.5–19.0 years) ages. The correlation values were $+.453$ for endomorphy, $+.501$ for mesomorphy, and $+.708$ for ectomorphy (ponderal index). Hence, agreement was best for the ponderal index and poorest for the visually assessed first and second components, suggesting perhaps relatively low predictive relationships between pre- and post-pubescence. Herron (1960) evaluated the constancy of various physique ratios over a 5-year period in 26 adolescent boys. Indices associated with linearity showed greater constancy than others, e.g., the correlation for the ponderal index was .718 and for the Rees–Eysenck index was .852. In general, ratios related to specific body parts showed less constancy over adolescence. Sinclair (1969) reported relatively high inter-age correlations between somatotype components in a longitudinal series of boys followed from 12 to 17 years. Correlations between adjacent ages ranged from $+.80$ to $+.85$ for endomorphy, $+.84$ to $+.94$ for mesomorphy, and $+.86$ to $+.90$ for ectomorphy. Significant variation in somatotype ratings, however, occurred between 12 and 17 years, the "percentages of significant differences" being greater for mesomorphy and ectomorphy (53%) than for endomorphy (40%). Comparing somatotype ratings made at 12–17 years and at 33 years of age in a series of males and females, Zuk (1958) found reasonable stability of somatotype from early adolescence through adulthood.

Relative to somatotype variation during adolescence, the observations of Tanner (1962, p. 104) are particularly appropriate:

> . . . it would certainly be wrong to leave any impression that the adolescent spurt, whether late or early, causes any radical change in body build: it certainly does not. It adds only the finishing touches to a physique which is recognizable years before. Anyone who has looked at serial pictures of children followed from infancy to adulthood must be impressed chiefly by the similarity the child shows from one age to another. So great is this that there is little doubt that someone used to looking at children's photographs could predict with accuracy the adult somatotype from a picture taken at age 5 or even earlier.

V. Sex Differences in Physique

Sex differences in body size, proportion, and composition, though apparent, are generally minor during the preadolescent years. Girls, however, are advanced in maturation rate at all ages, the difference becoming greater with increasing age up to adolescence. Girls also enter adolescence earlier and stop growing earlier than boys.

The effect of the adolescent spurt on sex differences in body size is well documented. Although girls are temporarily taller and heavier than boys in early adolescence, males attain greater adult stature and weights. Males enter adolescence later and thus grow over a longer period. Sex differences in body proportions also manifest themselves at this time. The sitting height–height ratio (an index of the contribution of the trunk and legs to total height), for example, becomes slightly higher in girls with the onset of the adolescent spurt and remains so throughout the teen-age years into adulthood (Simmons, 1944). This, of course, indicates relatively shorter legs in girls, or alternatively, relatively longer trunks. For equal stature, therefore, girls have shorter legs. Similarly, the shoulder–hip ratio (bicristal breadth–biacromial breadth) shows sex differences in laterality of build during adolescence. Preadolescent sex differences are minor, but from approximately 10 years of age on, there is a consistent sex difference which persists into adulthood—girls develop relatively broader hips and boys develop broader shoulders (Simmons, 1944). It is interesting to note that the adolescent spurt in bitrochanteric breadth in females is as great as that observed in males, although in most other dimensions (e.g., biacromial breadth) males have greater adolescent spurts (Tanner, 1962).

Somatotype comparisons between males and females during adolescence are lacking, perhaps for cultural reasons. By comparing somatotype ratings made between 12 and 17 years of age to ratings made at 33 years of age in males and females, Zuk (1958) found primary growth in the mesomorphic component for males and in the endomorphic component for females.

The foregoing in conjunction with changes in primary and secondary sex characteristics as well as body composition changes during adolescence (see Malina, 1969c) result in marked sexual dimorphism in physique at the cessation of growth. Males are taller and heavier, have relatively longer legs, narrower hips and broader shoulders, and have less subcutaneous fat and more muscle and bone (greater lean body mass) than females. These sex differences in body form primarily result from the effects of the hormones of adolescence acting upon the individual's preadolescent physique (Tanner, 1962).

VI. Maturation and Physique

In a small longitudinal series of boys, Barton and Hunt (1962) found prepubertal somatotype ratings (made at 11.5 years) to be poor predictors of the events of adolescence. Further, subadult somatotype ratings (made between 16 and 18 years) were less related to the chronology of adolescence. Rather, somatotype extremes at prepubertal and subadult ages showed marked contrasts in the events of the male adolescent cycle.

Relating adult somatotype ratings to events of male adolescence, Hunt *et al.* (1958) reported mesomorphy associated with early maturation of secondary sex characteristics, ectomorphy associated with later maturation, and endomorphy associated with an early onset and late completion of maturational events. Acheson and Dupertuis (1957) related adult somatotypes to skeletal maturation of the hip and pelvis with similar results. Mesomorphy and endomorphy were higher among faster maturers, while ectomorphy was more pronounced in slower maturers. Comparisons of extreme ectomorphs and mesomorphs suggested variation in time pattern of epiphyseal maturation associated with physique. Finally, Dupertuis and Michael (1953) related adult somatotype ratings to longitudinal height and weight records. Young adult males classified as mesomorphs were generally shorter (from 5 to 6 years of age on) and heavier (from 2 years of age on), and closer to adult size at all ages from 2 through 17 years than young adult ectomorphs. The mesomorph also reached his peak adolescent spurt earlier.

VII. Ethnic Differences in Physique

Ethnic boundaries, be they social or racial, have been maintained to a large extent, producing variations between different populations. Ethnic variation in body build has been reported, but studies of children of different ethnic origins are lacking (see Malina, 1969b). Rauh *et al.* (1967) and Krogman (1970), using height and weight measurements, noted a greater percentage of Negro boys (5–18 years) in the thin range of the scale. The data for girls showed greater percentages of Negro girls tending toward the stocky to heavy end of the physique continuum, especially during the adolescent years.

Sheldon *et al.* (1940) reported prominent development of the mesomorphic component among a series of 400 Northern Negroes. In an attempt to predict somatotypes from anthropometric dimensions, Damon *et al.* (1962) found that the regression equations for the white sample did not apply to their Negro subjects. The young adult Negro male sample was less endomorphic ($\overline{X} = 2.92$ compared to 3.86) and more mesomorphic ($\overline{X} = 5.14$ vs. 4.29), with little difference in the ectomorphic component ($\overline{X} = 2.99$ vs. 2.89). Robbins (1962) noted similar physique trends among a small sample of college women. Using Parnell's method, Robbins found higher muscularity and linearity in the Negro women, the difference more marked in the mesomorphic component.

Using a modified version of the Sheldonian somatotype method, Danby

(1953) described the physique of East Africans (primarily Kikuyu and Luo) as low in endomorphy, dominant but moderately high in mesomorphy, and variable in ectomorphy. In a study of Nilotic body build, Roberts and Bainbridge (1963) noted physiques low in endomorphy and mesomorphy but extremely high in ectomorphy (indeed so high, they had to extend the third component scale to 9 in some cases).

Kraus (1951) and Heath *et al.* (1961) reported a dominant second component in samples of men of Japanese ancestry. Hawaii-born Japanese women had somewhat higher first and second somatotype components than American white women (Heath *et al.*, 1961).

Ethnic variations in physique are apparent. The role of environmental factors in mediating the phenotypic expression of variation in physique is difficult to establish. Occupational and nutritional factors must be considered. Danby (1953) noted higher mesomorphy in a prison sample of Kikuyu compared to tribal Kikuyu, and related the difference to the physical work required of the prison sample (see below). Lasker (1947) reported marked somatotype alterations after 24 weeks of partial starvation. Roberts and Bainbridge (1963) and Heath *et al.* (1961) suggested true ethnic differences. Roberts and Bainbridge (1963, p. 362), for example, concluded that the Nilotic physique ". . . seems to have been developed primarily as an adaptation to the extreme climate, and mainly of genetic origin. Undernutrition is excluded as a primary cause." It is of interest to note that Nilotic children show a marked linearity of build (Roberts, 1960). Eveleth (1966), on the other hand, noted generally more linear (weight–height index) and less stocky build (calf girth relative to lower limb length) in American children growing up in the hot and wet climate of Rio de Janiero than in their Temperate Zone American counterparts. Eveleth attributed the body form differences (at least those reflected in the indices used) to heat stress during the growing years. Hence, to say the least, further study of occupational, activity, nutritional, and climatic influences on physique development are essential.

VIII. Exercise and Physique

It is difficult to assess the influence of physical exercise during the growing years on physique because of the lack of longitudinal data and adequate controls (see Rarick, 1960; Malina, 1969a). However, several short-term studies involving the exercise variable are available. Simon (1961) reported an acceleration of growth in body bulk as a result of a 6-month training program in 14–15 year old children. Length measurements, as would be expected, were not affected. Jokl *et al.* (1941) examined the effects of a

6-month training program on the physiques of "substandard" European recruits, 16 through 21 years of age. The most striking observation was an increase in body weight, which was attributed to muscular development. Similar results were also reported by Craven and Jokl (1946) for "substandard" adolescent boys 15 through 17.5 years of age. Seltzer (1946) found significant increases in chest circumference as a result of an 8-week physical training program in Air Force cadets (mean age, circa 21 years). The increase was attributed to the development of thoracic musculature.

The foregoing short-term studies of the adolescent and late-teen years indicate the beneficial effect of short-term physical training on muscular development as well as localized effects of specific activities. The persistence of exercise–induced effects, however, is questionable, particularly in the late teens and early adulthood. Tanner (1952), for example, found increases in muscularity with training; however, 4 months after the training program was discontinued, almost all measurements had reverted to pretraining values.

Steinhaus (1933) reported increased muscle girths in children, youth, and adults but noted a possible exception in the youth (presumably teen agers). During periods of rapid growth in body length (i.e., the adolescent height spurt), exercise had less effect on muscle growth than in periods when the tendency to grow in diameter predominated. In college age subjects, on the other hand, Steinhaus (1933) reported that "middle-sized" individuals showed the greatest increases in muscle girths with exercise, whereas the "thinnest aesthenic" types responded the least. Hence, training effects on muscularity appear to be related to stage of growth and possibly to physique type. Further, one may inquire as to the relationship between body type and exercise pursuits. Are certain physiques predisposed to certain types of physical activity pursuits and habits during the years of active growth as well as in adulthood? Experience would suggest this, and the observation that athletes tend to gravitate to sports suited to their body build reinforces this point of view.

Further examination of the role of physical activity in influencing physique, especially the second component, mesomorphy, is warranted. The beneficial effect of exercise in reducing fatness and possibly affecting endomorphy ratings is obvious. In Tanner's (1952) short-term study (4 months) of weight training and physique in young adult males, however, the training–induced changes were not sufficient to alter somatotype ratings, although regional somatotype estimates occasionally changed by one unit. Thus, at least in young adults, exercise does not appreciably alter physique.

Sufficient data on the above are lacking, however, during the growing years. Parízková (1968) reported one of the few longitudinal training studies from preadolescence into adolescence. Four groups of boys exposed to different degrees of participation and training in sports were followed over 4 years (11 through 15). All had similar socioeconomic and educational backgrounds, and similar maturity status at the beginning of the study. While basic anthropometric dimensions and skeletal maturity did not differ after 4 years, there were, as would be expected, striking differences in body composition—the more active boys had greater lean body mass. In addition, Parizkova noted that the most active boys developed relatively narrower pelves in relation to stature and shoulder breadth at the end of 4 years even though all four groups had similar values at the start of the study. Hence, Parizkova (1968, p. 222) commented: "It seems that the skeleton could be influenced by physical exercise in this special age period without remarkable change in absolute values of most traditionally measured indicators."

In an earlier study, Godin (1920) reported results of observations on the growth of gymnasts and nongymnasts from 14 through 18 years of age. The gymnasts were slightly taller, but they were heavier and especially larger in thoracic and forearm measurements. Godin attributed the positive differences to participation in gymnastic exercises over a long period.

As indicated at the outset, the effects of exercise and training on physique are not clear. Some data are suggestive, but adequate longitudinal observations in childhood are lacking. Many comments associated with short-term studies are presently merely speculative. For example, Jokl *et al.* (1941), on noting marked body weight and muscularity changes in their "substandard" recruits, commented: "Had the training which our recruits received, been applied eight or ten years earlier, the effect would have been more marked; they would have developed a better physique and with it a different personality" (p. 62). And further, de Wijn's (1968) comment on the interrelationships of growth, exercise, nutrition, and physique:

> We wonder what the pattern of chemical growth of the so-called "ectomorphic" child, with his poor muscular development and rapid growth, would be when his father had given him, at an early stage at school age, a pair of boots to play football with or hockey all days, and when—with the hockey stick from the father in one hand—his mother would care for an extra bottle of milk in the other hand. Maybe he would grow up as a medium build muscular type. (p. 311)

IX. Physique and Strength

The concept that muscular strength is related to body size is supported by the experiences of daily life. Children at an early age learn to accept the concept "big and strong." That body size and strength are associated is not surprising since strength is roughly proportional to the body's muscle mass and the latter constitutes approximately 40% of the body weight. Yet the relationship between body weight and strength is not as close as it might seem, for most studies, including the research of Jones (1949), show that in adolescent and in young adult males the correlation between weight and static strength measurements is within the range of .50 to .60. Similarly, the correlations between height and measurements of strength is substantially lower, approximating .30. Thus, the variance in strength attributable to differences in body weight is only 30–35%, whereas the variance accounted for by differences in stature amounts to less than 10%. Clearly, factors other than height and weight must be operating. While the quality of the muscle tissue and the ability and the willingness of the individual to mobilize his neuromuscular mechanism in strength tests probably account for a major part of the differences in strength, constitutional factors are believed to be of considerable significance in the development of muscular strength. What, then, do we know about the impact of physique upon muscular strength in childhood and adolescence?

A. CHILDHOOD

Information on the relationship between physique and strength in childhood is limited. That children show marked differences in body build early in life is well known, such differences tending to remain within well-defined limits throughout the growing years. Some children show a preponderance of endochondral bone formation, in which the linear component of growth is accentuated, resulting in relatively long slender bones. In other children there is greater appositional bone formation which results in heavier but shorter bones. The latter have essentially a mesomorphic physique with a sturdy, heavy boned skeleton, one well equipped to support a well-muscled body. In contrast, the child with light bones and slender skeleton, the child of ectomorphic build, is by nature less well endowed with muscle tissue than his heavily boned counterpart. Hence, one would expect the heavily boned child to be stronger than the child of ectomorphic build. From the scanty data available this would seem to be true, although if strength per unit of body mass is used the difference in strength between the heavily boned and light boned child is less well defined.

The above is illustrated by data from the Wisconsin Growth Study (Bohm, 1959) in which static strength measures were obtained annually on the upper and lower extremities of a group of 33 boys as they progressed from the first through the fifth grade. The correlations between strength and body weight were positive, but low, ranging for the most part between .20 and .45. As the boys advanced in age the correlation between weight and strength increased slightly, seldom going beyond .45. Using the ponderal in-

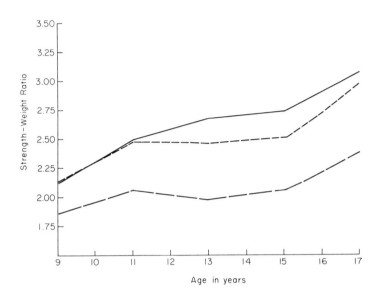

Fig. 1. Longitudinal strength–weight ratios of adolescent males grouped by physique categories (strength scores: sum of right and left grip, pull, and thrust). Key: endomorph (— —), mesomorph (——), and ectomorph (– –).

dex as an estimate of body build, the correlations between strength and the ponderal index were without exception low, ranging at all age levels between .05 and .40. While the child of stocky build tended to be the stronger, when appropriate allowances were made for differences in weight, the strength advantage of the stocky over the linear-framed child was slight. The evidence to date indicates that differences in static dynamometric strength in young boys can be attributed more to differences in body size and to yet-unidentified factors than to differences in physique.

B. ADOLESCENCE

The relationship of physique to strength is more clearly defined in adolescence than in childhood, particularly in adolescent males. Jones (1949) in the Adolescent Growth Study found that the correlation between strength (composite of right grip, left grip, pulling and thrusting strength) and mesomorphy was .61 with height and weight partialled out. The multiple correlation between the composite strength scores and an appropriately weighted combination of mesomorphy, ectomorphy, endomorphy, height, and weight was .886. Thus, 75% of the variance in strength was in this case attributed to these five variables, leaving only a small percentage to other variables. On the basis of this Jones holds that intrinsic factors, factors that determine body build, play a much more important role in individual differences in strength than does exercise or functional training. The adolescent growth study (Jones, 1949) also clearly demonstrated that the physique of the male changes little during adolescence and that the strength of the ectomorph lags well behind that of the mesomorph throughout the adolescent years.

The above is illustrated in Fig. 1 from data by Tuddenham and Snyder (1954) in which mean strength scores of adolescent males grouped according to physique are plotted for ages 9, 11, 13, 15, and 17. Throughout this age range the mean strength scores of the mesomorphs are substantially above that of the other two physique groups, even though their average weight at no point exceeded that of the endomorphs. The ectomorphs, as would be expected, were consistently the weakest of the three groups. While the strength superiority of the mesomorph increases with advancing age it is substantial in the preadolescent years. In terms of strength per pound of body weight the mesomorphs were as a group stronger on the average than ectomorphs and the endomorphs at 11 years of age and older (Fig. 2). It is clearly evident that strength, muscularity, and mesomorphy go hand in hand in males throughout the preadolescent and adolescent years.

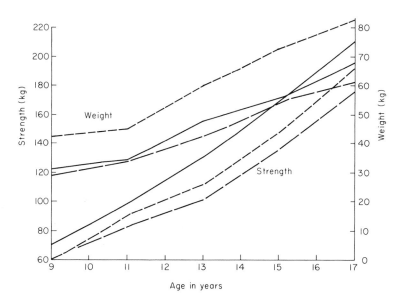

FIG. 2. Longitudinal strength and weight scores of adolescent males grouped by physique categories (strength scores: sum of right and left grip, pull, and thrust). Key same as in Fig. 1.

Only limited study has been made of the relationship between physique and strength of adolescent girls. Espenschade (1940) in the California Adolescent Growth Study reported that when comparing the group of girls with the best overall gross motor performance scores with those having the poorest records, the high group was slightly superior in manual and shoulder girdle strength to the poor group through the 3½-year period that their growth was followed. Differences throughout this period in body structure were evident, the high group being taller, lighter in weight, and having less subcutaneous tissue than the low performing group. While somatotyping procedures were not used, the high group tended to be slightly more slender in build with proportionately narrower hips than the low group. This would suggest that while the differences in strength between the extreme categories were not great, the girl with the build of the ecto-mesomorph tended to be stronger than the endomorph in spite of the fact that her body weight was substantially less.

There is some evidence that differences in strength in adolescents can in part be attributed to differences in the degree of masculinity and feminity within a sex. This is borne out by the research of Bayley (1951) who reported that when standards of somatic androgeny were used with a sample of 79 adolescent boys and 83 adolescent girls, and the adolescents grouped into somatic androgeny classifications, substantial differences in hand and shoulder girdle strength were found among the groups (see Fig. 3). While the mean strength index (sum of right grip, left grip, pull, and thrust divided by height in centimeters times weight in kilograms) of the boys was substantially greater than that of the girls, the girls in the bisexual feminine classification were on the average stronger than the boys in the bisexual masculine classification. Similarly, as one moves from the hyperfeminine category to the hypermasculine somatic androgeny category, the strength index becomes progressively higher. These data indicate that body morphology as reflected by assessments of somatic androgeny may well be a factor of some consequence in accounting for individual differences in strength within and between sexes.

X. Physique and Motor Performance

A. INFANCY AND CHILDHOOD

Motor development early in life is related to a variety of morphological features suggestive of physique-associated effects. Muscular and small-boned infants (Shirley, 1931) and those of linear frame (Norval, 1947) tend to walk at an early age. Large leg muscle mass (radiographically determined from calf x-rays) is associated with early standing and walking, while

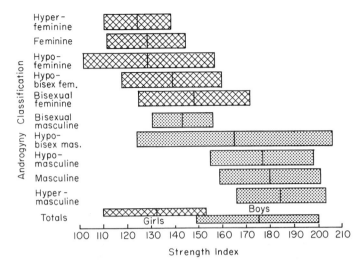

FIG. 3. Strength index scores according to somatic androgeny classification (strength scores: sum of right and left grip, pull, and thrust). (Reprinted with permission, *Child Development,* from Bayley, 1951).

relatively small muscle mass tends to delay the attainment of these developmental milestones (Garn, 1963). The work of Garn (1966) indicates that leg muscle mass at 6 months of age is predictive of walking at 1 year, which suggests a developmental rather than an activity-mediated relationship.

The foregoing observations, though limited, emphasize the need to extend physique–performance–body composition relationship studies to infancy and the preschool ages. Subcutaneous fat, for example, decreases from about 1 year of age through the preschool years (Tanner, 1962), the years when most motor skills are developing at perhaps their most rapid rate. Unfortunately, we know very little about the impact of the body's changing composition on the development of motor skills early in life. There is, however, evidence to indicate relative stability of physique from approximately 6½ years of age to maturity (Petersen, 1967). Hence, we have reason to believe that the physique of the man is established early in life. What this means in terms of stability of growth in motor performance is not known.

Similarly, we have only limited information on the role that physique plays in the motor development of children in early childhood. Some inferences can, however, be drawn from the work of Walker (1962) who used standard photographic procedures to somatotype 125 nursery children, 2, 3, and 4 years of age. In relating the somatotype of these children to teacher-assessed behavior characteristics, Walker concluded that there appear to be multiple determiners of behavioral traits which arise from primary body conditions, i.e., from strength, energy, and sensory thresholds. There was a tendency for mesomorphic girls to channel their energies into social activities, the mesomorphic boys into gross motor activities. While, according to Walker (1962), variations in physical energy and in body sensitivity seem to be important mediating links between physique and behavior, it would appear that at this early age the expectations of our culture are already having an impact on what children sense is appropriate behavior for each sex.

As was previously noted with strength, performance in motor skills during the elementary school ages is largely unaffected by body build and constitutional factors, except at the extremes of the continuum. Many factors, either related or unrelated to physique, are operating which affect motor performance during these formative years. With advancing age, a broadening range of experience, and an increase in size, motor performance improves. Boys perform better than girls in most gross motor activities in childhood. Performance is moderately related to skeletal maturation (Seils, 1951), though the maturity effect is of little consequence when chronological age, height, and weight are held constant (Rarick and Oyster, 1964). In the latter study, however, the extremes of the performance continuum, i.e., best vs. poorest, showed maturity differences, best performers being accelerated in skeletal maturity.

Considerable research has been done on the relationship between measurements of body build and measurements of motor performance in later childhood and early adolescence. For the most part these studies have shown only low to moderate relationships between physique and motor performance. For example, in a factor analytic study of preadolescent boys 7–11 years of age, Barry and Cureton (1961) reported practically no effect of morphological variables (physique) upon motor performance when general body size was controlled by rotation procedures.

Similarly, Eynon (1958), in relating physique to a variety of motor items in boys 6 through 12 years of age, noted that agility was positively related to mesomorphy ($r = .311$), negatively related to endomorphy ($r = -.332$), and not related to ectomorphy ($r = .132$). Otherwise, correlations of somatotype and motor variables were generally low ($r < .30$). Likewise, Voisard (1954) reported low correlations between the Iowa Brace Test and somatotype in boys 10–13 years. Ectomorphy was positively related to the Iowa Brace scores ($r = .434$), endomorphy was negatively related ($r = -.499$), and mesomorphy was not related ($r = .153$).

Using Wetzel physique ratings, Bookwalter (1952) noted that boys classified as obese were the poorest performers. Large and small boys in the thin and medium Wetzel physique channels performed equally well when compared within their respective size categories. In the "average" size category, thin boys performed better than boys in the medium channels. Although the data suggest physique–performance relationships, note that the Wetzel physique ratings are derived from height and weight with their inherent limitations as physique indicators.

In a later study using the Wetzel grid with junior high school boys, Wear and Miller (1962) reported that boys in the heavy physique channels performed poorly, while boys classified as medium in physique and "normal" in development performed best. It should be noted that Wetzel's developmental levels (accelerated, normal, and retarded) are not maturity categories in the traditional sense (i.e., skeletal, menarcheal, etc.). Developmental level is essentially an age–size rating. Further, Wear and Miller's sample comprised an age range from 11 years, 11 months to 16 years, 9 months, an age during which many boys experience the adolescent spurt, an age during which physique shows some instability, and also an age during which normal children, with no developmental problems or skeletal age deviations, are known to deviate as much as two or more Wetzel channels (Krogman, unpublished data). Hence, though Wear and Miller's (1962) data are suggestive, they must be interpreted within the context of the limitations imposed by cross-sectional studies and the Wetzel grid.

B. ADOLESCENCE

As in childhood the quest for high relationships between physique and gross motor performance has not met with much success. In comparing high and low performing groups in the Adolescent Growth Study, Espenschade (1940) reported that body build had little influence upon average performance but was a factor of some consequence in exceptional achievement. High performing boys tended to have a medium physique with relatively short legs and narrow hips as compared to low performers who had a stocky build, wide hips, and relatively long legs. The high performing girls were characterized by a slender physique and short legs in contrast to the broad and heavy build of the poor performing girls. More recently, Espenschade and Eckert (1967) reported Sheldonian somatotype ratings for high and low performing adolescents. The ratings represent a composite for each subject based upon longitudinal observations. High performing boys were only slightly higher in mesomorphy ($\overline{X} = 5.2$ vs. 4.5) and considerably lower in endomorphy ($\overline{X} = 3.0$ vs. 4.6), with little difference in ectomorphy ($\overline{X} = 3.1$ vs. 2.9). Similar somatotype trends were apparent for high performing girls. They were slightly higher in mesomorphy ($\overline{X}$ = 3.7 vs. 3.0) and lower in endomorphy ($\overline{X}$ = 4.9 vs. 5.6), with essentially identical ectomorphy ($\overline{X}$ = 2.7 vs. 2.8). It appears that in both sexes high endomorphy is related to poor motor performance during adolescence.

In a comprehensive study of Ceylonese subjects between 10 and 20 years, Cullumbine *et al.,* (1950) noted greater capacity for severe exercise in subjects of linear build than in stocky and obese types. On the other hand, response to moderate exercise was unrelated to physique. Those of midrange, normal physiques performed better in the dash and in the performance of moderate exercise carried to fatigue. An examination of the physiques of the high school males in the endurance study of Cureton *et al.* (1945) showed that mesomorphy was associated with the best scores and endomorphy with the poorest.

Studies of the young adult of college age clearly show the relationship between somatotype components and performance. In marked endomorphy excess fat (i.e., dead weight) is a handicapping factor, while in marked ectomorphy deficient strength is a handicap. Mesomorphic males perform better than either ectomorphs or endomorphs in agility, speed, and endurance tests, while ectomorphs perform better than endomorphs in the same items (Sills and Everett, 1953). In a factor analysis study, Sills (1950) found positive loadings of motor ability on mesomorphy and negative loadings on endomorphy. Interestingly, ectomorphy was not identified as a discrete factor; rather, it appeared to be reciprocal to endomorphy and mesomorphy.

In college women, Perbix (1954) reported similar observations. Mesomorphy was positively related to strength and power items, while endomorphy was negatively related to power and agility. Ectomorphy did not appear as a significant factor, perhaps because endomorphy was the dominant component of this female sample. Garrity (1966) reported better performance levels in mesomorphic-ectomorphic college women. Endomorphic-ectomorphs consistently performed at low levels. Marked development of the endomorphic component appeared to be a limiting factor in performance. Interestingly, women with balanced somatotypes performed somewhat more consistently and at higher levels than most other somatotype categories.

The studies cited throughout this section would seem to confirm the observation made at the outset that physique does not markedly influence performance except at the extremes of the physique continuum. There obviously is more to performance than physique. For some, physique clearly is a limiting factor in physical performance. Yet it is only one of many factors that affects motor performance of most individuals. Its significance, except in a few extreme instances, is far less than numerous physiological, kinesiological, and psychological factors.

References

Acheson, R. M., and Dupertuis, C. W. (1957). *Hum. Biol.* **29**, 167.

Anastasi, A. (1958). "Differential Psychology," 3rd ed., Macmillan, New York.

Bayley, N. (1951). *Child Develop.* **22**, 47.

Barry, A. J., and Cureton, T. K. (1961). *Res. Quart.* **32**, 283.

Barton, W. H., and Hunt, E. E., Jr. (1962). *Hum. Biol.* **34**, 254.

Bohm, G. E. (1959). Unpublished Master's Thesis, University of Wisconsin, Madison.

Bookwalter, K. W. (1952). *Res. Quart.* **23**, 271.

Carter, J. E. L. (1970). *Hum. Biol.* **42**, 535.

Comas, J. (1957). "Manual of Physical Anthropology." Thomas, Springfield, Illinois.

Craven, D., and Jokl, E. (1946). *Clin. Proc. (Cape Town)* **5**, 18.

Cullumbine, H., Bibile, S. W., Wikramanayake, T. W., and Watson, R. S. (1950). *J. Appl. Physiol.* **2**, 488.

Cureton, T. K. (1951). "Physical Fitness of Champion Athletes." Univ. of Illinois Press, Urbana.

Cureton, T. K. (1964). *Mon. Soc. Res. Child Develop.* **29**, Ser. No. 95, N. 4.

Cureton, T. K., Huffman, W. J., Welser, L., Kireilis, R. W., and Latham, D. E. (1945). *Mon. Soc. Res. Child Develop.* **10**, Ser. No. 40, N. 1.

Damon, A. (1970). *In* "Anthropology and the Behavioral and Health Sciences" (O. von Mering and L. Kasdan, eds.), pp. 179–205. Univ. of Pittsburgh Press, Pittsburgh, Pennsylvania.

Damon, A., Bleibtreu, H. K., Elliot, O., and Giles, E. (1962). *Amer. J. Phys. Anthropol.* [*N.S.*] **20**, 461.

Danby, P.M. (1953). *J. Royal Anthropol. Inst.* **83**, 194.

de Wijn, J. F. (1968). *Acta Paedopsychiat.* **35**, 301.

Domey, R. G., Duckworth, J. E., and Morandi, A. J. (1964). *Psychol. Bull.* **62**, 411.

Dupertuis, C. W., and Michael, N. B. (1953). *Child Develop.* **24**, 203.

Espenschade, A. S. (1940). *Mon. Soc. Res. Child. Develop.* **5**, Ser. No. 24, N. 1.

Espenschade, A. S., and Eckert, H. M. (1967). "Motor Development." Merrill, Columbus, Ohio.

Eveleth, P. B. (1966). *Ann. N. Y. Acad. Sci.* **134**, 750.

Eynon, R. B. (1958). Unpublished Master's Thesis, University of Illinois, Urbana (cited by Cureton, 1964).

Florey, C. du V. (1970). *J. Chronic Dis.* **23**, 93.

Garn, S. M. (1961). *In* "De.Genetica Medica, Pars II" (L. Gedda, ed.), pp. 415–434. Gregor Mendel Inst., Rome.

Garn, S. M. (1962). *Mod. Probl. Paediat.* **7**, 50.

Garn, S. M. (1963). *Ann. N. Y. Acad. Sci.* **110**, 429.

Garn, S. M. (1966). *In* "Review of Child Development Research" (L. W. Hoffman and M. L. Hoffman, eds.), pp. 529–561. Russell Sage Found., New York.

Garn, S. M., Clark, A., Landkof, L., and Newell, L. (1960). *Science* **132**, 1555.

Garrity, H. M. (1966). *Res. Quart.* **37**, 340.

Godin, P. (1920). "Growth During School Age" (transl. by S. L. Eby). Gorham Press, Boston, Massachusetts.

Hammond. W. H. (1953). *Hum. Biol.* **25**, 65.

Hammond, W. H. (1957). *Hum. Biol.* **29**, 40.

Heath, B. H. (1963). *Amer. J. Phys. Anthropol.* [*N.S.*] **21**, 227.

Heath, B. H., and Carter, J. E. L. (1966). *Amer. J. Phys. Anthropol.* [*N.S.*] **24**, 87.

Heath, B. H., and Carter, J. E. L. (1967). *Amer. J. Phys. Anthropol.* [*N.S.*] **27**, 57.

Heath, B. H., Hopkins, C. E., and Miller, C. D. (1961). *Amer. J. Phys. Anthropol.* [*N.S.*] **19**, 173.

Herron, R. E. (1960). Unpublished Master's Thesis, University of Illinois, Urbana (cited by Cureton, 1964).

Huber, N. M. (1969). *Amer. J. Phys. Anthropol.* [*N.S.*] **31**, 171.

Hunt, E. E., Jr., and Barton, W. H. (1959). *Amer. J. Phys. Anthropol.* [*N.S.*] **17**, 27.

Hunt, E. E., Jr., Cocke, G., and Gallagher, J. R. (1958). *Hum. Biol.* **30**, 73.

Jokl, E., Culver, E. H., Goedvolk, C., and de Jongh, T. W. (1941). *Publ. S. Afri. Inst. Med. Res., Reprint* No. 303.

Jones, H. E. (1949). "Motor Performance and Growth." Univ. of California Press, Berkeley.

Kraus, B. S. (1951). *Amer. J. Phys. Anthropol.* [*N.S.*] **9**, 347.

Krogman, W. M. (1970). *Mon. Soc. Res. Child. Develop.* **35**, Ser. No. 136, N. 3.

Krogman, W. M. Unpublished data.

Lasker, G. W. (1947). *Amer. J. Phys. Anthropol.* [*N.S.*] **5**, 323.

Lindegard, B. (1953). *Acta Psychiat. Neurol. Scand., Suppl.* **86**.

Lindegard, B., ed. (1956). "Body Build, Body Function and Personality." Lund, Gleerup.

McCloy, C. H. (1940). *Child Develop.* **11**, 249.

Malina, R. M. (1969a). *Clin. Pediat.* **8**, 16.

Malina, R. M. (1969b). *Clin. Pediat.* **8**, 476.

Malina, R. M. (1969c). *Clin. Orthop. Related Res.* **65**, 9.

Malina, R. M., Harper, A. B., and Holman, J. D. (1970). *Res. Quart.* **41**, 503.

Malina, R. M., Harper, A. B., Avent, H. H., and Campbell, D. E. (1971). *Med. Sci. Sports* **3**, 32.

Munroe, R. A., Clarke, H. H., and Heath, B. H. (1969). *Amer. J. Phys. Anthropol.* [*N.S.*] **30**, 195.

Norval, M. A. (1947). *J. Pediat.* **30**, 676.

Osborne, R. H., and De George, F. V. (1959). "Genetic Basis of Morphological Variation." Harvard Univ. Press, Cambridge, Massachusetts.

Parízková, J. (1968). *Hum. Biol.* **40**, 212.

Parnell, R. W. (1954). *Amer. J. Phys. Anthropol.* [N.S.] **12**, 209.

Parnell, R. W. (1958). "Behaviour and Physique. An Introduction to Practical and Applied Somatometry." Arnold, London.

Paterson, D. G. (1930). *In* "The Measurement of Man," pp. 117–170. Univ. of Minnesota Press, Minneapolis.

Perbix, J. A. (1954). *Res. Quart.* **25**, 84.

Petersen, G. (1967). "Atlas for Somatotyping Children." Royal Vangorcum Ltd., Publ., Assen, The Netherlands.

Pryor, H. B. (1940). "Width-Weight Tables," 2nd ed. Stanford Univ. Press, Palo Alto, California.

Pryor, H. B. (1966). *J. Pediat.* **68**, 615.

Rarick, G. L. (1960). *In* "Science and Medicine of Exercise and Sports" (W. R. Johnson, ed.), pp. 440–465. Harper, New York.

Rarick, G. L., and Oyster, N. (1964). *Res. Quart.* **35**, 523.

Rauh, J. L., Schumsky, D. A., and Witt, M. T. (1967). *Child Develop.* **38**, 515.

Rees, L. (1960). *In* "Handbook of Abnormal Psychology: An Experimental Approach" (H. J. Eysenck, ed.), pp. 344–392. Basic Books, New York.

Reynolds, E. L. (1949). *Hum. Biol.* **21**, 199.

Reynolds, E. L., and Asakawa, T. (1950). *Amer. J. Phys. Anthropol.* [N.S.] **8**, 343.

Robbins, L. M. (1962). *Proc. Indiana Acad. Sci.* **72**, 63.

Roberts, D. F. (1960). *Symp. Soc. Study Hum. Biol.* **3**, 59.

Roberts, D. F., and Bainbridge, D. R. (1963). *Amer. J. Phys. Anthropol.* [N.S.] **21**, 341.

Segraves, R. T. (1970). *Brit. J. Psychiat.* **117**, 405.

Seils, L. G. (1951). *Res. Quart.* **22**, 244.

Seltzer, C. C. (1946). *Amer. J. Phys. Anthropol.* [N.S.] **4**, 389.

Seltzer, C. C. (1966). *N. Engl. J. Med.* **274**, 254.

Sheldon, W. H., and Stevens, S. S. (1942). "The Varieties of Temperament." Harper, New York.

Sheldon, W. H., Stevens, S. S., and Tucker, W. B. (1940). "The Varieties of Human Physique." Harper, New York.

Sheldon, W. H., Dupertuis, C. W., and McDermott, E. (1954). "Atlas of Men: A Guide for Somatotyping the Adult Male at All Ages." Harper, New York.

Shirley, M. M. (1931). "The First Two Years; A Study of Twenty-five Babies," Vol. I. Univ. of Minnesota Press, Minneapolis.

Sills, F. D. (1950). *Res. Quart.* **21**, 424.

Sills, F. D. (1960). *In* "Science and Medicine of Exercise and Sports" (W. R. Johnson, ed.), pp. 40–53. Harper, New York.

Sills, F. D., and Everett, P. W. (1953). *Res. Quart.* **24**, 223.

Simmons, K. (1944). *Mon. Soc. Res. Child. Develop.* **9**, Ser. No. 37, N. 1.

Simon, E. (1961). *In* "Health and Fitness in the Modern World" (L. A. Larson, ed.), pp. 31–41. Athletic Institute, Chicago, Illinois.

Sinclair, G. D. (1969). Thesis, University of Oregon, Eugene (abstr.).

Steinhaus, A. H. (1933). *Physiol. Rev.* **13**, 103.

Tanner, J. M. (1952). *Amer. J. Phys. Anthropol.* [N.S.] **10**, 427.

Tanner, J. M. (1953a). *In* "Anthropology Today" (A. L. Kroeber, ed.), pp. 750–770. Univ. of Chicago Press, Chicago, Illinois.

Tanner, J. M. (1953b). *In* "Clinical Genetics" (A. Sorsby, ed.), pp. 155–174. Butterworths, London.

Tanner, J. M. (1962). "Growth at Adolescence," 2nd ed. Blackwell, Oxford.

Tanner, J. M. (1964a). "The Physique of the Olympic Athlete." George Allen and Unwin, London.

Tanner, J. M. (1964b). *In* "Human Biology," pp. 297–397. Oxford Univ. Press, London and New York.

Tuddenham, R. D., and Snyder, M. M. (1954). "Physical Growth of California Boys and Girls." Univ. of California Press, Berkeley.

Vandenberg, S. G. (1962). *Amer. J. Phys. Anthropol.* [N.S.] **20**, 331.

Voisard, P. P. (1954). Unpublished Master's Thesis, University of Illinois, Urbana (cited by Cureton, 1964).

Walker, R. N. (1962). *Mon. Soc. Res. Child Dev.* **27**, Serial No. 84, 3.

Wear, C. L., and Miller, K. (1962). *Res. Quart.* **33**, 615.

Zuk, G. H. (1958). *J. Genet. Psychol.* **92**, 205.

CHAPTER

7

Age Changes in Motor Skills

Helen M. Eckert

Within the context of this chapter, the term "motor skills" refers to all movements of any part of the body which are the result of the contractions of skeletal muscles within the individual. This interpretation includes the neuromuscular responses involved in reading, speaking, and writing as well as the "large muscle" activities of walking, running, jumping, and throwing. The former, which are ontogenetic in development and frequently used in learning experiments, will not be discussed here.

The central focus of this chapter will be directed toward tracing the development of large muscle, phylogenetic activities. Consideration will also be given to the development of motor skill components such as strength, balance, and coordination, which are required in various combinations and to different degrees in the motor activities of children. Measurements of motor skill should ideally involve the use of motor tests which assess a particular component to the exclusion of others. Rarely, if ever, is this the case, for human movement and single motor tasks invariably involve more than one basic component. The simplest of movements requires a certain amount of force, is executed at a certain speed, and requires the coordination and integration of certain body parts. Thus, while it is impossible to measure directly the basic components of motor skills, such as strength, speed, balance, and coordination, tests have been designed which by their nature logically require a greater proportion of one component than of others. Therefore, measurement procedures have been developed through which we can observe the changes in motor performance of children and at the same time, by inference, learn something about the basic components which underlie age changes in motor skill attainments.

I. Early Patterns of Motor Behavior

The reflexes of the newborn infant may be considered as well-established unitary motor skills, and there is general agreement that the establishment of reflex responses is an important aspect in the development and acquisition of additional motor skills. Piaget (1952) considered the reflexes of the newborn infant to be the building blocks for the development of more mature types of behavior. Gardner (1965) urged the utilization of reflexive behavior to expedite the learning of motor skills, yet also indicated that some motor skills require the inhibiiton of reflexive behavior for their proper execution.

, A number of reflexes and neonatal responses may be categorized as having locomotory implications. The "trotting" or alternate stepping reflex of the supported newborn infant when its feet are resting on a solid surface and the momentary raising of the head when it is placed in a prone position have been identified by McGraw (1946) as neonatal responses which have implications for standing and bipedal locomotion. In addition, McGraw (1940) has reported flexion of the knees and hips resulting in an up-and-down motion of the body when the neonate is suspended by the ankles. This response is suggestive of the up-and-down motion of the body in the two-footed hop of the young child. The assumption of the tonic neck reflex with the turning of the infant's head to its side suddenly when it is lying on its back (Pratt, 1954) may have implications for the alternate limb flexion and extension of bipedal locomotion. Hellebrandt et al. (1956) have associated the tonic neck reflex with postural alignment and facilitation during heavy resistance and stressful exercise.

The grasping reflex of the newborn infant would seem to be a vital precursor of all skills involving prehension. Moreover, in terms of the strength component of motor skills the grasping reflex would appear to be proportionately closer to adult grasping strength than is the strength of the legs in the trotting reflex in proportion to adult leg strength. When pulled toward suspension with both hands infants are generally able to support more than 70% of their body weight. Halverson (1937) found that 27 of 97 infants examined under 24 weeks of age were able to support their entire weight with one of the infants, aged 4 weeks, having the strength to support its weight with the right hand alone. The average length of suspension time with two hands for the neonate has been recorded by Richter (1934) as 60 sec with the longest time being 128 sec. No comparable figures for weight support with the legs are available at present, but McGraw (1946) indicated that genuine upright ambulation is impossible in the neonate because of lack of strength and undeveloped equilibratory apparatus. No records are currently available indicating either support or propulsion of the body by the arms of the neonate when in the prone position.

Although the responses of the newborn infant include unilateral and bilateral limb movements in addition to the selectively cited reflexes, the neonate lacks the strength to overcome the effects of gravity on the body as a whole. Furthermore, the neonate sleeps approximately 20 hr out of 24 (Pratt *et al.,* 1930) so that there is little opportunity or incentive for movement. The neonatal period is generally considered to be one of adjustment from aqueous to aerobic environment and of perfection of the newly activated physiological and sensory functions.

The period of infancy is highlighted by the refinement of prehension and the acquisition of upright, bipedal locomotion. The comparatively high strength component of the hand grasp of the young infant and the great change that occurs during the first month after birth in the area controlling hand movements of the gyrus centralis in the brain (Conel, 1939, 1941) are undoubtedly factors contributing to the early refinement of prehension.

The grasping reflex of the neonate is elicited by touch and has been identified by Halverson (1937) as having two phases; first, gripping or clinging which is considered to be a proprioceptive response to pull on the finger tendons and, second closure of the fingers to light pressure on the palm. Although the young infant will grasp and hold an object of appropriate size, no attempt is made to reach for an object until after control has been attained over the oculomotor muscles, at about the sixteenth week. After this time, the sight of objects will result in arm activation that is ineffective until approximately 4 weeks later when a "corralling" action of the arms and hands working together will sweep the object closer rather than pushing it out of reach. At approximately 24 weeks, the infant is able to reach for and secure an object with one hand; however, the procedure involves the distinct movements of raising the hand, a circuitous thrusting forward of the hand, and the final lowering of the hand to grasp the object. Reaching and grasping become coordinated into a single continuous movement by approximately the fortieth week (Halverson, 1931; Gesell and Ilg, 1946).

The grasping reflex of the neonate, which involves palm and finger flexion without thumb opposition, begins to disappear at about 4–6 months and is gradually replaced by the voluntary, adult grasp with opposing thumb (Pratt, 1954). This shift of thumb position in the grasp coincides with ulnar-radial shift in the positioning of the grasped object in the palm of the hand. The end product of the changes in thumb position and of object position is the precise control afforded by the pincherlike opposition of thumb and forefinger at approximately 52 weeks (Halverson, 1931). The achievement of the adult mode of prehension does not mean that it will be consistently used by the infant in the grasping of all objects. The size, shape, weight, and texture of the object are determining factors in the mode of prehension employed. Increased control and independence of digital action occurs with increasing age with the rate and extent of increase being dependent upon environmental opportunities and heredity.

TABLE I

DEVELOPMENT OF LOCOMOTION[a]

Stage	Time of onset	Activity
1. Postural control of upper body	Before 20 weeks	Chin up[b] Chest up[b] Stepping[c] Sit on lap[b]
2. Postural control of entire trunk and undirected activity	25–31 weeks	Sit alone momentarily[b] Knee push or swim Rolling Stand with help[b,c] Sit alone 1 min[b]
3. Active efforts at locomotion	37–39.5 weeks	Some progress on stomach Scoot backward
4. Locomotion by creeping	42–47 weeks	Stand holding to furniture[b,c] Creep Walk when led[c] Pull to stand[b]
5. Postural control and coordination for walking	62–64 weeks	Stand alone[b,c] Walk alone[c]

[a] Adapted from Shirley (1931).
[b] Progression in upright posture. See text.
[c] Progression in attainment of bipedal locomotion. See text.

II. Locomotor Development

The aqueous enviornment of the uterus supports the developing fetus in comparative weightlessness so that there is no fetal necessity or demand for strength other than that required for changes in position. Moreover, the equilibratory apparatus is shielded from gravitational force prior to birth and the motor areas of the gyrus centralis associated with trunk and legs lag in cortical development in comparison with the areas which mediate the movements of the neck and shoulders (Conel, 1939, 1941). Thus, the prenatal environment offers the developing organism limited einvironmental stimuli to activate the mechanisms needed for developing the postural and locomotor mechanisms. It is not surprising, then, that the acquisition of upright, bipedal locomotion is such a long and involved procedure.

In dealing with the mass of data from studies of the development of upright posture and bipedal locomotion, investigators have found it convenient to identify phases in the development of these behavioral patterns and the factors which are associated with their onset. While different investigators do not describe these developmental changes in exactly the same way, an examination of two such approaches clearly indicates that the sequence of events in the development of upright locomotion is the same in both. Shirley (1931), as shown in Table I, combines a listing of 17 activities into five stages which are indicative of progressive steps in the acquisition of upright posture and of walking. The activities marked by superscript *b* are classified by Shirley as those which indicate progression in upright posture and those identified by superscript *c* are indicative of progression in the attainment of bipedal locomotion. The unmarked activities listed in Table I are considered to provide a temporary means of locomotion and merely overlap the other two categories without aiding their development other than through the incidental increase in strength.

Similarly, Gesell and Ames (1940) identified 23 stages of locomotor behavior and then compressed these into four cycles, each of which are considered to be a "continuum of closely related stages." The following cyclic descriptions illustrate fluctuations in flexor and extensor dominance as well as the integration of bilateral, unilateral, and crossed lateral movements into more complex movements. In general, the most mature expression of flexor dominance occurs in creeping, whereas upright locomotion eventually involves a permanent perponderance of extensor action. This cyclic approach to the development of upright locomotion led Gesell (1954) to the promulgation of his principle of reciprocal interweaving.

1. In the first cycle (stages 1–10, covering the first weeks of life) the dominant pattern of bilateral flexion of arms and legs which is present at

birth gradually gives way to more mature unilateral flexion of the extremities. During this cycle the trunk remains in contact with the supporting surface, although the extremities are used with limited success for circular locomotion.

2. During the second cycle (stages 11–19, the 30th through the 42nd week) the infant is able to elevate the trunk above the table surface. This requires a temporary reversion of the arms to the more elemental position of bilateral flexion. The predominate motor patterns of this cycle include bilateral extension of arms and bilateral extension and flexion of the legs as observed in the backward crawl, the low creep, the high creep and rocking. In Stage 19 the creep involves a high level of movement with rather involved coordinations, for not only is there alternate extension of the arms and alternate flexion of the legs, but these movements occur with the members of the opposite side moving alternately.

3. The third cycle (stages 21a and 21b, the 49th through 56th week) entails a temporary reversion to an immobile bilateral extension in maintaininig the plantigrade stance. Shortly thereafter, arms and legs extend forward alternately in plantigrade progression, the right arm and leg moving simultaneously.

4. The fourth cycle (stages 22 and 23, and 50th through 60th week) finds the infant capable of full trunk extension and assuming an upright posture. Walking follows as the arms and legs extend bilaterally, but with the movements occurring alternately.*

The cyclic aspect of Gesell's analysis (1954) is indicated by an increase in the maturity gradient within each cycle and a temporary regression (or reversal) in the maturity gradient at the beginning of each subsequent cycle so that the total progression toward upright, bipedal locomotion is of a spiral nature. Shirley (1931), however, explained periods of little or no gain in the progression toward upright locomotion in terms of the development of functional equilibrium. The infant must be able to control its body in any new postural alignment in a static position before it can undertake the shifting postural sets which accompany movement. The transition from the creeping to the bipedal position involves not only a raising of the center of gravity but also a greatly reduced base of support from four relatively broadly spaced points to two comparatively close ones. Thus, the base of support at any one point in time with a shifting center of gravity changes from three points to one point as the child assumes the bipedal manner of locomotion.

Strength may also be a factor contributing to apparent delays between new postural alignments and movement. Reduction in the number of points of support means that a proportionately greater weight of the body must be supported on the remaining points. The infant undergoes a very marked period of growth especially in the length of the limbs, and the gains in strength must be sufficient to support the previously acquired body weight in addition

* From Rarick (1954). By permission of the author.

to the weight from recent growth. Accounts of the swimming ability of infants indicate that locomotion is greatly facilitated when the supported body weight is reduced by immersion in water. Analysis of the "dog paddle" type of movement used by infant swimmers may prove to be an avenue for exploration of the persistence of the trotting reflex as equilibratory and strength requirements are minimized during immersion in water.

The primitive modes of locomotion, such as rolling, pivoting, rocking, scooting, crawling and creeping, were considered by Shirley (1931) as transient motor skills that did not contribute to either upright posture or bipedal locomotion whereas Gesell (1954) regarded these activities as functional expressions of necessary stages in the organization of the neuromuscular system. Delecato (1963) has proposed a theory of neurological organization which attempts to explain learning difficulties in terms of faulty neuromuscular patterning during the developmental sequence leading to upright locomotion and recommended a remedial training program to overcome these difficulties. In view of the contradictory theories of Shirley and Gesell and in light of more recent findings of little or no relationship between locomotory developmental training programs and learning difficulties (Skinner, 1968), questions may be raised as to the effectiveness of such training programs.

The various stages and cycles in the progressive development of upright locomotion identified by investigators such as Shirley (1931), Gesell and Ames (1940), and Bayley (1935) are based upon averages obtained from observations of groups of children. Therefore, it is possible that all phases do not appear in all children nor that their duration is in any sense the same for all children. Variations also occur in the listed times for the onset of various phases; for example, Bayley (1935) recorded 13 months as the median age of walking alone whereas Nicolson and Hanley (1953) reported the mean age for walking of 114 boys to be 13.4 months while that of 123 girls was 13.6 months.

Physique, environment, and heredity would certainly appear to influence the onset of upright locomotion. The recent trend toward larger and heavier babies at birth may have produced some discrepancy in the normative age of the onset of walking for today's infants in comparison with previously recorded averages. However, the sequence of events in the development of upright, bipedal locomotion recorded by early definitive investigations remains valid today.

The achievement of the skill of "walks alone" is only one phase in the progression toward an adult walking pattern. The increase in functional equilibrium during walking is reflected by a decrease in the use of the arms for balance and by a reduction in the base of support through a reduction in the degree of toeing out of the foot and a decrease in the width of the step.

With increasing balance and concomitant confidence, the child is quickly able to increase the speed of his gate and, in the process of doing so, increases the length of his stride, decreases its width, and gradually reduces the angle at which the leg is raised. The rate of walking tends to stabilize at approximately 170 steps/min between the ages of 18 months and 2 years. In comparison, the rate for briskly walking adults has been found to be 140–145 steps. Since the rate of stride of the child does not increase as rapidly as the amount of distance covered within any period of time, the length of stride must increase from 18 months to 2 years to account for the noted increase in the traversed distance (Shirley, 1931).

Foot action in walking changes from the out-toed, full sole step of the beginning toddler to the two-year-old's stride in which the alignment of the foot is parallel to the direction of the walk and the transfer of weight, upon contacting the ground, is from heel to toe. The changes in arm, leg, and foot action that occur with increasing age result in a transition from the toddler with outstretched arms, wide stance, and uneven jerky steps to the child with a uniform length, width, and relative smoothness of stride at 2 years of age. In addition to walking in a forward direction, children soon experiment with walking sideways at about 16.5 months and walking backward at 16.9 months. The more complicated feat of walking on tiptoes is not normally attempted until after 30 months of age (Bayley, 1935). At 3 years of age the child has developed a certain amount of individuality in the way he holds his head and trunk and walking has become so automatic that little attention needs to be given to it even if the surface is slightly uneven. The adult pattern of the walk is approximated by age 4 with the assumption of a smooth transfer of weight together with a rhythmical, swinging style.

III. Development of Phylogenetic Skills

A. CLIMBING

At about the time when an infant learns to creep, he also begins to pull himself to a stand position in his crib, playpen, or nearby furniture. It is not surprising, then, that the infant will creep upstairs approximately 2 months before he learns to walk alone (Shirley, 1933) using movement patterns almost identical to those used in creeping on a level floor (Ames, 1937). Descent of stairs is not mastered as early as ascent, however, and it is not unusual for a child to creep up a short flight of stairs and then cry indignantly because he does not know how to get down. With support, the young child will usually ascend stairs in an upright position between 18 and 20 months of age. The foot pattern in this ascent, however, is not the alternate step pattern of walking. Rather, the child employs a "marking time" pattern in

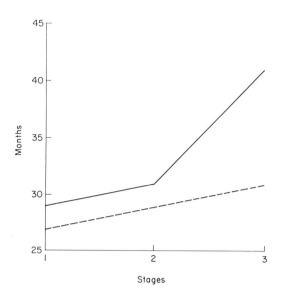

FIG. 1. Ascent of stairs. Stage 1: mark time, without support; stage 2: alternate feet, with support; and stage 3: alternate feet, without support. (———) Long flight, 11 steps; (- -) short flight, 3 steps. (Adapted from data by Wellman, 1937.)

which the child advances the same foot for each riser with the trailing foot placed on the riser beside the lead foot before the next step upward is taken.

The alternate foot pattern is used by the child of approximately 29 months when climbing a short flight (3 steps) of stairs with the support of an adult's hand or a railing (Fig. 1). Unsupported ascent of the same length of stairs using the alternate foot pattern is not accomplished until approximately 2 months later. The ascent of longer flights of stairs occurs later than that of shorter flights with the extent of delay being dependent upon the length of the stairs and the level of achievement of the child. Similarly, as noted for creeping, the descent of stairs is delayed at each level of achievement and is also dependent upon the length of the stairs (Wellman, 1937). Facility in climbing stairs is influenced by the height of the risers with lower risers being mastered first (Gesell and Thompson, 1934). The opportunity for stair climbing and the age at which initial attempts are made affect the progress children will make in acquiring this skill. In general, by the time the child is approximately 41 months old he is able to climb a long flight (11 steps) of stairs unassisted and with an alternate foot placement. Descent of the same flight of stairs in the same manner usually occurs at 55 months (Wellman, 1937).

Proficiency in ladder climbing lags behind that in stair climbing but involves the same sequence in achievement levels with respect to ascent, descent, and foot patterns. The height of the ladder and the angle of inclination are also determinants of the age at which ladder climbing is first attempted. An examination of general climbing ability with equipment such as packing boxes and jungle gyms indicates that 90% of children can be classified as proficient climbers at 6 years (Gutteridge, 1939).

B. RUNNING

The basic limb movements of running are similar to those of walking. Therefore, the neuromuscular adjustments necessary in translating the walking pattern to the run are not great. The increased tempo of step which is achieved shortly after the child begins to walk alone is soon increased to a modified run at approximately 18 months. Although the rate of step in this modified run is faster than in the walk, there is no period at which the body is not supported by one foot and the style of the modified run is similar to the stiff-legged, full-soled, uneven stride of the walk. The true run, however, has a period of momentary nonsupport; hence, the forward and upward thrust of the body mass and the absorption of force in landing requires greater leg strength than in the walk. Moreover, dynamic equilibrium must be maintained at this accelerated tempo.

Between the ages of 2 and 3, the transfer of weight from heel to toe in foot action results in a smoother transfer of body weight and shortly thereafter the movement develops into a true run as more and more force is exerted by the toes at the end of the step. However, control of force and equilibrium is still such that the child is limited in his ability to stop and turn quickly. At the ages of 4 and 5 years, there is continued improvement in equilibrium, power, and the form of the run which results in the gradual development of the controls necessary for stopping, starting, and turning. The adult manner of running becomes reasonably well established by the time the child is 5 and 6 years of age, and at this time the child is able to use his running skill effectively in most play activities.

After achievement of the adult pattern of running, the speed of the run, which is basically a function of the length and frequency of the stride, increases. The steady increase in body size with age and the concomitant increase in muscular strength and lever length allows for increased stride length and tempo in the run. With advancing age there is also an increase in endurance as reflected in the ability to run greater distances at near maximum speed. Therefore, in tests of running speed, older children are usually measured over distances of 50–60 yards and younger ones over 30–35 yards. Since the start used in the run has less effect on the total time of the

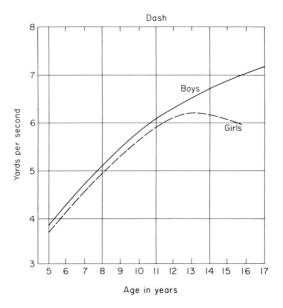

Fɪɢ. 2. Running. (From Espenschade, 1960. Reprinted by permission of author and publisher.)

dash in longer distances than in shorter ones, the starting time in the older age groupings listed in Fig. 2 has less influence on their yards-per-second averages than in the younger age groupings.

Level of achievement of skill in running as expressed in terms of speed in the dash is similar for the sexes and increases with age until adolescence when there is a leveling off and slight decline for girls whereas boys continue to improve at approximately the same rate. Factors which are considered to contribute to the noted sex difference in skill level at adolscence will be discussed in a separate section.

C. Jumping

Jumping, like running, requires that the body mass be projected off the ground, but the period of elevation is longer and the deceleration more abrupt in the jump than in the run. In the latter, forward momentum is transferred by the rocker action of the foot to add force to the next stride, whereas in the jump the landing action must absorb all the momentum that has been generated in the jump. Obviously, the almost instantaneous acceleration of the body mass in the takeoff of the jump followed by the sudden deceleration in landing places great demands upon the functional equilibrium of the organism and necessitates the application of tremendous muscular force at exactly the proper moment. It is not surprising, then, that jumping activities are not undertaken by the child until some degree of facility has been attained in basic locomotor skills.

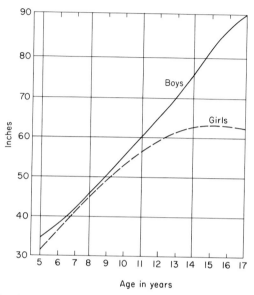

FIG. 3. Standing broad jump. (From Espenschade, 1960. Reprinted by permission of author and publisher.)

The exaggerated step down that occurs when the child descends stairs in an upright position marks the onset of the pattern of movement required in jumping. A one-foot jump or downward leap occurs when the child's feet are momentarily unsupported during his downward progress. In this instance, the child must have achieved enough balance and strength to accommodate the gravity-generated force of his descent to immediate deceleration on a narrow one-foot base. With help, a child will undertake a one-foot jump from a height of 18 in. at approximately 27 months (Wellman, 1937). At approximately the same time the child also jumps up and down on the floor with both feet (Bayley, 1935). Although the two-foot landing provides a wider base, all of the child's previous upright locomotor techniques have been based upon single, not bilateral, limb propulsion. Therefore, it is not surprising that the early patterning in the standing broad jump is a one-foot takeoff and a two-foot landing. The standing broad jump with its two-foot takeoff and landing and with the arms acting as augmenters of the momentum generated by the thrust of the legs is a highly complex skill and, along with the running long jump, constitutes an advanced level of attainment in a sequence of behavioral modifications of phylogenetic origin (Hellebrandt *et al.*, 1961). Jumping over objects and for height is attempted after the onset of jumps from height and for distance with proficiency in this area being subject to ontogenetic refinements such as the straddle jump, western roll, and "Fosbury flop."

Observations by Gutteridge (1939) indicate that 80% of children have reasonably good mastery of jumping skills at 5 years of age. Following the acquisition of the adult pattern in the standing broad jump, jumping proficiency, as measured by the distance jumped, increases in both sexes with increasing age through the childhood years. While the increase in distance jumped with each succeeding age level is approximately linear between the ages of 5–17 years for boys, the performance of girls levels off at the onset of adolescence (Fig. 3). Factors which may influence sex difference in the performance of this skill will be discussed in another section.

D. Throwing

The beginnings of the throwing pattern may be considered to originate in the first release of an object held by an infant. Many children of approximately 6 months are able to perform a crude and unrefined throw which involves limited and isolated use of the throwing arm when the child is in a sitting position. Such a throw usually occurs when the child is rejecting an undesirable object. Shortly before the first year reasonably well defined direction is given to the object. Both distance and direction improve during the second year, but the throwing pattern still remains immature consisting of stiff, jerky movements of the arm with little or no effective use being made of trunk or foot movements (Gesell and Thompson, 1934).

Analysis of the cinematographic records of boys and girls aged 2 through 7 years indicates four distinct types of throws which are closely associated with definite age groupings (Wild, 1938). The least mature throw, seen at 2–3 years of age, is identified by movements of the arm and body which are confined mainly to the anteroposterior plane. For the preparatory phase, the arm is drawn up either obliquely or frontally with a corresponding extension of the trunk until the object to be thrown is at a point high above the shoulder. During delivery, the trunk straightens with a forward carry of the shoulder as the arm comes through in a stiff, downward motion. Both feet remain firmly in place with the body facing in the direction of the throw during both the preparatory phase and delivery of the projectile.

The major change in pattern that occurs in the second type of throw, typical of the 3½–5-year age group, is the shift to horizontal arm and body movements. The more horizontal arm action combined with trunk rotation provides a greater arc for generating velocity thus permitting additional momentum to be imparted to the projectile. The feet, however, continue to remain together and in place during the entire throw.

A great change in the throwing pattern occurs during the fifth and sixth years. This involves the transfer of weight during ball delivery from one foot to the other. In the right-handed child, the step forward is made with the right foot as the ball is delivered with the right hand. The initial transfer of weight from the rear left foot to the leading right foot during the delivery phase of the throw combined with the resulting body rotation and horizontal movement of the arm during forward projection increases the arc and the power of the throw.

Achievement of the mature overhand throwing pattern is marked by a readjustment of the foot patterning to increase the extent (range) of body rotation. For the right-handed individual, this is accomplished by the transference of weight to the rear right foot during the preparatory phase followed by a step forward onto the leading left foot with delivery. This additional trunk rotation coupled with the horizontal adduction of the arm during the forward swing provides the maximal utilization of the body leverage for the attainment of velocity at the most distal segment, the hand.

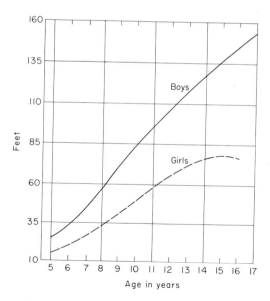

FIG. 4. Distance throw. (From Espenschade, 1960. Reprinted by permission of author and publisher.)

Gutteridge (1939), using a number of trained observers to rate children in throwing ability, noted that 84% were considered to be proficient in this skill at 6 years of age, whereas Wild (1938) reported that none of the girls in her cinematographic analysis of the overhand throw progressed beyond the third type of throw in either arm or foot action. Although data are not available for adult women at present, personal observation indicates that mastery of the mature overhand throwing pattern has not been achieved by many high school girls and college women. Therefore, the marked sex differences in average performance in the overhand distance throw that occurs at all age levels (Fig. 4) would appear to be a reflection of sex differences in throwing pattern which in turn may be influenced by cultural forces. Although boys do on the average, have larger length and girth of forearm (Dupertuis and Hadden, 1951; Meredith and Boynton, 1937) which gives them a mechanical and strength advantage in the propulsion of projectiles, the morphological and strength differences between the sexes are not great enough to account for the extremely large sex differences in this skill. In any event, the proficiency difference in throwing, as measured by the overhand distance throw, widens with advancing age (Fig. 4). This is particularly true after 15 years when the average girl's performance levels off whereas that of the average boy continues to improve.

IV. Bases for Sex Differences in Skill Level

The preceding figures for average performances in the dash (Fig. 2) and the standing broad jump (Fig. 3) indicate slightly higher performance levels for boys than for girls until the onset of adolescence when the differences favoring the boys become more marked. Performance differences may be, in part, a reflection of the greater height and weight of boys from birth to maturity with the exception of early adolescence when the earlier sexual maturity and concomitant accelerated growth of girls tend to make them slightly taller and heavier than boys. The obvious connection between height and limb length together with the contribution of the latter as levers in the mechanics of execution of motor skills would seem to indicate that the slight differences in average performance may be accounted for by slight differences in body size. However, such an inference is complicated by many considerations, one of which is sex differences in body proportions. Prior to the age of 5 years, proportional limb length is slightly greater for girls than for boys although the average actual limb lengths may be the same because of slightly greater height for boys. Bayley's Scale of Motor Development (1936) does not differentiate between the sexes prior to 5 years of age. Between the ages of 5 and 11 years the proportional limb length of the sexes is approximately the same (Bayer and Bayley, 1959); thus, the average actual limb lengths of boys is slightly greater than for girls giving the former a mechanical advantage in most physical skills performance. In addition, sex differences in the structure of the pelvis (Reynolds, 1945, 1947) and the insertion of the head of the femur into the pelvis may contribute to observed sex differences in running and jumping, particularly at adolescence. After 11 years of age the proportional limb length becomes greater for boys than girls (Bayer and Bayley, 1959). This factor together with the greater increment in height that occurs in boys at adolescence gives the adult male a very decided advantage over the female in the performance of phylogenetic physical activities. Furthermore, the greater increase in shoulder width for boys after 11 years in comparison with a reduction in the shoulder growth for girls (Simmons, 1944) accentuates the sex difference in the arc of the shoulder–arm action in the overhand throw for distance.

Overt changes in body proportions reflect proportional changes in body tissues with increasing years. Before 6 years of age, subcutaneous tissue measurements show a similar pattern for both sexes (Boynton, 1936) but shortly thereafter sex differences in mean gain in various body tissues during preadolescent and adolescent years result in different proportions of bone, muscle, and fat for adult males and females. Figure 5 indicates that the mean gain in breadth of skin and subcutaneous tissue (fat) in the calf of girls is generally higher and of longer duration than the mean gain of the same tissues in boys. The mean gain in breadth of bone and muscle in girls declines slightly from 6 to 9 years after which there is a sharp increase through the twelfth year with a gradual decline until the fifteenth year when there is a sharp decline which results in a negative gain at 16 years. On the other hand, the profile of mean gain for the boys indicates a fairly consistent gain from 6 through 11 years which marks the onset of a rapid period of gain extending through the fifteenth year followed by a year of less marked mean gain in bone and muscle.

Increases in bone and muscle tissues should be reflected in increases in strength and such is indeed the case upon comparative analysis of Lombard's data (1950) in Fig. 5 with isometric strength data collected by Rarick (Espenschade and Eckert, 1967). Girls registered their greatest gain in mean isometric strength in four of seven muscle groups tested between the ages of 9 and 10 years, while the greatest gains in mean strength of all seven muscle groups occurred between 11 and 12 years for the boys. Similarly, the rapid gain in height and weight and concomitant increases in muscle tissue during the adolescent growth spurt is reflected in a pronounced increase in grip strength in boys whereas the increment in strength is much less for girls at this time (Jones, 1949). Boys tend to double their grip strength between

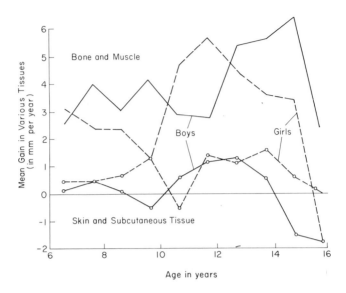

FIG. 5. Gains in breadth of bone, muscle, skin, and subcutaneous tissue. (From Espenschade and Eckert, 1967. By permission of the publisher. Adapted from data by Lombard, 1950.)

the ages of 6 and 11 while an increase of 359% is shown between 6 and 18 years of age (Meredith, 1935). On the other hand, girls increased in grip strength by only 260% between the ages of 6 and 18 with the discrepancy between the sexes being largely attributed to the girls' lessened increase in strength during the adolescent years (Metheny, 1941). A close association has been made between strength in males and male hormone secretion (Simonson et al., 1941; Burke et al., 1953) which would seem to account in part for the pronounced increase in strength in males during adolescence.

Rapid increments in bone and muscle tissue and in strength may require periods of readjustment and reorientation to the increases in body size and strength. Heath (1949), in a study of children ranging in age from 6 to 14 years, and Goetzinger (1961), in a similar study of children aged 8–16 years, reported continuing improvement for both sexes in rail walking over the entire age range with the exception of a slackening off of performance level for females from 12 to 14 years. On the other hand, Keogh (1965) claimed that there is rather clear indication that performance in beam walking does not increase steadily with age in childhood. He reported a consistent change with age in boys except for a leveling off from 7 to 9 whereas girls have a marked increase from 7 to 8 followed by a leveling off from 8 to 10 years. In addition to the reduced increment noted for girls in rail walking during the adolescent years, boys are noted to have steady improvement in beam walking until the age of 11 years when there is a subsequent decline (Seashore, 1947), a reduced rate of change from 13 to 15 years (Espenschade et al., 1953), or greater variability in adolescence (Wallon et al., 1958).

Beam walking tests are classified as dynamic balance tests, while general coordination tests such as the Brace test include items which are believed to measure control, agility, and flexibility in addition to balance. The composite score of the 20 stunts in the Brace test results in performance curves similar to those for the phylogenetic activities of running, jumping, and throwing with increased age (Espenschade, 1960). Analysis of individual items within the test, however, revealed widely divergent trends with item five of the Brace test, which consists of three push-ups, showing the most striking sex differences whereas item seven, in which the individual sits down and then stands again with the arms folded and the feet crossed, produced the least amount of sex difference and the least amount of variation with age (Espenschade, 1940). Factor analyses of numerous items believed to measure coordination have resulted in the identification of different factors by different investigators (Cumbee, 1954; Fleishman, 1964). A comparison of the factors identified from data collected on college women with the factors identified from data for third and fourth grade girls led Cumbee et al. (1957) to recommend that consideration should be given to a different definition of motor coordination at different age levels.

In addition to the sex differences that have been noted in increments in body size and body tissue during adolescence, other anatomical and physiological changes occur which tend to contribute to the marked performance superiority of the male following adolescence. Boys develop a larger thoracic circumference and girls have relatively larger thighs (Boynton, 1936). In addition to having a greater vital capacity and higher maximum breathing capacity (Ferris et al., 1952; Ferris and Smith, 1953) which allows for a greater exchange of gases, boys also have higher mean hemoglobin values and a greater number of red blood cells per cubic centimeter of blood (Mugrage and Andresen, 1936, 1938) for the transport of gases within the body. Blood volume is greater in the male (Sjöstrand, 1953), and the greater increase in heart size which results in the establishment of a greater basal stroke volume (Nylin, 1935) probably accounts for the increased pulse pressure and systolic blood pressure in males (Shock, 1944). The lower resting heart rate and body temperature of adolescent males (Iliff and Lee, 1952) coupled with a greater alkali reserve which allows for the absorption of larger quantities of lactic acid without a change in the pH of the blood (Robinson, 1938) would seem to better equip the male with greater re-

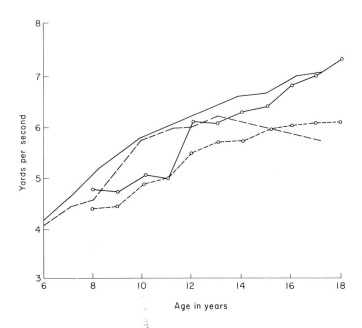

Fig. 6.   Curves of running performance for Bulgarian and American boys and girls. Boys: (○—○) Bulgarian and (——) American. Girls: (○- -○) Bulgarian and (- - -) American. (From Espenschade and Eckert, 1967. By permission of the publisher. Adapted from Bulgarian data by Mangarov, 1964. Adapted from American data by Espenschade, 1960.)

serves for strenuous muscular activity in comparison with the adolescent female.

It is logical to attribute the noted sex differences in phylogenetic physical skills to sex differences, body size, anatomical structure, and physiological functioning. However, it is also possible that social and cultural factors may influence mean performance in running, jumping, and throwing in such a way that the sex differences in performance are magnified. A comparison of Bulgarian mean scores (Mangarov, 1964) with American mean scores (Espenschade, 1960) in the distance throw and dash indicates that performance levels off for the American girls at 15 and 13 years for the respective skills whereas that of the Bulgarian girls continues to improve through 16 and 18 years, respectively. The curves of running performance for Bulgarian and American boys and girls in Fig. 6 indicate that the Bulgarian girls exceed the American girls in yards per second after 15 years of age. The depressed scores for the Bulgarian boys and girls from the ages of 8 through 11 and the rapid gain in yards per second from 11 to 12 years for the Bulgarian boys may be a function of the distance used for the dash. The 40-meter dash was used to test the Bulgarian children through age 11 and the 60-meter dash thereafter. On the other hand, the American data are based upon runs of varying lengths and the depressing effect of the start on yards per second scores may have been reduced by longer runs. In any event the profile of increments is similar for Bulgarian and American boys whereas the performance of Bulgarian girls does not show the marked adolescent lag of American girls. It would appear, therefore, that differences in cultural attitudes toward strenuous physical activity for girls may account to some extent for performance differences. Certainly, the negative increment in bone and muscle for American girls at 16 years (Fig. 4) is attributable to a reduction in muscle tissue rather than bone which implies atrophy of muscle tissue because of lack of maximal use.

Although the performance curves for the phylogenetic activities are very similar, a review of studies reporting intercorrelations between these skills shows that the correlation coefficients are low (Espenschade, 1960). These results may be considered to be at variance with the report of Glassow and Kruse (1960) that girls, aged 6 through 14 years for whom data were available for at least three consecutive years, tended to remain in the same relative position in their group, especially in the run and standing broad jump. However, a number of investigators (Seashore, 1947; Jenkins, 1930; Jones, 1949) also reported a high degree of variability within any age group for specific activities. Such results suggest that different statistical procedures and/or more sophisticated measuring techniques may provide more definite information on the apparent communality of the relationship of phylogenetic physical skills and age. Longitudinal studies using electromyographic, cinematographic, and electroencephalographic techniques in addition to the tests that have been somewhat standardized in the past would certainly add materially to our knowledge and understanding of age changes in motor skills.

References

Ames, L. B. (1937). *Genet. Psychol. Monogr.* **19**, 409.

Bayer, M., and Bayley, N. (1959). "Growth Diagnosis." Univ. of Chicago Press, Chicago, Illinois.

Bayley, N. (1935). *Monogr. Soc. Res. Child Develop.* **1**, No. 1.

Bayley, N., (1936). "The California Infant Scale of Motor Development." Univ. of California Press, Berkeley.

Boynton, B. (1936). *Univ. Iowa Stud. Child Welfare* **12**, No. 4.

Burke, W. E., Tuttle, W. W., Thompson, C. W., Janney, C. D., and Weber, R. J. (1953). *J. Appl. Physiol.* **5**, 628.

Conel, J. L. (1939). "The Postnatal Development of the Human Cerebral Cortex," Vol. I Harvard Univ. Press, Cambridge, Massachusetts.

Conel, J. L. (1941). "The Postnatal Development of the Human Cerebral Cortex," Vol. II Harvard Univ. Press, Cambridge, Massachusetts.

Cumbee, F. Z. (1954). *Res. Quart.* **25**, 412.

Cumbee, F. Z., Meyer, M., and Peterson, G. (1957). *Res. Quart.* **28**, 100.

Delecato, C. H. (1963). "The Diagnosis and Treatment of Speech and Reading Problems." Thomas, Springfield, Illinois.

Dupertuis, C. W., and Hadden, J. A. (1951). *Amer. J. Phys. Anthropol.* [N.S.] **9**, 15.

Espenschade, A. S. (1940). *Monogr. Soc. Res. Child Develop.* **5**, No. 1.

Espenschade, A. S. (1960). *In* "Science and Medicine of Exercise and Sports" (W. R. Johnson, ed.), pp. 419–439. Harper, New York.

Espenschade, A. S., and Eckert, H. M. (1967). "Motor Development." Merrill, Columbus, Ohio.

Espenschade, A. S., Dable, R. R., and Schoendube, R. (1953). *Res. Quart.* **24**, 270.

Ferris, B. G., and Smith, C. W. (1953). *Pediatrics* **12**, 341.

Ferris, B. G., Whittenberger, J. L., and Gallager, J. R. (1952). *Pediatrics* **9**, 659.

Fleishman, E. A. (1964). "The Structure and Measurement of Physical Fitness." Prentice-Hall, Englewood Cliff, New Jersey.

Gardner, E. B. (1965). *JOPHER* **36**, 61.

Gesell, A. (1954). *In* "Manual of Child Psychology" (L. Carmichael, ed.), pp. 335–373. Wiley, New York.

Gesell, A., and Ames, L. B. (1940). *J. Genet. Psychol.* **56**, 247.

Gesell, A., and Ilg, F. L. (1946). "The Child from Five to Ten." Harper, New York.

Gesell, A., and Thompson, H. (1934). "Infant Behavior: Its Genesis and Growth." McGraw-Hill, New York.

Glassow, R. B., and Kruse, P. (1960). *Res. Quart.* **31**, 426.

Goetzinger, C. P. (1961). *J. Educ. Res.* **54**, 187.

Gutteridge, M. V. (1939). *Arch. Psychol., N. Y.* **244**, 1.

Halverson, H. M. (1931). *Gent. Phychol. Monogr.* **10**, 107.

Halverson, H. M. (1937). *J. Genet. Psychol.* **56**, 95.

Heath, S. R. (1949). *Motor Skills Res. Exch.* **1**, 34.

Hellebrandt, F. A., Houtz, S. J., Partridge, M., and Walters, C. E. (1956). *Amer. J. Phys. Med.* **35**, 144.

Hellebrandt, F. A., Rarick, G. L., Glassow, R., and Carns, M. L. (1961). *Amer. J. Phys. Med.* **40**, 14.

Iliff, A., and Lee, V. A. (1952). *Child Develop.* **23**, 237.

Jenkins, L. M. (1930). "A Comparative Study of Motor Achievements of Children Five, Six and Seven Years of Age." Teachers College, New York.

Jones, H. E. (1949). "Motor Performance and Growth." *Univ. of California Press,* Berkeley.

Keogh, J. (1965). "Motor Performance of Elementary School Children." Univ. of California, Los Angeles.

Lombard, O. M. (1950). *Child Develop.* **21**, 229.

McGraw, M. B. (1940). *Amer. J. Dis. Child.* **60**, 1031.

McGraw, M. B. (1946). *In* "Manual of Child Psychology" (L. Carmichael, ed.), pp. 332–369. Wiley, New York.

Mangarov, I. (1964). *Bull. Inform., Bulg. Olympic Comm., Year IX* **5**, 22.

Meredith, H. V. (1935). *Univ. Iowa Stud. Child Welfare* **11**, No. 3.

Meredith, H. V., and Boynton, B. (1937). *Hum. Biol.* **9**, 366.

Metheny, E. (1941). *Univ. Iowa Stud. Child Welfare* **18**, No. 2.

Mugrage, E. R., and Andresen, M. I. (1936). *Amer. J. Dis. Child.* **51**, 775.

Mugrage, E. R., and Andresen, M. I. (1938). *Amer. J. Dis. Child.* **56**, 997.

Nicolson, A. B., and Hanley, C. (1953). *Child Develop.* **24**, 3.

Nylin, G. (1935). *Acta Med. scand.* **69**, Suppl.

Piaget, J. (1952). "The Origins of Intelligence in Children." International University Press, New York.

Pratt, K. C. (1954). *In* "Manual of Child Psychology" (L. Carmichael, ed.), pp. 215–291. Wiley, New York.

Pratt, K. C., Nelson, A. K., and Sun, K. H. (1930). *Ohio Univ. Stud. Psychol.* No. 10.

Rarick, G. L. (1954). "Motor Development during Infancy and Childhood." Univ. of Wisconsin Press, Madison.

Reynolds, E. L. (1945). *Amer. J. Phys. Anthropol.* [N.S.] **3**, 321.

Reynolds, E. L. (1947). *Amer. J. Phys. Anthropol.* [N.S.] **5**, 165.

Richter, C. P. (1934). *Amer. J. Dis. Child.* **48**, 327.

Robinson, S. (1938). *Arb. Physiol. Angem. Entomol. Berlin-Dahlem* **10**, 251.

Seashore, H. G. (1947). *Res. Quart.* **18**, 246.

Shirley, M. M. (1931). "The First Two Years: A Study of Twenty-five Babies," Vol. I. of Minnesota Press, Minneapolis.

Shirley, M. M. (1933). "The First Two Years: A Study of Twenty-five Babies," Vol. II. Univ. of Minnesota Press, Minnesota.

Shock, N. W. (1944). *Amer. J. Dis. Child.* **68**, 16.

Simmons, K. (1944). *Monogr. Soc. Res. Child Develop.* **9**, No. 1.

Simonson, E., Kearns, W. M., and Enzer, N. (1941). *Endocrinology* **28**, 506.

Sjöstrand, T. (1953). *Physiol. Rev.* **33**, 202.

Skinner, P. H. (1968). *In* "Psychological and Physiological Aspects of Reading" (A Report of the Twenty-fourth Annual Conference and Course on Reading), p. 149–161. Univ. of Pittsburgh Press, Pittsburgh, Pennsylvania.

Wallon, H., Evart-Chmielniski, E., and Sauterey, R. (1958). *Enfance* **11**, 1.

Wellman, B. L. (1937). *Child. Educ.* **13**, 311.

Wild, M. R. (1938). *Res. Quart.* **9**, 20.

CHAPTER

8

Motor Learning as a Function of Age and Sex

Robert N. Singer

I. Introduction

Motor learning, or the experiencing and mastering of movement behaviors, is perhaps the child's most prominent means of self-expression. Any attempt to separate cognitive, affective, and psychomotor behavior, while convenient, is in a sense unreal, because behavior is a complex phenomenon and is the result of the interaction of many factors. Nonetheless, behavior primarily realized through movement-oriented actions is generally classified in the psychomotor domain. These actions will be the concern of this chapter.

Much of the research in the psychomotor domain has been associated with the identification of particular behaviors at given ages. Such research has provided valuable information from cross-sectional and longitudinal data on age and sex differences in motor performance, but it has given us little insight as to the true nature of the learning phenomena. Such normative data will receive only minor attention here. Representative sources, e.g., Stone and Church (1968), Smart and Smart (1967), and Crow and Crow (1962), can be referred to if this information is desired. Rather, the focus of this chapter will be on research bearing on motor learning as a function of age and sex. Generally speaking, those concerned with learning in children are interested in their responsiveness to various environmental incentives and experiences.

For ease of presentation, the material will be categorized under the general headings "Infancy," "Early Childhood," "Middle Childhood," and "Adolescence." Infancy is usually considered to encompass the first 18 months of life. Near the end of this period the child has established reasonable proficiency in bipedal locomotion and the beginnings of speech. The Latin derivation of the term "infancy" means "not speaking." Early childhood, the period from 18 months to 5 years, includes the preschool years, a time of rapid sensorimotor development. Middle Childhood (5–11 years of age) includes the school age and encompasses the elementary school period, a time of slow and relatively even growth. Finally, adolescence ushers in the period of sexual maturing, covering the age range from 12 to 18 years. General and unique motor learning phenomena associated within and among these classifications are discussed. As will be evident throughout the chapter, one of the major problems is to differentiate the effects of maturation and learning on motor behavior. Quality research is sparse, especially in the early and formative years of the child's life. This is true for the apparent reason that it is much easier to control and investigate subhuman motor activities and make inferences to human behavior than it is to conduct experiments on humans. This chapter has been written within the framework of these and other limitations.

II. Infancy

A. GENERAL MOTOR CHARACTERISTICS

A number of the neonatal responses and those of early infancy are reflex in origin. Some of them are lost as maturation takes place while others are built into complex movement patterns in the normal processes of development. Of the former, those under the control of subcortical centers, such as the Babinski reflex and the tonic neck reflex, subside and are lost by the end of the fourth month. The latter are built into such primitive movement patterns as reaching, crawling, creeping, standing, and walking. The exact role of maturation and learning in the development of these basic abilities is not known. However, there is good reason to believe that maturation is a powerful force in the development of body control and the readiness to learn. Dennis and Dennis (1940) have demonstrated that Hopi infants bound to a cradleboard and allowed little movement during the first year were no slower in beginning to walk than those babies who were not restrained. The evidence strongly indicates that the motor sequences which infants undergo are genetically and maturationally centered, for normative data document the stages and ages when certain behaviors may be expected to merge. Children go through orderly steps in which the maturation process is important for generalized responses. Later complex skills reflect specific practice, not general experiences. From birth to later infancy, to early school years and adolescence, motor behavior changes from the expression of primarily reflexive and generalized movement patterns to more differentiated, specialized, and integrated movements (Breckenridge and Murphy, 1969).

More recent research indicates that the newborn is not passive but instead possesses numerous learning potentialities. Learning can occur very shortly after birth. Capacity and rate of acquisition of learning matter increases with age after birth and during childhood. Interestingly enough, earlier learning is more likely to be forgotten than learning that occurs later in life. Retention is better in adulthood than childhood. This probably results from the immaturity of the child's nervous system. The paradox of the situation is that the influence of early experience on adult behaviors is recognized and yet the long-term retention of learned behaviors increases with age.

Lorenz (1958) has reported on the differences in innate motor patterns among species of birds and ducks. His research in hybridization in which new combinations of motor patterns are produced indicates that such characteristics are "dependent on comparatively simple constellations of genetic factors."

Apparently, cultural enrichment can hasten some behavioral processes while extremely restrictive and barren conditions will delay or prevent the appearance of certain patterns of behavior. The infant undergoes a series of changes which enable him to execute more complex acts. He learns eye–hand coordination tasks by integrating visual and neuromuscular components. Raising the head, recognizing a bottle or nipple, reaching for objects, and even playing games such as pat-a-cake are indicative of maturationally-induced but environmentally influenced activities. Thus the infant, as he develops, demonstrates the means of learning how to manage toys and objects and to master simple neuromuscular skills.

However, a number of questions pertaining to motor development and the acquisition of effective movement patterns remain unresolved. Perhaps the most prominent of these are the questions of maturation versus learning and the nature of the interrelationship of human abilities within and between stages of development. Throughout the century, at least three major theoretical positions bearing on these and other questions have influenced research in motor development (Crowell, 1967): (a) behaviorism, (Watson, 1919), (b) the maturational approach (Gesell, 1940), and (c) the interaction hypothesis (Piaget, 1952). Gagné (1968) has proposed a cumulative learning model, where "the child progresses from one point to the next in his development, not because he acquires . . . new associations, but because he learns an ordered set of capabilities which build upon each other in a progressive fashion through the process of differentiation, recall, and transfer of learning."

The popularity of the above approaches has generally corresponded with the approximate years surrounding their publication. Each position holds interesting implications for those who are concerned with motor development, and each attempts to explain and predict the emergence of motor responses. Unfortunately, space does not permit further elaboration of these points of view.

Brackbill and Koltsova (1967) suggested that gross motor skill learning in the first few years of life actually competes with language development. For example, learning to walk impedes the development of speech. These authors summarized and analyzed the research on this and related aspects of development, most of which was published years ago in the 1930's and 1940's.

The actual process of skill acquisition in children and its impact on other aspects of behavioral development have over the years commanded the attention of psychologists. Skilled performance encompasses image representation with the corresponding appropriate action. It requires the coordination of both excitatory and inhibitory controls, the inhibitory controls of the higher centers playing a prominent role. According to Greenwald (1970), infants 2 years of age or younger can show some voluntary inhibition of performance. Luria (1961) has documented the voluntary regulation of performance in an experiment in which verbal direction was given to a child squeezing a rubber balloon. The developmental sequences of actions from ages 18 months to 5 years are described. The child shows increasing control over actual movements through the thought and mediation process. At about age 5, the child can effectively transfer behavioral control from overt verbalization to self-verbalization (thought) over task execution. In contemporary Soviet child psychology. El'konin (1969) reports that much attention is given to the study of the development of movements and motor habits.

Considerable data are available on the relationship of race, sex, and socioeconomic conditions to the motor development and skill level of young children. Girls usually reveal a slightly higher level of motor development at birth, e.g., girls sit and walk earlier than boys. Girls mature faster than boys and show a more advanced skeletal development at all ages up to maturity. The boy's skeletal development is on the average only 80% of a girl's at each age from birth to maturity (Tanner, 1961).

It has been documented (Williams and Scott, 1953) that Blacks are more advanced in skeletal maturation than white children. As early as the first year of life, Black children possess a greater motor development. This is because they are probably allowed more freedom than white children. But even within groups of Black children, as distinguished by family socioeconomic levels, such differences exist, for lower socioeconomic Black children are more advanced in motor development than are upper socioeconomic class Black youth. They reveal a significant gross motor acceleration, apparently because of permissive and less exacting rearing practices among lower socioeconomic families. For instance, they are no doubt less restricted to the crib. Williams and Scott (1953) concluded in their study that motor acceleration is more dependent on child rearing practices associated with socioeconomic classes than upon racial affiliations.

B. CRITICAL LEARNING PERIODS

It is now recognized that there are critical periods in the acquisition of many aspects of behavior. Experimental work on a variety of organisms has clearly shown that the development of behavioral responses is dependent not only on maturation but also on the opportunity for the animal to respond actively to environmental stimuli. If such opportunities are not afforded at the period when these behaviors would usually be expressed, they are not likely to develop normally. Furthermore, that which is learned in one period of development may be crucial in later learning. Scott (1962) has summarized the research on behavioral development and relationships to critical periods with regard to children and animals. He wrote:

> A great deal of needed information regarding the optimum periods for acquiring motor and intellectual skills is still lacking. These skills are based not merely on age but on the relative rate of maturation of various organs. Any attempt to teach a child or animal at too early a period of development may result in his learning bad habits, or simply in his learning "not to learn," either of which results may greatly handicap him in later life.

With regard to lower forms of life, research is becoming more plentiful concerning behavioral introduction, deprivation, and enrichment. For example, Melzack and Thompson's (1956) data revealed that Scottish terriors raised in competitive situations were more aggressive at maturity than the controls. The restricted dogs were inept in competitive situations, curiosity, and social behavior. Ravizza and Herschberger (1966) have shown that rats raised in an inhibited environment where no climbing was permitted differed in adulthood from rats allowed to climb. The restricted group was less active in exploration tasks, more emotional in novel situations, and inferior on an intelligence test. Sluckin and Salzen (1961) discussed the process of imprinting in ducklings and chicks as a function of a critical period. Research methods, problems, and parameters in studies dealing with the effect of early experiences on adult animal behavior have been reviewed by King (1958). Articles on stimulus restriction in young organisms, particularly as this relates to the behavior of lower species, can be found in Newton and Levine (1968).

Returning to skill acquisition and critical periods of learning in humans, the study most often cited is that of McGraw (1935) and her research with the twins, Johnny and Jimmy. Johnny was the experimental subject and was provided very early in life with numerous gross motor experiences such as creeping up an incline, swimming. and roller skating. His initial training began the latter part of the first month of life. Jimmy served as a control and received his practice and instruction in the same activities but later (at 22 months) and in a condensed version.

McGraw concluded from the findings that the so-called phyletic (racial) activities are under infracortical control and are subject to very limited modification during postnatal development. Such skills as creeping, crawling, sitting, and walking are under the control of innate developmental forces and their onset is not speeded up by special training. On the other hand, ontogenetic activities (those more recently acquired in man's cultural evolution) can be more easily modified by training, but only within the limits set by the child's level of maturity.

McGraw's study supports the timeliness of instruction and implies that children should be introduced to many skills at earlier ages than generally assumed and that introduction to other skills should be delayed.

In Germany, at the present time, Professor Liselott Diem (1970) is investigating means of introducing swimming techniques to babies as young as 2 months of age. Long-term effects will be noted, and the data collected on controls will provide the basis for comparative analyses. Her initial efforts have resulted in babies being able to swim on the front or back without any assistance at about 5 months of age. Earlier studies by McGraw (1935) and other researchers, as well as empirical observations, attest to the fact that neonates can perform coordinated swimming patterns. When placed in water they utilize the arms and legs in a swimming motion until about 4 months of age, at which time the pattern becomes disorganized.

C. SPECIAL TRAINING

Educators and researchers, interested in facilitating learning, have tried various training programs that might provide an earlier-than-usual acquisition of skills. Investigations focusing attention on critical learning periods have on occasion included special training programs for given learning experiences for subjects of different ages. On the other hand, research has been developed where the prime concern was not the search for a critical learning period but instead a knowledge of the benefits in providing special programs at specified early points in life.

Special motor skill instruction of young children given ample opportunity to practice has provided inconsistent results from study to study (Gesell and Thompson, 1941; Hilgard, 1932; Ketterlims, 1931). These studies have shown that maturation is a highly significant factor in skill acquisition in the young child. The time required to master a simple skill can be markedly reduced if practice is withheld until the child is mature enough to profit from the practice. Hicks (1930a,b), Hicks and Ralph (1931), and Wenger (1936) noticed that those that were trained were generally not superior to controls in certain perceptual motor tasks. Thus, attempting to teach a child a motor task before he shows evidence of readiness is of little use.

Some studies indicate that certain activities can be readily learned by the very young child. Unfortunately, as McGraw (1970) warned, "we have not as yet learned how to develop to the maximum all potentials of the growing child." She further stated:

> The pressure is on learning, early learning. It seems clear that the infant and toddler are capable of learning a great deal, *if* the opportunities for learning are properly presented. It also seems evident that the principles of learning derived from laboratory studies of animals or college students are inadequate when it comes to dealing with rapid behavior development of the human infant.

Apparently, a number of physiological responses can be conditioned to various kinds of stimuli in infancy and early childhood. Steinschneider (1967) summarized studies showing physiological changes and adaptions to sensory stimulation. For instance, a change in skin resistance to indifferent stimuli can be elicited, thus showing that infants can be physiologically conditioned.

Staats (1968) believed, as has been expressed by others, that coordination, balance, and other behavioral patterns develop similarly in all children.

However, he felt that individual differences are a function of dissimilarities in training techniques. More specifically, he wrote that many of these behaviors "involve complex stimulus controls acquired through the individual's reinforcement history." He placed great import on reinforcement opportunities; i.e., what is reinforced and how it is reinforced. Staats favored a systmatic reinforcing training program for more rapid motor development. This view coincides with current research on learning in infants which is focused on two considerations: the usage of operant learning in newborns and the application of effective reinforcers in the learning situation (Elkind and Sameroff, 1970).

D. Abilities

Data from a number of sources reveal the "going-togetherness" of abilities in the infant. A greater independence of motor abilities is found at the preschool level than in infancy, and this independence grows more noticeable with advancing age. Beyond infancy, the concpet of "general motor ability" has been largely refuted. It would seem more appropriate to speak of motor abilities, no one of which might be considered as unifying. In other words, learning becomes more specific with age.

In school-aged children and adolescents, performance in motor skills has been found to have a low relationship with intelligence test scores. The relationship between cognitive and motor abilities in preschool children shows a more meaningful relationship (Bayley, 1935). Bayley administered the California Infant Scale of Motor Development to 61 infants each month from birth to 15 months, and at 3-month intervals from 15 to 36 months. A correlation of .50 was obtained between the motor and cognitive items given during the first 15-month period. A correlation of .39 was found between the age of walking and age of talking. While these correlations are too low to be of predictive value, they nevertheless are indicative of a meaningful relationship.

Furthermore, a number of significant human abilities and characteristics develop quickly in the first 5 years of life (Bloom, 1964). The implication from Bloom's research is that the preschool and early years are highly important in the development of learning patterns and general achievement. Research with lower forms of animal life as well as with infants points to the need for a good learning environment early in life. A solid background of early learning experiences which provides the opportunity to develop a repertoire of movement patterns lays the foundation for more complex skill learnings later in life. The mastery of elementary skills assists the learner in acquiring new activities involving these skills. Staats (1968) provided the following example:

. . . the child who has acquired skilled throwing responses will learn to serve in tennis more easily than a less skilled child; for the tennis serving stroke involves a throwing movement. In general it may be suggested that a child who has acquired the basic skills involved in running, dodging, and sliding, and so on, is in a better position to learn new games involving these basic skills.

The implications for the early introduction of a wide variety of movement experiences are fairly clear. Skills and abilities are developed which are necessary for later proficiencies in other more sophisticated skills. Furthermore, achievement in motor skills is directly related to achievement in cognitive efforts. Staats (p. 425) suggested that cognitive and motor skills are interrelated; e.g., "the child must acquire a vast repertoire of sensory-motor skills under the control of verbal stimuli."

Abilities are determined by both genetics and experience. Obviously, capacities can only be developed within the framework which genetics has laid down. To what extent does inheritance determine success in motor skills? Twin resemblance was studied by McNemar (1933) in the following motor skills tests: Koerth pursuit rotor, Whipple steadiness, Miles speed drill, Brown spool packer, and card sorting. Ninety-eight sets of twins of junior high school age, 48 pairs being fraternal and 47 pairs being identical, provided the data. The identical twins showed a degree of resemblance in performance which was higher than in the fraternals. McNemar concluded that the hereditary hypothesis was the most plausible explanation of individual differences in motor skills. In Scarr's (1966) study, identical twins had higher correlations for five measures of reaction time than did fraternal twins. Genetics and situational factors interact an determine achievement in performance measures.

III. Early Childhood

A. MOTOR DEVELOPMENT

During the preschool years, motor development occurs at an accelerated pace. The type of motor skills the child is able to master involves greater perception and cognition. As Kay (1969) wrote: "Skills such as block building which depend upon visual-spatial perception, motor manipulation, positioning, and release show marked improvement from three to five years and . . . the whole task is made up of many subskills." Thus, the child is able to indulge in more complex activities to a reasonably satisfactory extent.

In an early study, Gutteridge (1939) administered a battery of motor tests to almost 2000 children from 4 to 6 years of age. Some of the highlights of the findings are as follows: In jumping, 40% of the children were judged as proficient at 3½ years of age and 85% at 6 years of age. Sixty-three percent demonstrated success in tricycling at 3 years of age, while 100% of the children were able to perform this task at age 4. In ball throwing, 85% of the 6-year-olds were appraised as being reasonably skilled. The more complex skills, such as throwing and catching balls, and hopping and skipping, were usually performed adequately at ages 4 or 5. Boys were judged to be better in some activities, girls in others.

A series of five motor skills requiring speed of movement was recently given by Schulman *et al.* (1969) to 375 children between ages 3 and 8. As might be expected, age was found to be the most important variable influencing performance scores, although performance was not a linear function of age. No consistent patterns emerged when boys and girls were compared in performance.

Although a number of researchers are concerned with normative data as reflected by quantitative scores, others are interested in analyzing movement patterns. Such techniques as a computerized photographic system (Herron and Frobish, 1969) go beyond traditional observational records in providing refined data. Observational information on movements involved in motor activities tend to reveal maturational levels and capabilities to acquire skills.

B. SKILL ACQUISITION

With each year, problem solving and performing skills become more a matter of transfer and less a situation of new learning. Skill proficiency, in many cases, is shown to be fairly related to chronological age. In Dusenberry's (1952) investigation of 56 subjects 3–7 years of age, learning to throw a ball for distance occurred over and above the effects of maturation and general practice. She concluded that throwing ability was associated with age.

In a study of two groups of nursery school children matched on various performance factors, sex and age, Hilgard (1932) found that an experimental group receiving special training in cutting with scissors, buttoning, and climbing for a period of 12 weeks performed these skills no better than the control group that received only 1 week of training during the last week of the experiment. In another investigation, one pair of twins 4½ years of age was tested by Hilgard (1933). The learning activities included performance on a walking-board, cutting, digit memory, ring tossing, and object memory. After 8 weeks of practice, the experimental twin was superior on all test measures except the largest walking-board time score. However, at 3- and 6-month intervals after practice, performance between the twins was as similar to each other as at the beginning of the study.

Hicks (1930b) collected data from 60 children ranging in age from 3 to

6 on strength, a perforation test, a tracing test, and a task involving throwing at a moving target. Hicks concluded that skill improvement on these tests was primarily the result of maturation and general practice at these ages.

Fowler (1962), on the other hand, has questioned the assumption that skills are basically acquired through normal maturation and in definite sequences corresponding to developmental stages. In his summary of related research he reported the usual finding that training in specific skills improves performance but the training is more effective if it is not given too early. Maturation is usually cited as being more important than experience. Fowler proposed that this conclusion is erroneous: (a) It is mainly observed because the skills tested are too simple; a peak level is reached too fast. Training is of greater worth with complex tasks; hence, better results would be shown with them. (b) The typical control group in the investigations analyzed has had other related experiences in the tasks tested. (c) The control group may have had preexperimental experience in the tasks of interest. Thus, Fowler would support early specialized training in complex skills, especially those involving a high component of cognitive and perceptual learning.

In the McGraw study reported previously, Johnny and Jimmy were tested again 4 years later at age 6 to determine the effects of the earlier specialized training of Johnny with the more routine program Jimmy experienced (1939). Johnny, the experimental twin, was generally reported to display at this later period greater motor coordination and daring in physical performance.

Learning considerations, such as reinforcement. knowledge of results, and transfer of training, are usually investigated with college students. One of the exceptions is the investigation of Pumroy and Pumroy (1961) of reinforcement effects in a motor learning situation with children. The subjects were 48 nursery school children of both sexes, from 35 to 65 months of age. Different amounts and percentages of reinforcement were provided to determine their effects on task extinction. As might be expected, the more reinforcements the children received the more responses they made during the extinction period. More studies are needed on all age groups to determine the ubiquity of the conclusions reached on various learning phenomena since they are primarily determined on the basis of college subjects' performance.

Appropriate practice and reinforcing schedules help to promote motor learning. But perhaps what should be given greater attention is an environmental setting that encourages intrinsic motivation for youngsters to participate in desired motor activities and to achieve reasonable levels of skill. An apparently successful application of this principle is reported by Pronko

(1969), who described the work of Sinichi Suzuki, a Japanese violin teacher. Suzuki specialized in instructing very young children. His achievements with them were remarkable since the violin is not associated with the Japanese culture. His philosophy was based on the concept that the motivation to play must come from the child's desire to play for playing's sake. External rewards and punishment were omitted. The reward came from the satisfaction of playing well. Suzuki's program was based on the child's identifying with his parents and peers in appreciating good music, a listen-and-play method. Many of his pupils were only 4 years of age.

C. Preschool Education

That young children can learn much more than they are given credit for is well accepted today by educators and psychologists. The belief in the significance of early learning experiences has brought about an increased demand for nursery schools. However, the form that preschool education should take has raised many heated issues. Since intelligence develops most rapidly before the age of 4, it becomes apparent that the environment in which the child finds himself is a major consideration in his future development. Exactly what the environmental stimuli should be to provide the proper incentives for educating nursery school children is a matter on which specialists are not in complete agreement. Some are inclined toward a highly motivating, intellectually oriented environment whereas others feel that it is unnecessary or even harmful to "push" children in this direction too rapidly.

Regardless of the approach that is recommended the major emphasis is on the intellectual development of the child. Few individuals today are concerned with special training programs oriented to enhancing the motor proficiency of young children per se. In a society where intellectual achievements are valued highly, it is not surprising that where motor development is emphasized, it is with the intent of facilitating the cognitive processes. The belief in the influence of successful sensorimotor experiences upon later accomplishments in the cognitive domain is evident by the current wave of specially designed perceptual-motor or sensorimotor programs that have been designed for preschool children.

Popular programs initiated by Kephart, Doman and Delacato, Barsch, Frostig, and Getman are based on the premise of the existence of a strong relationship between sensorimotor activities and cognitive pursuits in the young child. These programs are based on the assumption that experiencing various sensory inputs and learning to move the body effectively will positively influence the cognitive learnings that follow.

It is not the function of this chapter to treat the validity of these programs nor their ramifications other than to show that there is not general agreement on their value. It might be well to point out here that generalized movement patterns, motor abilities, and motor skills have usually been studied to (a) collect normative data or (b) determine relationships with abilities and achievements in the cognitive domain. Rarely are motor activities in the young investigated (a) to understand the learning process involved in their acquisition or (b) the implications for future successful psychomotor performance.

Certainly, indirect and circumstantial evidence would indicate the need for experiencing and learning a wide variety of motor activities in early childhood. Generalized movement patterns serve as the basis for developing highly refined motor tasks. The more varied the movement experiences and proficiencies that are acquired, the greater the range of movement patterns that are available for transfer to future motor situations. Timely, enriched programs may benefit a child in many ways. They will no doubt have an impact on the adolescent's and adult's attainment of skills.

IV. Middle Childhood

A. DEVELOPING LEARNING CAPACITIES

In middle childhood, the child receives a somewhat formal introduction to numerous physical activities. Organized programs in and out of school encompassing a variety of activities substitute for spontaneous play. Music and dance lessons, organized athletic teams, industrial art programs, and the like provide outlets for children to develop their learning capacities in the psychomotor domain. At this stage of life the nervous system is sufficiently mature to provide the controls for refined coordination of the body parts and for the proficiency to engage in reasonable complex motor tasks.

In spite of the moderately high level of neuromuscular coordination attained by most children as they enter the period of middle childhood it would be a waste of time and an experience in frustration to introduce certain tasks before the organism is mature enough to handle them. Scott (1968) wrote:

> The results . . . are more satisfactory than those obtained by trying to teach complex physical skills directly at an early age. Most children are not able to perform activities requiring good coordination of the whole body much before the ages of 7 or 8, and introducing them too early to such activities only results in unskilled performance or failure.

It should, however, be reemphasized that many activities can be *modified* to meet the child's maturational level. Thus, activities designed for children of a given maturity level can, with modifications, be made appropriate for less mature children. Practice or special training is effective only when the organism is maturationally ready.

Learning can be encouraged or discouraged by the nature of the teacher. For example, the teacher's sex is evidently an important variable. Boys do more poorly than girls in kindergarten. It has been suggested that this may be because female teachers dominate this level of schooling. Girls conform better, are quieter, and more obedient. Some recent experiments indicate that boys instructed by a male teacher progress much faster than boys in coed classes. There is every reason to believe that performance in physical activities will be similarly affected.

Not only is the instructor important in learning. "Maturation, individual ability, practice, and tuition (or instruction) affect the acquisition and learning of motor skills," stated Smith (1956). Yet, the instructor sets the environment in which learning can best occur. He can stimulate learning in various ways. Ragsdale (1950) offered some cogent observations about the child's learning process, e.g., (a) there is no one best form for the execution of a skill; (b) the motor learning process is active and thoughtful, not blind and a matter of mimicry; (c) trial and error or repetitive behavior is wasteful; and (d) there are various methods of attempting to achieve a goal— models are to be observed but not necessarily copied.

Individual abilities develop within the genetic structure laid down with age and experience. Although Brace (1930) has designated the 9–12-year period as the time when most abilities emerge for the performance of complex skills, individuals continue to refine their execution of skills. Highly skilled performance in motor tasks is normally not achieved until the late teens or early-to-midtwenties.

The basis for motor aptitude of boys in the 10–12-age bracket has been determined by Ismail and Cowell (1961). Twenty-five test items were administered that were believed to contribute to motor aptitude. Five factors emerged when the data were submitted to factor analysis. They were speed, growth and maturity, kinesthetic memory of the arms, body balance on objects, and body balance on the floor. According to these investigators, these factors are most important for success in a variety of motor tasks for this age group.

Learning capacities may be influenced by the child's prior movement experiences and by his concepts of movement and his attitudes toward physical activity. Since transfer occurs more effectively when one situation closely resembles another, the broader the child's range of motor experiences the greater the chance for transfer. It has been proposed that special rhythm experiences will improve overall athletic coordination. The benefits of such practices are yet to be established. Groves (1969) attempted to answer the question: Does rhythmic training affect children's abilities to synchronize bodily movements with rhythmic stimuli? One-hundred sixty-four first, second, and third graders served as subjects. Children specially trained in rhythmic training performed no better than the nontrained in their ability to attain the objective. Evidently, favorable transfer effects were not obtained under the conditions of the study. Boys and girls obtained fairly similar scores, with the boys being slightly higher. Of all the variables studied, age and grade level contributed most to synchronization ability. Furthermore, a follow-up study seemed to indicate that age and maturation were of greater relevance to rhythmic-synchronization ability than instruction. Children improved without the benefit of instruction.

In the elementary school years, body size and motor performance are reasonably highly correlated. Weight, age, and strength are fairly good predictors of athletic accomplishments. The relative influence of maturational level and body build on performance ability as contrasted with learning capabilities is difficult to assess. Many of the tasks undertaken by children are fairly gross in nature, requiring less refined bodily organization in movement than those skills performed in later years. Strength, jumping ability, speed, and the like exert a profound influence on achievement.

Physical maturity is important in the athletic accomplishments of boys in later childhood (Espenschade and Eckert, 1967). Hale (1956), in testing participants in the 1955 Little League World Series, reported that their average chronological age was 12.53 years, whereas 50% of the athletes were skeletally as mature as the average 14 or 15 year old. Clarke (1967), after studying boys representing basketball, football, track and field, and wrestling, concluded that the successful athletes in elementary and junior high school possess greater body size, strength, endurance, power, and maturity than the average child.

We might conjecture that learning capacities for complex motor tasks are best expressed within a mature body framework. Learning is dependent on past experiences, motivation, and a host of other variables. Thus, superior athletes and others successful in motor skills reflect the presence of the best combination of these factors, but the relative importance of each is not known, probably varying considerably among individuals.

B. Motor Characteristics

A fair amount of literature is available concerning the physical characteristics and skill levels of boys and girls of elementary school age. For instance, Dohrmann (1964) noted that 8-year-old boys were superior to girls in throwing and kicking skills. Apparently, boys also learn the skills involved in ball-throwing accuracy and in a novel ball-bounce test faster than girls in the 6–9 age bracket (Smith, 1956).

There is some evidence to indicate that with girls, age changes in motor performance in middle childhood is relatively constant. In examining the performance of girls, ages 6–14 years, in a 30-yard run, standing broad jump, and the overarm throw, Glassow and Kruse (1960) concluded that there was a tendency for the girls to remain in the same relative position within the group. They conjectured: "It may be that early development of motor coordination is essential for later success or that an inherent nature of motor ability may determine the limit of achievement during the growing years." In the tasks measured, performance generally improved with age.

With outstanding young male athletes, Clarke's (1967) data provide a different picture. He found that outstanding elementary school athletes were not necessarily outstanding junior high school athletes and vice versa. Only 5 of 20 athletes studied were superior at both school levels. Additional evidence for individual differences is reported by Clarke in that the pattern of common characteristics varied from athlete to athlete.

When viewing performance differences in various age groups, trends became apparent. Age plays an important role in determining proficiency in a variety of tasks. Reaction time improves in childhood from 2½ to 11½ years (Goodenough. 1935) and from childhood (6 years) to 19 years (Hodgkins, 1962). Discrimination reaction time improves from 8 years to the late teens with no sex differences until the latter time when the boys exceed the girls and maintain an advantage for a number of years (Noble *et al.*, 1964). Complex coordination skill, as measured on the Toronto Complex Coordinator, is positively related to age when comparing children ranging in age from 4 to 10 (Humphries and Shephard, 1959). Some age and sex differences occur in psychomotor tests (Stachnik, 1959). Reaction time and speed of movement in both males and females increase most between ages 6 and 12, although not until age 12 do males show a superiority over females (Hodgkins, 1963), and balance on a stabilometer task and a vertical ladder climb improves from ages 7 to 17, with no appreciable sex differences (Bachman, 1961). In the last study, it is interesting to note that the amount of improvement, or rate of learning, was found to be *independent* of age and sex in the 6–26-year range.

Data from Alderman (1968) and Henry and Nelson (1956) concur with Bachman's results. Alderman compared 10- and 14-year-old boys and girls in performance on the rho motor task. The older group was superior in initial performance and final performance. No sex differences were noted, nor was three any dissimilarity between groups in the amount of learning. Henry and Nelson compared the performance of 10- and 15-year-old boys in a motor skill task, the results clearly indicating superior performance of the older boys, although the younger group improved more with practice than the older boys. Buckellew (1968) reported progressive although not necessarily significantly better performance in grades five, six, seven, and eight in measurements of physical fitness.

In a laboratory situation, Singer (1969) tested third and sixth grade boys and girls in an assortment of motor tasks. When measures of performance were determined, the sixth grade boys and girls were generally superior to the third grade boys and girls. Also, the boys were better (although not necessarily significantly) than their counterparts in such tasks as (a) stabilometer balance, (b) discrimination reaction time, (c) the Figure Reproduction Test; and (d) pursuit rotor tracking. The subjects were not afforded any instruction. An examination of the learning trials indicates that (a) six trials on the stabilometer yielded similar incremental learning curves for each of the four groups, although the sixth grade boys performed at a higher rate than the girls on trials four and five; (b) performance on five trials in discrimination reaction time, in which stimulus–response associations had to be acquired, yielded similar profiles among the groups; (c) in pursuit rotor performance, an analysis of four trials revealed continual positive increments from trial to trial for the sixth grade boys while the other groups fluctuated; and (d) the seven subtasks on the Figure Reproduction Test yielded similar performance patterns for the groups. It should be reemphasized that performance levels were generally best for the sixth grade boys, followed in order by the sixth grade girls, third grade boys, and third grade girls. Nevertheless, the skill acquisition process, as detected by performance scores on each trial for each task, was generally similar among the groups.

In studying pursuit rotor performance in young children Davol et al. (1961) reported findings on third grade children similar to those reported by Singer (1969). Even though the boys became increasingly superior to the girls, no significant sex differences in performance were reported. Davol et al. examined the performance of kindergarten children through third grade, noting that the older children made greater gains in proficiency with practice than the younger children.

The Ammons et al. (1955) study was designed to investigate the relationship of age and sex to pursuit rotor performance in children in grades three six, nine, eleven, and twelve. It was found that there was a marked overall improvement in proficiency accompanying each age period, with boys showing an increasing superiority over girls. The proficiency of girls declined from grade nine to grade twelve.

In most studies concerned with motor performance and age relationships, laboratory tasks are not of prime interest. Gross motor skills, typically athletic, have been reported with both sexes in various age brackets. For instance, Seils (1951) found performances in running, throwing, balancing, striking, catching, and jumping to be greater at succeeding age levels in the primary grades. Sex differences in the 7–9-year bracket in favor of boys in ball-rolling skills (Witte, 1962) and in the standing broad jump (Kane and Meredith, 1952), in various athletic events from ages 5 to 7 (Jenkins, 1930), and in other activities from ages 9 to 11 (Latchaw, 1954) have been reported. Sex differences in motor performance, of course, may result from a number of factors. The noted differences do not imply that sex is a variable in learning per se. Sociocultural factors, play patterns, motivation and interest, and anatomical and physiological factors play a large role in determining performance levels in assorted activities.

Keogh (1965) tested 1171 kindergarten through sixth grade children on a number of motor tasks. Age changes in motor performance on most tasks were generally linear. Boys were more skilled in throwing and girls were better in hopping, while unique patterns of sex and age interactions existed with the other activities. In another investigation, Keogh (1969) examined 5–7-year-old boys and girls in limb control tasks and body control tasks. Age improvements were demonstrated, and girls were better performers than boys at each age.

It may be surmised that most gross motor activities are learned to a higher degree and performed at a greater performance level in boys as compared to girls in the childhood years. These distinctions become more apparent with age. But in simple, fine motor tasks, females usually surpass the males. In reviewing the pertinent literature, Broverman *et al.* (1968) concluded:

> . . . evidence exists that females exceed males in tasks that require rapid, skillful, repetition, articulation, or coordination of "lightweight," over-learned responses (perceptual responses, small muscle movements, simple perceptual-motor coordinations). (p. 25)

Research is reported showing that girls exceed boys in tasks of fine manual dexterity, verbal functions (speech and reading do involve perceptual-motor behaviors), clerical functions, speed of color naming, and other tasks involving rapid perception and frequent shifts of attention.

C. INSTRUCTION

There is a great difficulty in ascertaining the age at which children can significantly benefit from special instruction. The issue has been continuously raised as to the benefits, if any, of encroaching on the maturational process. Conflicting research evidence and theory has confused the situation. In Miller's (1957) experiment, first grade children were taught the overhand throw for 26 20-min periods. Although this group performed better than a control group, the difference was not significant. It is possible, however, that longer practice might have resulted in more meaningful distinctions between the groups. Using another approach to the problem, namely, analyzing motion pictures and the overhand throwing patterns of children from ages 2 to 7, Wild (1938) concluded that maturational factors are most important until the age of 6. After age 6, learning is very important in skill acquisition.

McDonald's (1967) thesis refuted this notion. Eight-year-olds, both male and female, were subjects. The experimental group learned to throw a tennis ball at a vertical wall target, 10 trials per day, 2 days per week for 8 weeks. The experimental and control groups were pretested and posttested. It was found that the practice afforded the experimental group did not lead to significant improvement. The boys were more accurate than the girls, but they did not profit more from practice than the girls.

The value of instruction in a supplementary physical education program in improving motor skills was investigated by Olson (1968) with primary school children. Children deficient in motor skills were found to benefit from the supplementary program as compared to a control group. In fact, some experimental children reached a level of proficiency comparable to normal primary grade school children who did not participate in the supplemental physical education program. A pertinent point in the controversy over instruction vs. no-instruction is that Olson found that when two groups were compared in performance as a result of their experience in the supplementary program, one with instruction and one without, there did not appear to be any differences in performance. However, a retention test, administered 2 months later, suggested that retention may be aided by initial instructional procedures.

Unfortunately, too few studies have actually been designed to measure the effects of special instruction and training in motor learning with children as subjects. The age at which such instruction becomes most meaningful is probably dependent on the nature of the task, the maturity and readiness level of the child, and the instructional process utilized.

V. Adolescence

A. MOTOR PERFORMANCE

It has been shown that boys accelerate markedly in motor performance during the adolescent years whereas girls level off and even show decreases in performance in comparison to the preadolescent or early adolescent years. In other words, sex differences in relation to motor performance become more apparent with increasing age from later childhood to the end of adolescence. It should be reemphasized that innate learning capacities do not necessarily differ between the sexes but rather bodily structure, socio-cultural influence, interest and motivation, and need-satisfaction can probably explain performance distinctions for the most part. Prior learning experiences influence rate and level of skill acquisition as well.

Only limited research has been done on the factors affecting motor learning prior to late adolescence and early adulthood (the college years). Most research presents performance data per se, rather than changes due to instruction, training, or merely as a result of practice. For example, Espenschade (1947) administered the Brace Motor Ability Test to 325 girls and 285 boys, ranging in age from 10½ to 16 years. Among the conclusions, boys were found to increase in ability to perform the varied activities. Girls improved in agility up to 14 years, then their performance declined. Minor changes across age were observed in the test items requiring control, flexibility, and balance for the girls. Prior to 14 years, the boys and girls were quite similar in performance scores, but after that, the boys' superiority increased rapidly at each successive age level.

A number of authorities and researchers have stated that physical fitness factors and athletic abilities improve in boys throughout adolescence whereas they decline in girls after the age of 13 and 14. One of the few contrary findings is reported by Vincent (1968), who observed that girls from 12 to 18 years of age evidenced improvement with age in five of the eight activities tested. Furthermore, the achievements of female Olympic athletes indicate the potentiality of proficiency for that sex in many athletic activities. Although females will probably not equal males in activities involving strength and power because of physiological and anatomical limitations, they are *capable* of learning and performing motor skills at a much higher level than has been generally recorded in the literature. Tanner (1961) has stated that power, athletic skill, and physical endurance, when considered absolutely, improve progressively and quickly during the adolescent years. This statement has definitely been confirmed by research on males and can be shown perhaps to a lesser degree to be true with females.

B. The Attainment of Skill

Since each year blends together the maturational aspects of the organism with its prior experiences, it is not surprising that until growth is stabilized at about age 19 or 20, performance in most skills improves. McGeoch and Irion (1952), in summarizing a number of investigations undertaken in the early portion of this century, stated: "practice effects on a wide range of perceptual-motor acts increase with age from about chronological age 8 to the years of late adolescence or early maturity." In very complex skills involving a high degree of refined coordination of movement responses to appropriate cues, improvement in skill may occur continuously in the years that follow.

Bachman (1961), after studying motor learning rates of children 6–16 years of age, raised the point that simple skills are typically taught to younger children because of the belief that they are limited in learning ability, as contrasted with older children and adolescents. Using two balance tasks in which practice was provided, Bachman concluded that rate of learning was not affected by age or sex. However, the amount of learning, as measured by the improvement between initial and final levels of skills, increased through the adolescent period. Thus, it can be seen that much depends on the criterion measurement of learning and practice effects as to those conclusions that may be reached. McGeoch and Irion (1952) stated:

> The measures of practice effect have usually been composites of initial performance and gain from specific practice. Only the latter is strictly a measure of rate of learning, since initial status is probably more heavily weighted by transfer from prior learning, done at unknown rates, to the first or early trials of the activity of the experiment. It is often a question, therefore, whether to call the results measures of rate of learning or of transfer plus rate of learning, and usually the latter is the more accurate.

Skill achievement in the adolescent period, as is the case in other years, is influenced not only by environmental, situational, and maturational variables but also by genetic factors. Although it has been found that similar activities are performed by the families of outstanding athletes, evidence has been produced that point to the strong impact heredity makes on athletic and other psychomotor accomplishments (Gedda *et al.*, 1964).

The development of discrimination-reaction skill, where speed of movement, response learning, and stimulus–response associations are important, was studied by Noble *et al.* (1964). Performance relationships with sex and age were reported. The performance of the males improved from 8 until 20 years, then declined; that of the females improved from 8 to 16 and then became poorer. It is difficult to determine why the peak years of performance differed between the sexes. The investigators claim that the findings result from "a combination of sensorimotor maturation and transfer of (nonspecific) training. . . . generalized experiences in discriminating spatial cues and reacting promptly in selected directions . . . could account for differences."

In this investigation the two sexes were similar in performance level until the age of 16. Hodgkins (1963), on the other hand, found that in reaction time and speed of movement the performance of the two sexes was similar up to the age of 12 years, but thereafter the males were superior. For both sexes, reaction time improved until the approximate age of 19 and then decreased. Peak movement time speeds were reached for both sexes at about age 15. Findings in these and other studies, in which the males's skills and abilities are shown to surpass those of the female at certain ages, provide some insight as to performance differences between the sexes when tasks are introduced requiring these skills and abilities.

C. THE BENEFITS OF INSTRUCTION

With increasing age, maturity, and comprehension, the ability to learn more complex tasks becomes realized. At advanced skill levels, directions can be more detailed and lengthy, thus leading to further learning. Not only do the adolescent years bring the body and its repertoire of responses under greater control, but the organism is more capable of intellectualizing the task. Attention, perception, imagery, and the like are developed corresponding to the increased maturity of the individual during adolescence. Such personal qualities are necessary for the skilled execution of complex motor skills.

Furthermore, positive transfer probably occurs more easily during youthful years rather than in advanced years. Previous learnings affect later learnings, especially the more the tasks are specifically related. At older ages it becomes increasingly difficult to reorganize patterned behavioral responses. This is supported by the experiment of Ruch (1934) in which three age groups, 12–17, 34–59, and 60–82, were involved in learning a pursuit rotor task with direct vision and with mirror vision. The oldest group performed the poorest on the mirror vision task, the youngest group the best. Apparently, until early maturity, the amount of positive transfer increases with age. Later in life, learning abilities may decline because of competing fixed responses.

When one scans the literature on sex differences in learning many diverse tasks, representing the psychomotor, cognitive, and affective domains, the general trend is to find small, if any, dissimilarities. Naturally, the nature of the material may favor one sex or the other. Prior related experiences may also contribute to sex differences in performance. But assuming sufficient practice, motivation and interest, and minor inequities in anatomical structure or physiological functions the two sexes perform comparably in most learning situations. There is probably no known biological factor that should favor one sex in ability to learn most motor skills. Any obtained differences mostly result from differences in past experiences and the nature of the present task and the context in which it is learned.

References

Alderman, R. B. (1968). *Res. Quart.* **39**, 428.

Ammons, R. B., Alprin, S. I., and Ammons, C. H. (1955). *J. Exp. Psychol.* **49**, 127.

Bachman, J. C. (1961). *Res. Quart.* **32**, 123.

Bayley, N. (1935). "The Development of Motor Abilities During the First Three Years," Monogr. Soc. Res. Child Develop., Washington, D. C.

Bloom, B. S. (1964). "Stability and Change in Human Characteristics." Wiley, New York.

Brace, D. K. (1930). "Measuring Motor Ability: A Scale of Motor Ability Tests." Barnes, New York.

Brackbill, Y., and Koltsova, M. M. (1967). *In* "Infancy and Early Childhood" (Y. Brackbill, ed.), pp. 207–286. Free Press, New York.

Breckenridge, M. E., and Murphy, M. N. (1969). "Growth and Development of the Young Child." Saunders, Philadelphia, Pennsylvania.

Broverman, D. M., Klaiber, E. L., Kobayashi, Y., and Vogel, W. (1968). *Psychol. Rev.* **75**, 23.

Buckellew, W. F. (1968). Unpublished Doctoral Dissertation, University of Arkansas, Fayetteville.

Clarke, H. H. (1967). *Proc. 8th Nat. Conf. Med Aspects Sports*, **8**, 49.

Crow, L. D., and Crow, A. (1962). "Child Development and Adjustment." Macmillan, New York.

Crowell, D. H. (1967). *In* "Infancy and Early Childhood" (Y. Brackbill, ed.), pp. 125–203. Free Press, New York.

Davol, S., Hastings, M., and Klein, D. (1961). *Percept. Mot. Skills* **21**, 351.

Dennis, W., and Dennis, M. G. (1940). *J. Gen. Psychol.* **56**, 77.

Diem, L. (1970). Personal communication.

Dohrmann, P. (1964). *Res. Quart.* **35**, 464.

Dusenberry, L. (1952). *Res. Quart.* **23**, 9.

Elkind, D., and Sameroff, A. (1970). *In Annu. Rev. Psychol.* **21**, 191–238.

El'konin, D. B. (1969). *In* "A Handbook of Contemporary Soviet Psychology" (M. Cole and I. Maltzman, eds.), pp. 163–208. Basic Books, New York.

Espenschade, A. S. (1947). *Res. Quart.* **18**, 30.

Espenschade, A. S., and Eckert, H. M. (1967). "Motor Development." Merrill, Columbus, Ohio.

Fowler, W. (1962). *Psychol. Bull.* **59**, 116.

Gagné, R. M. (1968). *Psychol. Rev.* **75**, 177.

Gedda, L., Milani-Comparetti, M., and Brenci, G. (1964). *In* "International Research in Sport and Physical Education" (E. Jokl and E. Simon, eds.), Thomas, Springfield, Illinois.

Gesell, A. L. (1940). "The First Five Years of Life." Harper, New York.

Gesell, A. L., and Thompson, H. (1941). *Genet. Psychol. Monogr.* **24**, 3.

Glassow, R. B., and Kruse, P. (1960). *Res. Quart.* **31**, 426.

Goodenough, F L. (1935). *J. Exp. Psychol.* **25**, 431.

Greenwald, A. G. (1970). *Psychol. Rev.* **77**, 73.

Groves, W. C. (1969). *J. Res. Music Educ.* **17**, 408.

Gutteridge, M. V. (1939). *Arch. Psychol.* **244**.

Hale, C. J. (1956). *Res. Quart.* **27**, 276.

Henry, F., and Nelson, G. (1956). *Res. Quart.* **27**, 162.

Herron, R. H., and Frobish, M. J. (1969). *J. Exp. Child Psychol.* **8**, 40.

Hicks, J. A. (1930a). *Child Develop.* **1**, 90.
Hicks, J. A. (1930b). *Child Develop.* **1**, 292.
Hicks, J. A., and Ralph, D. W. (1931). *Child Develop.* **2**, 156.
Hilgard, J. R. (1932). *J. Genet. Phychol.* **41**, 31.
Hilgard, J. R. (1933). *Gen. Psychol. Monogr.* 14, 6.
Hodgkins, J. (1962). *J. Gerontol.* **17**, 385.
Hodgkins, J. (1963). *Res. Quart.* **34**, 335.
Humphries, M., and Shephard, A. H. (1959). *Percept. Mot. Skills* **9**, 3.
Ismail, A. H., and Cowell, C. C. (1961). *Res. Quart.* **32**, 507.
Jenkins, L. M. (1930). *Teach. Coll. Contr. Educ.* No. 414.
Kane, R., and Meredith, H. (1952). *Res. Quart.* **23**, 198.
Kay, H. (1969). *In* "Principles of Skill Acquisition" (E. A. Bilodeau, ed.), pp. 33–58. Academic Press, New York.
Keogh, J. F. (1965). "Motor Performance of Elementary School Children." Dept. Phys. Educ., University of California, Los Angeles.
Keogh, J. F. (1969). "Analysis of Limb and Body Control Tasks," USPHS Grant HD 01059. V.S. Pub. Health Serv., Washington, D.C.
Ketterlinus, E. (1931). *Child Develop.* **2**, 200.
King, J. A. (1958). *Psychol. Bull.* **55**, 46.
Latchaw, M. (1954). *Res. Quart.* **25**, 439.
Lorenz, K. Z. (1958). *Sci. Ameri.* **199**, 67.
Luria, A. R. (1961). "The Role of Speech in the Regulation of Normal and Abnormal Behavior." Pergamon, New York. Oxford.
McDonald, C. L. (1967). Unpublished Master's Thesis, Pennsylvania State University, University Park.
McGeoch, J. A., and Irion, A. L. (1952). "The Psychology of Human Learning." McKay, New York.
McGraw, M. B. (1935). "Growth: A Study of Johnny and Jimmy." Appleton, New York.
McGraw, M. B. (1939). *Child Develop.* **10**, 1.
McGraw, M. B. (1970). *Ameri. Psychol.* **25**, 754.
McNemar, Q. (1933). *J. Genet. Psychol.* **42**, 70.
Melzack, R., and Thompson, W. R. (1956). *Can. J. Psychol.* **10**, 82.
Miller, J. L. (1957). *Res. Quart.* **28**, 132.
Newton, G., and Levine, S., eds. (1968). "Early Experience and Behavior." Thomas, Springfield, Illinois.
Noble, C. E., Baker, B. L., and Jones, T. A. (1964). *Percept. Mot. Skills* **19**, 935.
Olson, D. M. (1968). *Res. Quart.* **39**, 321.
Piaget, J. (1952). "The Origins of Intelligence in Children." International University Press, New York.
Pronko, N. H. (1969). *Psychol. Today* **2**, 52.
Pumroy, D. K., and Pumroy, S. (1961). *J. Genet. Psychol.* **98**, 55.
Ragsdale, C. E. (1950). *In* "Learning and Instruction" (N. B. Henry, ed.), pp. 69–91. Univ. of Chicago Press, Chicago, Illinois.
Ravizza, R. J., and Herschberger, A. C. (1966). *Psychol. Rec.* **16**, 73.
Ruch, F. L. (1934). *J. Gen. Psychol.* **11**, 261.
Scarr, S. (1966). *Child Develop.* **37**, 663.
Schulman, J. L., Buist, C., Kaspar, J. C., Child, D., and Fackler, E. (1969). *Percept. Mot. Skills* **29**, 243.
Scott, J. P. (1962). *Science* **138**, 949.

Scott, J. P. (1968). "Early Experience and the Organization of Behavior." Brooks/ Cole, Belmont, California.

Seils L. (1951). *Res. Quart.* **21**, 244.

Singer, R. N. (1969). *Res. Quart.* **40**, 803.

Sluckin, W., and Salzen, E. A. (1961). *Quart. J. exp. Psychol.* **8**, 65.

Smart, M. S., and Smart, R. C. (1967). "Children: Development and Relationships." Macmillan, New York.

Smith, J. A. (1956). *Res. Quart.* **27**, 220.

Staats, A. W. (1968). "Learning, Language, and Cognition." Holt, New York.

Stachnik, T. J. (1959). Unpublished Master's Thesis, Illinois State University, Normal.

Steinschneider, A. (1967). *In* "Infancy and Early Childhood" (Y. Brackbill, ed.), pp. 3–50. Free Press, New York.

Stone, L. J., and Church, J. (1968). "Childhood and Adolescence." Random House, New York.

Tanner, J. M. (1961). "Education and Physical Growth." Univ. of London Press, London.

Vincent, M. F. (1968). *Res. Quart.* **39**, 1094.

Watson, J. B. (1919). "Psychology From the Standpoint of a Behaviorist." Lippincott, Philadelphia, Pennsylvania.

Wenger, M. A. (1936). *Univ. Iowa Stud. Child Welfare* **12**, No. 1.

Wild, M. (1938). *Res. Quart.* **9**, 20.

Williams, J. R., and Scott, R. B. (1953). *Child Develop.* **24**, 103.

Witte, F. (1962). *Res. Quart.* **33**, 476.

CHAPTER

9

Stability and Change in Motor Abilities

G. Lawrence Rarick

I. Prediction of Developmental Change

The primary purpose of the study of human growth and development is to advance knowledge of the developmental processes and to learn how genetic and environmental factors affect the course of human growth. Any attempt to predict the magnitude of change in human characteristics requires an understanding of the nature and individuality of human growth and an insight into the effects that different environments have on the development of human traits and abilities.

It has been well established that the blueprint for development is laid down at the time of conception and that innate forces shape the general pattern of development during the prenatal period and through the early years of life. In the sense that changes in most physical traits and many behavioral characteristics are initially and primarily controlled by genetic forces, it is apparent that developmental changes will follow well-defined sequences, although these changes will not necessarily occur at the same ages for all children. Thus, one can predict with considerable accuracy the nature and sequence of the changes, although the age at which these will occur will often vary considerably among children. With advancing age and with a greater impingement of diverse environments on children, the prediction of developmental change obviously becomes more difficult. As Bloom (1964) pointed out, the early years of life are of utmost significance in the development of the child because variations in the environment at this time shape those characteristics that are in the process of formation.

Much of our current knowledge of human growth has come from longitudinal investigations in which repeated observations have been made on the same children from infancy to maturity (Jones and Bayley, 1941; Tuddenham and Synder, 1954; Ebert and Simmons, 1943). These studies have provided a wealth of information on the interrelationship of traits and abilities at different points in development, but perhaps more important they have enabled scientists to learn something about the magnitude and velocity of changes in human characteristics as children grow to maturity. The longitudinal method has also made it possible to determine the accuracy of predicting developmental changes over short and long periods of time. Similarly, this approach has also given researchers insight into the impact of different environments upon the development of physical and behavioral characteristics of children of similar and of different genetic stock.

Much of the early work on the prediction of human growth involved the use of anthropometric measurements. This was done partly because of the ease and accuracy with which these measurements could be taken and partly because at that time they were widely used to reflect the nutritional status of children. More recently, the interest of developmental psychologists has centered on the behavioral aspects of development. Educational researchers have for some time been interested in predicting educational achievement and the impact that various educational programs have on future academic success.

The purpose of this chapter is to bring to the attention of the reader information on stability and change in the development of certain physical abilities. With advances in precision of measurement, behavioral scientists have centered their interest on the study of individual differences and the extent to which these differences change with age and environmental circumstances.

The age-old question of the respective roles of heredity and environment in accounting for individual differences as children grow and develop is still largely unresolved. Bloom (1964) believed that much of the individual variability in behavioral characteristics can be explained by environmental variation. One cannot, however, ignore the role of genetic factors in the development of certain physical abilities, particularly those that are dependent on inherited physical attributes. Muscular strength, for example, is dependent to a large extent on body build (Jones, 1949). The robust mesomorphic boy, well-endowed by nature for an athletic career, may or may not choose to devote his growing years to athletic pursuits. The reasons underlying the decision would clearly be a function of the environment. On the other hand, those less well endowed physically learn early in life that they must choose physical activities compatible with their morphology or turn their interest in other directions.

While there may be some truth in the statement that athletes are born not made, this statement has not been put to the test and it is not likely that it will be in the foreseeable future. The problems involved in sorting out the genetic and environmental factors that bear upon this question and the difficulties involved in identifying and assessing the multitude of traits responsible for the athlete's success are many and varied. In view of our present state of knowledge, it is much simpler and perhaps more informative to measure specific performance capabilities and determine the extent to which these change in a systematic and predictable way over a period of years. In other words, it would seem appropriate to determine if those who performed well on tests of muscular strength and power and on certain motor performance tests in childhood are the ones most likely to maintain superiority on these tests as they grow older. Is there consistency among children in the development of specific physical abilities as they advance in age or is the variance in the growth of these abilities so great that prediction is impossible? This is the question this chapter seeks to answer.

There are several ways of assessing the consistency with which human characteristics develop, each appropriate for a particular purpose. One can, for example, observe the development of a given trait in groups of children who differ materially in respect to some criterion measurement. Shuttleworth (1939) used skeletal and sexual maturity as criterion measurements in his longitudinal study of the physical growth of children noting that, on the average, early maturing children were taller and heavier at each age than late maturing children. Jones (1949) and Espenschade (1940) in their longitudinal studies of the development of physical abilities in adolescents compared the strength and motor performance of early and late maturing boys and girls at 6 month intervals from early to late adolescence. While there may be individual differences within criterion groups in respect to the characteristic under observation, the variance will be minor as compared to the variance for the population as a whole, if the trait in question is related to the criterion measurement. The use of criterion groups in longitudinal studies has provided important information on stability and change in many physical and behavioral characteristics.

Correlation analysis is frequently used in determining the accuracy of assessing the consistency with which a characteristic changes over time. This provides a quantitative relationship between measurements of a growth variable obtained at one point in time with the same measurements taken at a later date. If, for example, one were to correlate the measured statures of a sample of boys at 10 years of age with their heights a year later and the boys have all grown relatively the same (kept their same position in the group), the correlation between their heights over this span of time would be $+1.0$. If there were no relationship whatsoever between their heights at these two points in time the correlation would be zero.

It is also possible to use gain scores as a means of studying change, although Cronback and Furby (1970) pointed out that this procedure has distinct limitations. In this case one is not so much concerned with the question of determining the extent to which individuals keep their relative position in the group but rather the extent to which the gains are related to the magnitude of the measurements at the time of initial measurement.

Rarely, if ever, is the development of a human trait unrelated to other physical or psychological characteristics. Hence, insight into the influence that other characteristics may have on the growth of a particular trait may be of interest. This may be determined by the use of multiple correlation. This procedure involves the correlation of the scores of a cluster of traits at one point in time with the scores of the trait in question at a later time. For example, one might use this procedure to determine if the correlation between performance on a strength test at two points in time would be improved by including in the computations the heights and weights of the subjects at the earlier age. While the major thrust of this chapter is to consider stability and change in specific physical abilities, some consideration will be given to predicting the relative magnitude of change over relatively long periods of time by using a combination of measurements as the basis for the prediction.

II. Stable and Unstable Growth Characteristics

Bloom (1964) defined a stable characteristic as "one that is consistent from one point in time to another." By this he meant that although a trait may differ quantitatively over time, the change can be predicted to some degree. He suggested that the minimum level of consistency is represented by a correlation of .50 or higher over a period of a year or more. According to this definition most measurements of physical growth can be classified as stable traits. For example, year-to-year correlations of the heights of boys and girls over 3 years of age and up to 17 years inclusive have been shown to be of the order of .94 or higher (Tuddenham and Snyder, 1954). The year-to-year correlations of weight and of many other anthropometric measurements are almost as high. Correlations of height over time periods of over more than 1 year while not as high as those given above are nevertheless substantial. There is a good reason for the high between age correlations of physical growth measures such as height, because the correlation involves correlating a part (the first year's height) with the whole (the first year's height plus the gain in height for the following year). Since ordinarily all children realize some growth in height each year and since this figure is relatively small in proportion to the height already achieved prior to the year's growth, the magnitude of the correlation is likely to be less a function of the year's gain in stature than of the height itself. Thus, much lower between year correlations are obtained by correlating height with the gain in stature the following year or by correlating between year gains in height and stature at maturity. As Bloom (1964) pointed out the correlations between yearly gains in height and mature stature range from .40 to as low as −.40. Thus, it is apparent that the high correlations obtained for height at yearly intervals are only slightly affected by annual gains. It is equally evident that growth variables in which the yearly changes are additive are invariably classified as stable traits. Those characteristics which fluctuate between annual gains and losses would be considered as unstable traits. Ordinarily one does not shrink in height except in old age, nor is one likely to show annual decreases in bone measurements.

Many traits and abilities may show little if any measureable change over extended periods of time. For example, skills or abilities once learned, may show no improvement if they are not practiced and quite often the performance regresses. Similarly, under adverse nutritional conditions or in illness the body weight of children may decline. Thus, developmental changes are subject to the environment and the way in which the growing child interacts with environmental forces. Clearly, the more homogeneous the environment, the more predictable will be the changes in both physical and behavioral characteristics.

III. Time, Maturity, and Prediction of Change

Time is a factor of considerable significance in the prediction of human performance. Consistency or reliability of performance (and of measurement) is frequently determined by test–retest trials, the trials often separated by only a few seconds. Correlations under these conditions are almost without exception higher than those obtained by correlating the scores of between day measurements, and the latter in turn show higher correlations than those obtained by correlating test scores with a week or longer between trials. It is logical that this would be the case, since the longer the time interval between trials, the greater will be the opportunity for intervening factors to have differing effects. The disparity in the magnitude of the correlations as a function of the time interval between testing is illustrated by the research of Jones (1949) who reported that within day reliability of the grip strength of boys of varying ages ranged from .933 to .959. With a week's time between trials the correlations dropped to .915 and with 6 months intervening between tests the correlations ranged from .794 to .904.

As mentioned earlier the age and maturity level at which developmental data are taken affects the stability or consistency of the growth of human characteristics. While the environment may affect characteristics at any time, its greatest effect both quantitatively and qualitatively is at times of most rapid developmental change. Hence, the first few years of life and early adolescence are the periods when traits and abilities are most sensitive to change.

IV. Hereditary Factors and Human Development

The influence of hereditary factors in determining the course of physical growth and development has been well established. It is well known that tall parents are likely to have tall children. Viewed more precisely the correlation between the heights of parents and the heights of their grown offspring has been shown to range between $+.50$ and $+.60$ (Elderton and Pearson, 1915; Sanders, 1934). The influence of hereditary factors on physical growth is more dramatically shown in the study of identical twins by Newman *et al.* (1937), where the correlation between the heights of identical twins reared in the same family was .98 and that for identical twins reared apart was almost identical, .97. Thus, it would seem the innate forces that control the growth mechanisms are so potent that even under different environmental circumstances identical twins show essentially the same physical growth characteristics.

It is interesting to speculate on the extent to which children inherit their physical abilities. Insofar as one's basic body structure is inherited and insofar as structure determines function, it would seem reasonable to believe that inheritance is a factor that cannot be ignored in the motor performance of children. Although there is limited data relative to this question, there is evidence to indicate that there is a well-defined parent size-related trend associated with the static muscular strength of children (Malina *et al.*, 1970). The trend was found to be more prominent in girls than in boys. This was attributed by the investigator to the impact of cultural expectations on young males which tend to overshadow the effect that the relatively small inherited differences in size has on strength. In girls, on the other hand, the genetic rather than the cultural effect assumes the more prominent role. In respect to motor performance, however, Malina found that parental size was of little consequence. Garn *et al.*, (1960; Garn, 1962), however, found that children born of large parents tended early in life not only to be taller and heavier but were more advanced in motor development in infancy (6–18 months) than children born of small parents.

In respect to educational achievement it is becoming increasingly clear that environmental factors play a more significant role than genetic factors in children of normal intelligence. As Bloom (1964) pointed out, the correlations between achievement measurements on identical twins reared together are high (.85 or higher) as compared to correlations of .70 for identical twins reared apart. Burt (1955) found correlations on achievement tests of .90 for identical twins reared together in comparison to correlations of .68 for those reared apart. Correlations of similar magnitude (.96 and .51) were reported by Newman *et al.* (1937). While data of these kinds are not available on measures of gross motor performance of identical twins reared in the same and in different environments, it is reasonable to believe the results would be similar.

While the blueprint for growth is determined by the genes, the final expression of development may be modified by environmental factors. Since physical attributes are less easily affected by environmental forces than by behavioral traits it is obvious that genetic influences would be more apparent in the former than in the latter. This is not meant to imply that physical growth is not affected by environmental influences, for it is indeed sensitive to both favorable and unfavorable conditions. The effects of an unfavorable environment on physical growth may not be lasting if the growing organism is not exposed to these factors for long periods of time, for according to Waddington (1957), growth is canalized. By this he meant that the growing organism vigorously persists in maintaining its normal rate of growth in spite of disrupting environmental factors. If, for some reason it is thrown off course, it will, as soon as the environment becomes favorable, accelerate its growth until it is back on its normal schedule of development. Prader *et al.* (1963) provided clear-cut evidence of what they call catch-up growth in children following illness and starvation. Olson (1949) held strongly to the belief that the growth of children tends to resist displacement from factors which tend to either accelerate or retard development. As Prader *et al.* (1963) pointed out some human traits can be pushed off the normal course of development more easily than others, and similarly some children have certain traits that are more sensitive to environmental factors than are others. There is, however, some kind of regulative force which tends to maintain development in its original channel. It is because of this tendency that it should be possible to predict with some accuracy the development of many human traits and characteristics, particularly if the effects of environmental factors can be assessed.

V. Growth in Strength

A. RELIABILITY OF STRENGTH MEASUREMENTS

The muscular strength of children has traditionally been assessed by measurements of static dynamometric strength usually employing either spring dynamometers or cable tensiometers. Spring dynamometers have been used in several cross-sectional and longitudinal studies in measuring the gripping, pulling, and thrusting strength of children. More recently, the cable tension method (Clarke, 1948, 1950), designed to measure the strength of limb movements, has been used with children and adolescents. Both methods produce highly reliable results when administered by trained personnel. Within day test–retest correlations of gripping, pulling, and thrusting strength, as reported by Jones (1949), range between .932 and .964 for boys and between .885 and .948 for girls. Clarke (1948), using the cable tensiometer, reported within day test–retest correlations for college age males ranging from .92 to .97. Test–retest correlations with the cable tensiometer when used with children are likewise high, usually well above .90 (Rarick and Mohns, 1955; Bohm, 1959). In view of the high reproducibility of these measurements, it is surprising that only a few growth studies have used tests of static dynamometric strength to reflect an important functional dimension of growing children, one that undergoes rather dramatic changes with advancing age. Data from the few longitudinal studies employing these measures have provided valuable information on the individuality and the predictability of growth in strength of children as they moved through childhood into the adolescent years.

B. CONSISTENCY IN GROWTH IN STRENGTH

Perhaps the first attempt to assess the consistency in the growth of strength of children over time was the research of Baldwin (1920) in which correlations of .65 for boys and .45 for girls were reported using measurements of gripping strength at 9 and 10 years of age, repeating the tests on the same children 6 years later. Correlations of a similar magnitude were found by Tuddenham and Snyder (1954) in their longitudinal study of California children. Using a composite strength score (sum of right grip, left grip, pull, and thrust), they obtained between year correlations, ages 9–18

TABLE I

CORRELATIONS BY SEX ACROSS AGES: UPPER EXTREMITY STRENGTH MEASUREMENTS[a]

Measurements (by sex)	Age-to-age correlations				
	Childhood				
	7–12	8–12	9–12	10–12	11–12
Boys					
Wrist flexion	.375	.550	.163	.486	.533
Elbow flexion	.279	.628	.825	.807	.733
Shoulder medial rotation	.202	.372	.605	.625	.702
Shoulder adduction	.523	.491	.315	.550	.676
Girls					
Wrist flexion	.361	.406	.553	.640	.655
Elbow flexion	.367	.627	.592	.763	.770
Shoulder medial rotation	.184	.300	.751	.774	.758
Shoulder adduction	.133	.608	.302	.597	.740
	Childhood to Adolescence				
Boys	7–17	8–17	9–17	11–17	12–17
Wrist flexion	.378	.493	.327	.244	.419
Elbow flexion	.235	.193	.602	.618	.634
Shoulder medial rotation	.258	.307	.568	.505	.674
Shoulder adduction	.491	.418	.485	.681	.662
Girls					
Wrist flexion	.387	.013	.454	.436	.457
Elbow flexion	.566	.308	.336	.350	.114
Shoulder medial rotation	.156	−.152	.327	.496	.372
Shoulder adduction	.260	.360	.622	.499	.185

[a] Adapted from Rarick and Smoll (1967).

years, of .63 for boys and .57 for girls. It is apparent that the consistency of performance of the subjects across this span of time was not great for the variance in strength in common between these two age levels was of the order of only 36% for the boys and approximately 30% for the girls.

Findings similar to those noted above have been reported by Rarick and Smoll (1967) using data from the Wisconsin Growth Study. In this study longitudinal anthropometric, strength, and motor performance data were obtained annually on 25 boys and 24 girls from 7 to 12 years of age and once again after an intervening period of 5 years. The cable tension method was used in securing four upper extremity and four lower extremity strength measurements. The results of the between age correlation of the four upper extremity strength measurements obtained on the two sexes are shown in Table I. As would be expected the size of the correlations decreased as the time interval between testing increased. Only one of the correlations across the span of 10 years (7–17 years) was above .50, the majority being in the .20's and .30's. The age-to-age correlations in childhood across the age span of 1 and 2 years, respectively were for the most part in the .60's and the .70's. They dropped progressively, however, as the time interval increased, and from age 7 years to age 12 the correlations were of a magnitude similar to those noted across the age span 7–17 years.

Data on the lower extremity strength scores taken from the Wisconsin Growth Study (Rarick and Smoll, 1967) are shown in Table II. While the between age correlations for these measurements are generally somewhat higher than those cited above, they are not high enough to be of predictive value. Perhaps the slightly higher between age correlations of the lower extremity strength measures in comparison to the upper extremity measurements is a reflection of the rather universal demands placed on the leg muscles of all children. It is reasonable to believe, on the other hand, that the extent of use of the arm muscles varies widely among children from one period of time to another. This would tend to accentuate the between age variance in the strength of the upper extremities.

The findings on between age correlations of static muscular strength reported above are supported by the research of Clarke (1971) in the Medford Growth Study. By using the cable tensiometer, Clarke obtained correlations over a 5-year age span in childhood (7–12 years) ranging from .217 to .583 for measurements of upper and lower extremity strength. Across a 5-year age span in adolescence (12–17 years) the correlations ranged from .013 to .465.

There are many factors that may account for the relatively low between age correlation in strength scores. While errors in measurement should not be a major factor in view of the high reliability of strength tests, individual

TABLE II

CORRELATIONS BY SEX ACROSS AGES: LOWER EXTREMITY STRENGTH MEASUREMENTS[a]

Measurements (by age)	Age-to-age correlations				
	Childhood				
	7–12	8–12	9–12	10–12	11–12
Boys					
Hip flexion	.353	.406	.668	.682	.813
Hip extension	.706	.481	.614	.674	.861
Knee extension	.430	.465	.277	.668	.732
Ankle extension	−.020	.634	.737	.794	.763
Girls					
Hip flexion	.477	.512	.774	.898	.890
Hip extension	.636	.512	.665	.769	.796
Knee extension	.763	.722	.661	.651	.754
Ankle extension	.035	.276	.533	.770	.670
	Childhood to Adolescence				
	7–17	8–17	9–17	11–17	12–17
Boys					
Hip flexion	.334	.393	.355	.402	.515
Hip extension	.344	.076	.430	.426	.488
Knee extension	.429	.503	.483	.735	.797
Ankle extension	.209	.416	.413	.480	.590
Girls					
Hip flexion	.506	.366	.530	.666	.712
Hip extension	.336	.394	.595	.378	.575
Knee extension	.673	.508	.730	.641	.646
Ankle extension	.267	.306	.523	.275	.265

[a] Adapted from Rarick and Smoll (1967).

variations in motivation from one test period to another may well be a factor of considerable consequence. Differences among individuals in the rate of strength development resulting from differences in regimens of physical activity would logically seem to be highly significant. In view of the positive relationship between muscular strength and body size, the low between age correlations may be a reflection of within and between subject variability in the tempo of physical growth and in the rate of sexual maturation. This becomes particularly important in early adolescence when the rate of physical growth varies widely among individuals because of rather wide age variations at the onset of sexual maturity. This has been dramatically illustrated by Jones (1949) who found that early maturing boys and early maturing girls are not only stronger but also show an earlier surge in strength development than their less mature counterparts of the same chronological age. The variability in these characteristics is accentuated in boys in the age range 13–15 years and in the girls from 11 to 14 years of age.

C. PHYSICAL GROWTH AND STRENGTH DEVELOPMENT

The above findings, and those from previously cited research, show clearly that the prediction of muscular strength of children from one point in time to another is hazardous. It will be recalled, however, that almost without exception the between age correlations were positive, which suggest

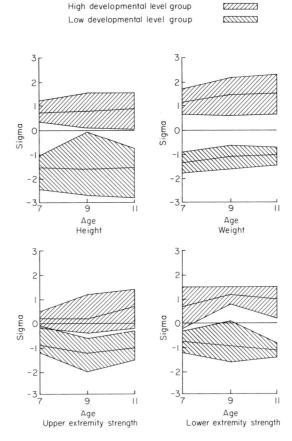

FIG. 1. Mean and range of standard scores of measurements of strength and physical growth of physically mature and physically immature boys plotted over a period of 5 years.

some tendency for strength to be a relatively stable quality with advancing age, although it cannot be accurately predicted over time. This tendency can perhaps be brought into focus more clearly if one looks at the development of strength over time in a group of children who differ markedly in physical maturity. This is illustrated by using the Wisconsin Growth Data and identifying by means of the Wetzel grid (Wetzel, 1941) the 5 most mature and the 5 least mature 7 year olds in the sample of 25 boys. The physical growth and strength development of these 10 boys were followed for a period of 5 years, from 7 through 11 years of age. The raw strength scores of all subjects on each of the eight strength tests were converted to standard scores, and from these values the average standard strength score of the four upper extremity measurements and of the four lower extremity measurements was calculated for each boy. Similarly, the heights and weights of subjects were converted to standard scores. The means and ranges of the standard scores of height, weight, upper extremity strength, and lower extremity strength of the 5 most mature and the 5 least mature boys are plotted against age in Fig. 1. This method of classifying the children clearly dichotomized the two groups not only in terms of the large and the small but also in respect to the strong and the weak. At no age level was there an overlap in strength scores of the two groups. It is evident that as a group, these large, physically advanced, and strong boys tended to retain their morphological and strength superiority throughout this 5-year period, although within each group there

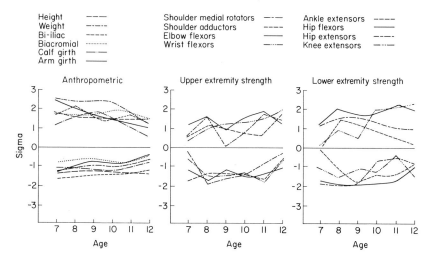

FIG. 2.   Standard score curves of the growth of 2 boys differing consistently in measurements of body size and strength from 7 to 12 years of age.

was probably considerable shifting of relative positions from year to year in both size and strength. Hence, while the correlation analysis emphasized the hazards of prediction, children do, within rather narrow limits, show considerable consistency in strength development in childhood. The consistent pattern of growth in strength and in body dimensions is further illustrated by measurements (converted to standard scores) taken on 2 children who were at opposite ends of the continuum on all growth measurements at age 7 (Fig. 2).

Findings similar to the above have been reported on adolescent boys. For example, Jones (1949) dichotomized a portion of his sample of adolescent males into two groups, one scoring high on measurements of arm and shoulder girdle strength throughout adolescence, the other scoring low on these measurements. The mean standard scores of four anthropometric measurements of the boys classified as strong in comparison to those classified as weak are shown in Fig. 3. The "strong" group was not only well above the mean of the total sample on all anthropometric measurements at 11 years of age, but it increased its advantage progressively to age 17. The "weak" group, on the other hand, was substantially below the general mean on all measurements except height, and it changed little in relative size from 11 to 17 years of age. The association between body size and strength in adolescent males is apparent when subjects are grouped into contrasting strength categories, as is the tendency for the groups to remain consistently different in measurements of both strength and body size.

VI. Motor Performance

Longitudinal studies of the development of fine and gross motor skills indicate that individual differences in these abilities are evident in infancy and remain remarkably constant through this period and into the childhood years. For example, Gesell (1937), using cinematographic observations of children engaged in routine motor tasks in the home, found that the rank order of the motor behavior of the 5 children under investigation was the same at 5 years of age as it had been in early infancy. Ames (1940), using similar techniques, reported constancy of individual differences in locomotor and manipulative behavior of 8 infants observed over a period of 3 years. Six of the eight retained the same relative rank in such behaviors as placing pellets in a bottle, spoon control, and in the attainment of successive stages of prone progression.

The above studies would lead one to believe there is an innate propensity in motor behavior that is evident early in life and a "psychomotor constancy" that is sufficiently persistent to guarantee continuation of the individuali-

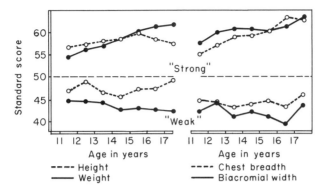

FIG. 3. Standard score curves of contrasting groups (physical measurements). Reprinted by permission from Jones (1949).

ty of these traits during infancy and early childhood. Further support of the above comes from a longitudinal study of 100 infants (Chess *et al.,* 1959) in which the investigator observed persistence of initially determined patterns of reactivity throughout the first few years of life. These findings are supported by Birns (1965) who reported that in the first 5 days of life there is considerable consistency of individual differences in response intensity. Furthermore, Edwards (1968) found that the correlations between the 1-min. Apgar score at the neonatal level and measures of fine and gross motor coordination at 4 years of age were .446 and .493, respectively, and that the muscle tone at birth accounted for more of the variance in the latter scores than any of the other four Apgar rating scores (heart rate, respiratory effort, reflex irritability, and color). It has been reasonably well established that infants who are markedly undersized at birth have developmental and learning problems in childhood (Robertson, 1964; Babson *et al.,* 1964). It would seem that there is a reasonable probability that difficulties at birth and health problems in the neonatal period have an unfortunate effect on later physical and behavioral development.

Those who work with preschool children frequently speak about the wide variation in motor aptitude at this age period, yet there is essentially no information of a longitudinal nature to indicate whether the less skilled and the highly skilled preschool children are destined to retain their respective levels of motor proficiency in the years ahead. There is, however, some evidence to show that in certain gross motor tasks, school age children tend to maintain their relative positions in the group over a period of several years. Glasgow and Kruse (1960), using longitudinal data, reported a moderately high correlation (.739) between broad jump scores obtained in grade one and broad jump scores on the same children 6 years later. Similarly, the correlation for the 30-yard dash across the same age span was .702. On these two gross motor tasks there was considerable stability of growth.

The above findings are supported by data from the Wisconsin Growth Study (Rarick and Smoll, 1967). It is clear from the size of the correlations in Table III that both the boys and the girls showed moderate between age consistency in broad jump performance in the age range 7–17 years, the majority of the correlations being above .65. The magnitude of the between age correlations of broad jump performance in the childhood years was also moderately high, being similar to those noted above. By Bloom's definition (1964) broad jumping performance would appear to be a reasonably stable trait.

The between age correlations for the 30-yard dash (Table III) were uniformly high for the girls in childhood (.784 to .933), less consistent for the boys (.386 to .780). From childhood to adolescence the correlations on the dash for the girls over the 10-year span (ages 7–17 years) were .562 for the girls but only .181 for the boys. It is evident that for this sample of boys, the prediction of running speed from 7 to 17 years was little, if any, better than chance.

The between age correlation of throwing velocity in childhood for both sexes, while positive, tended to be low (Table III). Similarly, the correlations from childhood to adolescence for both sexes on this measurement were low—the longer the time span between measurements, the lower the correlations. It is apparent that the power thrower in childhood is not likely to retain his superiority in the velocity of the throw in adolescence.

The foregoing indicates that although individual performance scores on these physical abilities over time show positive relationships, the correlations covering a span of more than a year or two are not high enough to be of predictive value. In other words, the results of the above study raises serious doubts that children retain their relative ranks among their peers in these abilities as they grow older. Differences in rates of physical growth and maturation probably account for some of the variance in performances with advancing age, but probably more important are the increasingly diversified interests and variations in patterns of living characteristic of children.

Correlations of the magnitude similar to those previously noted were reported by Espenschade (1940) with California adolescent children. Between year correlations (ages 13–16) on such events as the 50-yard dash, the standing broad jump, the jump and reach, the distance throw, the Brace test, and the throw for accuracy ranged in the boys from .29 in the dash to .66 in the Brace test and in the girls from .36 in the throw for accuracy to .84 in the distance throw. These findings are essentially in agreement with the results obtained in the Wisconsin Growth Study.

TABLE III

CORRELATIONS BY SEX ACROSS AGES: MOTOR PERFORMANCE MEASUREMENTS[a]

Measurements (by sex)	Age-to-age correlations				
	Childhood				
	7–12	8–12	9–12	10–12	11–12
Boys					
Broad jump	.484	.534	.663	.849	.780
30-Yard dash	.386	.424	.460	.694	.780
Velocity throw	.501	.308	.479	.580	.501
Girls					
Broad jump	.709	.705	.755	.807	.896
30-Yard dash	.924	.830	.784	.915	.933
Velocity throw	.115	.534	.462	.361	.552
	Childhood to Adolescence				
	7–17	8–17	9–17	11–17	12–17
Boys					
Broad jump	.596	.563	.694	.665	.728
30-Yard dash	.181	.138	−.073	.354	.517
Velocity throw	.278	.136	.378	.330	.404
Girls					
Broad jump	.502	.804	.704	.714	.661
30-Yard dash	.562	.699	.765	.710	.696
Velocity throw	.127	.252	.204	.290	.295

[a] Adapted from Rarick and Smoll (1967).

VIII. Growth Status and Incremental Changes in Strength and Gross Motor Proficiency

The question is frequently raised regarding the relationship between the presently achieved state of development and subsequent growth increments. The evidence in respect to measurements of physical growth is reasonably clear on this. Children, for example, who are tall for their years do not necessarily make the greatest gains in stature in future years, although they tend to retain their height superiority. In fact, most of the evidence points to low "status-gain score" correlations, some positive and some negative, depending on the ages across which the correlations are run. As Bloom (1964) pointed out up to the fourteenth year height gains correlate positively with the height at the age of original measurement. On the other hand, from age 13 to age 16 the gains are negatively correlated with height at age 18. The negative correlations during the latter period are a reflection of the wide individual differences in the age of onset of the adolescent growth spurt.

TABLE IV

Correlations Between Gains in Strength and Motor Performance, Ages 7–17 Years, with Strength and Motor Performance Scores Taken at Age 7

Measurements	Boys	Girls
Ankle extensor strength	.1395	−.1832
Hip flexor strength	.4621	.2311
Hip extensor strength	−.3010	−.1044
Knee extensor strength	.1089	.2870
Standing broad jump	.0610	−.3544
30-yard dash	.8188	.6264

A. STRENGTH

In view of the positive relationship between body size and most measurements of physical performance, one would logically expect that the findings reported above would also hold true for incremental gains in measurements of motor performance. Data from the Wisconsin Growth Study (Rarick and Smoll, 1967) are shown in Table IV. The correlations between the gain scores in strength over the span of 10 years (age 7–17 years) and the strength scores at age 7 are consistently low. This would support the contention that the strong children at 10 years are not necessarily the ones who will make the greatest gains in strength from childhood to adolescence.

B. MOTOR PROFICIENCY

The above seems likewise to hold true in the case of the standing broad jump (Table IV). On the other hand, in running speed as measured by a short dash the incremental gains in performance over this time span for both sexes are substantially related to the performance at age 7. In other words, it is evident that those who performed well on the dash at age 7 were the ones who showed the greatest improvement in this event over the span of 10 years. The reason for this apparent inconsistency is not entirely clear. Since running is basic to most of the sports and games of childhood and adolescence it would seem reasonable to believe that with advancing age the motorically more talented children would be the physically active ones and would tend to develop this ability to a greater degree than those who were initially the poorer performers. The standing broad jump, on the other hand, is a skill that is not routinely used in the daily physical activities of children; hence, it would not, as in the case of the run, be used as frequently by physically inactive children.

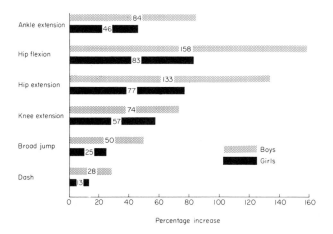

FIG. 4. Percentage increases in measurements of strength and motor performance of boys and girls from 10 to 17 years of age.

C. SEX DIFFERENCES IN INCREMENTAL GAINS IN STRENGTH AND MOTOR PROFICIENCY

It is a common observation that on most gross motor tasks involving strength and power, the performance of boys on the average is superior to that of girls at all ages, but the differences become greater in adolescence. To illustrate the relative sex differences in the increases in strength and motor performance from later childhood to adolescence the percentage increase in four strength measurements and two motor performance tests from 10 to 17 years of age (Wisconsin Growth Study) is shown in Fig. 4. It will be noted that in most instances the percentage gains in the measurements were approximately twice as great for boys as for girls.

D. ATHLETIC POTENTIAL

The extent to which boys of elementary school age who have been judged to have athletic potential do, in fact, realize this potential has received only limited study. It is worthy of mention that in the Medford Growth Study, Clarke (1967) found that many participants on interscholastic athletic teams in the Medford community, who possessed at an earlier age many of the characteristics of the interscholastic athlete, were not listed as outstanding athletes. Of those named as outstanding interscholastic athletes only one-fourth of them were considered outstanding in both elementary and junior high school. Forty-five percent were outstanding in elementary school but not in junior high school, whereas 30% were considered outstanding in junior high school but not in elementary school. Thus, while these boys as a group possessed many of the physical and behavioral attributes of the athletes in earlier years, any attempt to predict their relative athletic prowess over a span of 3 years or more was subject to considerable error.

VIII. Stability and Change in Physical Fitness

The long range effects of exercise programs on the growth and development of children and adolescents have received only limited attention. It is widely recognized that a vigorous program of physical exercise will produce substantial training effects on the muscular and cardiovascular systems of those undergoing training, but such effects are transitory. When the training program is stopped the individual's physiological capabilities return to the pretraining level in a relatively short time. Almost without exception these studies have been done on males who have completed or have almost completed their growth. Hence, we have little information on growth changes of children and adolescents in which physical training programs have been interspersed with relatively long periods of relative inactivity.

That adolescent boys and girls show substantial physiological changes as a result of vigorous training programs has been clearly demonstrated. Ekblom (1969) involved 6 11-year-old boys in a 6-month training program using a group of 7 untrained boys of the same age for controls. The aerobic working capacity of the trained group improved 15%, that of the untrained group remained unchanged. Five of the six boys in the training group remained in training for an additional 26 months with the result that their oxygen uptake increased a total of 55%, the heart volume increased by 45%, and the vital capacity was augmented by 54%. These changes were considerably more than would be expected from age-dependent increases in body dimensions. Furthermore, the training group had an accelerated rate of growth in body height when compared with growth estimates on samples of healthy untrained adolescents of the same age. Similar findings have been reported by Astrand et al. (1963) on adolescent girls who were involved almost continuously in a competitive swimming program over a period of some 4 years. It is apparent that highly significant changes in physiological functions can be ellicited in adolescent boys and girls who are subjected to vigorous programs of physical training.

The extent to which children and adolescents tend to retain their relative state of physical fitness over long periods of time when not involved in formal training regimens has received little attention. Data from the Denver Child Research Council suggest that physical fitness, as assessed by heart rate recovery from standard exercise on the bicycle ergometer, shows relatively high individual consistency from one age level to another regardless of the mode of life the child and adolescent has followed (McCammon and Sexton, 1958; and Sexton, 1963). The longitudinal data covering the ages from 8 to 20 years demonstrated clearly that the individuals tended to have their own level of fitness and that for the most part they remained at this level over extended periods of time. Approximately three-fourths of the population demonstrated this rather remarkable uniformity of heart rate response despite widely varying levels of physical activity. Some were involved in competitive athletics during the high school years without radically changing their heart rate response to the standard exercise.

IX. Factors Affecting the Prediction of Motor Performance

The moderately high relationship between measurements of body size and measurements of strength and motor performance in childhood and adolescence has been pointed out (McCloy, 1932; Neilson and Cozens, 1934; Cozens et al., 1936). In view of the relatively high between age correlations of measurements of physical growth (Bloom, 1964), it would seem reasonable to believe that the large, robust, and strong child would tend to become the adolescent who performs well in tests of motor performance. Data from

TABLE V

CORRELATIONS OF MEASUREMENTS OF PHYSICAL GROWTH, MUSCULAR STRENGTH, AND MOTOR PERFORMANCE OBTAINED AT AGE 10 WITH MEASUREMENTS OF MOTOR PERFORMANCE TAKEN AT AGE 17[a]

Measurements (age 10)	Boys		Girls	
	Broad jump (age 17) r	Dash (age 17) r	Broad jump (age 17) r	Dash (age 17) r
Height	.369	.320	.101	−.047
Weight	.084	.131	−.117	−.244
Ankle extension	.328	.512[b]	.206	.009
Hip flexion	.496[b]	.479[b]	.591[b]	.438[b]
Hip extension	.444[b]	.311	.665[b]	.442[b]
Knee extension	.531[b]	.479[b]	.397	.372
Standing broad jump	.788[b]	.233	.745[b]	.567[b]
30-Yard dash	.636[b]	.381	.671[b]	.718[b]

[a] Reprinted by permission from Rarick and Smoll (1967).
[b] Significant at the 5% level.

the Wisconsin Growth Study (Rarick and Smoll, 1967) indicate that while there is indeed a trend in this direction, the relationship between measurements of body size and of strength with motor proficiency is not high. As may be noted in Table V the correlation between height at age 10 and performance in the 30-yard dash and the broad jump was not high enough to indicate more than a chance relationship between these measurements. On the other hand, it is apparent from the correlations in Table V that strength at 10 years of age as measured was generally of more significance than height or weight as a factor associated with high level performance in the jump and the dash at 17 years of age. Of greater significance in determining performance in these two activities at age 17, particularly in the case of the boys, was the level of the performance in these activities themselves at the earlier age.

In order, however, to improve the prediction of performance in the dash and broad jump over the age span 10–17 years combinations of physical growth, strength, and motor performance variables taken at age 10 were used in multiple correlations. In the case of the boys 74% of the variance in broad jump performance at age 17 was accounted for using in a multiple correlation three measurements taken at age 10, namely, ankle extension, strength, knee extension strength, and broad jump scores. For the girls 64% of the variance at age 17 was accounted for by a combination of height, broad jump, and 30-yard dash performance at age 10.

The prediction of dash performance at age 17 by use of multiple correlation using data at age 10 was less precise than for the broad jump. In the case of the boys, six variables at age 10 (the dash, broad jump, weight, ankle extension, hip extension, and knee extension strength) were needed to account for 50% of the variance in dash performance at 17 years of age. For the girls performance data on the dash, broad jump, hip extension, and ankle extension strength were needed to account for 58% of the variance in the dash at age 17. It is evident from the above that a considerable proportion of the variance in the dash and broad jump at 17 years of age was attributable to factors other than those assessed at age 10. Nevertheless, it is clear from the above and from an inspection of the correlations in Table V that an appreciable part of the variance in the dash and broad jump at age 17 was attributable to individual differences in strength at age 10.

It is worthy of note that both Espenschade (1940) and Jones (1949) found that those who scored high in the various measures of motor performance in early adolescence tended on the whole to remain well above average throughout adolescence. Jones, for example, reported that the boys scoring highest on strength tests at 11 years of age, substantially improved their relative position in the group, whereas those who were rated as weak initially tended to retain this position (Fig. 5).

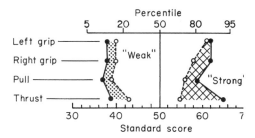

FIG. 5. Standard score profiles of contrasting groups (strength measurements). Key: (- - -) 11.5 and (———) 17.5 years. Reprinted by permission from Jones (1949).

The strong boys as a group were above average in height and weight and of mesomorphic frame. This group, according to Jones, also scored high on a personal-social inventory, particularly in the category "freedom from tensions" and "freedom from personal inferiority." Their adjustment to the scholastic demands of the school was not particularly noteworthy, but this in no way seemed to have a disturbing effect on them. The boys, so endowed, tended generally to be proficient in a variety of athletic endeavors, to have social prestige, and to be well adjusted. On the other hand, those in the weak group presented the opposite picture. Generally they were late maturers, of ectomorphic build, frequently ill, and as weak socially as they were weak physically. As Jones pointed out the above is not meant to imply a causal relationship between biological, social, and psychological traits. On the other hand, he was of the opinion that one cannot dismiss the possibility that factors related to vitality and fitness may predispose the individual to psychological health or that the social group in which he lives may accord a high value to motor proficiency.

References

Ames, L. B. (1940). *J. Genet. Psychol.* **57**, 445.

Astrand, P.-O., Engstrom, L., Eriksson, B., Karlberg, P., Nylander, I., Saltin, B., and Thoren, C. (1963). *Acta Paediat. (Stockholm), Suppl.* **147**, 1–75.

Babson, G., Kangas, J., Young, N., and Bramhall, J. L. (1964). *Pediatrics* **33**, 327.

Baldwin, B. T. (1920). *Univ. Iowa Stud. Child Welfare* **1**, 1–411.

Birns, B. M. (1965). *Child Develop.* **36**, 249.

Bloom, B. S. (1964). "Stability and Change in Human Characteristics." Wiley, New York.

Bohm, G. E. (1959). Unpublished Master's Thesis, University of Wisconsin, Madison.

Burt, C. (1955). *Brit. J. Educ.* **25**, 158.

Chess, S., Thomas, A., and Birch, J. (1959). *Amer. J. Orthopsychiat.* **29**, 791.

Clarke, H. H. (1948). *Res. Quart.* **19**, 118.

Clarke, H. H. (1950). *Res. Quart.* **21**, 399.

Clarke, H. H. (1967). *Proce. Nati. Conf. Medi. Aspects Sports, 8th, 1966* p. 49–57.
Clarke, H. H. (1971). "Physical and Motor Tests in the Medford Boy's Growth Study." Prentice-Hall, Englewood Cliffs, New Jersey.
Cozens, F. W., Trieb, M. H., and Neilson, N. P. (1936). "Physical Education Achievement Scales for Boys in Secondary Schools." Barnes, New York.
Cronbach, L. J., and Furby, L. (1970). *Psychol. Bul.* **74**, 68.
Ebert, E., and Simmons, K. (1943). *Monogr. Soc. Res. Child. Develop.* **8**, No. 2.
Edwards, N. (1968). *Genet. Psychol. Monogr.* **78**, 258.
Ekblom, B. (1969). *J. Appl. Physiol.* **27**, 350.
Elderton, E. M., and Pearson, K. (1915). "The Relative Strength of Nurture and Nature." Cambridge Univ. Press, London and New York.
Espenschade, A. S. (1940). *Monogr. Soc. Res. Child. Develop.* **5**, 1–126.
Garn, S. M. (1962). *Mod. Probl. Pediat.* **7**, 50.
Garn, S. M., Clark, A., Landkof, L., and Newell, L. (1960). *Science* **132**, 1555.
Gesell, A. (1937). *J. Genet. Phychol.* **47**, 339.
Glassow, R. B., and Kruse, P. (1960). *Res. Quart.* **31**, 426.
Jones, H. E. (1949). "Motor Performance and Growth." Univ. of California Press, Berkeley.
Jones, H. E., and Bayley, N. (1941). *Child Develop.* **12**, 167.
McCammon, R. W., and Sexton, A. W. (1958). *J. Amer. Med. Ass.* **168**, 1440.
McCloy, C. H. (1932). "The Measurement of Athletic Power." Barnes, New York.
Malina, R. M., Harper, A. B., and Holman, J. D. (1970). *Res. Quart.* **41**, 503.
Neilson, N. P., and Cozens, F. W. (1934). "Achievement Scales in Physical Education Activities." State Department of Education, Sacramento, California.
Newman, H. H., Freeman, F. N., and Holzinger, K. L. (1937). "Twins: A Study of Heredity and Environment." Univ. of Chicago Press, Chicago, Illinois.
Olson, W. C. (1949). "Child Development." Heath, Boston, Massachusetts.
Prader, A., Tanner, J. M., and von Harnack, G. A. (1963). *J. Pediat.* **62**, 646.
Rarick, G. L., and Mohns, M. J. (1955). *Res. Quart.* **26**, 74.
Rarick, G. L., and Smoll, F. L. (1967). *Hum. Biol.* **39**, 295.
Robertson, J. H. (1964). *Obstet. Gynecol.* **23**, 458.
Sanders, B. K. (1934). "Environment and Growth." Warwick & York, Baltimore, Maryland.
Sexton, A. W. (1963). *Pediatrics* **32**, 730.
Shuttleworth, F. K. (1939). *Monogr. Soc. Res. Child Develop.* **4**, No. 3.
Tuddenham, R. D., and Snyder, M. M. (1954). "Physical Growth of California Boys and Girls from Birth to Eighteen Years." Univ. of California Press, Berkeley.
Waddington, C. H. (1957). "The Strategy of the Genes." Allen & Unwin, London.
Wetzel, N. C. (1941). *J. Amer. Med. Ass.* **116**, 1187.

CHAPTER

10

Motor Performance of Mentally Retarded Children

G. Lawrence Rarick

Mental retardation is almost invariably accompanied by substandard levels of performance in both fine and gross motor skills. It is now well established that on standardized tests of motor performance educable mentally retarded children perform well below the average of intellectually normal children of the same age and sex (Francis and Rarick, 1959; Malpass, 1959, 1960). The more severe the retardation the greater is the deficit in motor proficiency (Tredgold *et al.,* 1956). Thus, the relationship between intelligence and motor behavior in the retarded would seem to be self-evident.

The difficulties which the retarded child has in the execution of simple movement patterns clearly have an adverse effect on the rate at which he acquires new and complex motor skills. However, the way in which the neuromuscular mechanism and the motor learning processes function in the mentally retarded is not as yet fully understood. Nor do we know precisely how the marginal intellectual processes in the mentally retarded limit the performance of physical skills. In fact, the magnitude of motor retardation in particular skills as a function of age and sex has only in recent years been a matter of systematic study.

The possibility that many dimensions of the retarded individual's life which are related to his intellectual development may be benefited by well-planned programs of physical activity and perceptual motor training has only recently received attention (Oliver, 1958; Corder, 1966; Cratty and Martin, 1970). The rationale in support of this is the underlying belief that the neuromuscular mechanism constitutes man's most primitive learning medium, one through which the infant and young child acquire a major part of their early learnings. This medium involves a broad range of perceptual-motor processes that are essential in the normal development of those aspects of intelligence upon which the child's concepts about his environment are built. The normal development of perceptual-motor functions is held by many to be fundamental in bridging the gap between the concrete and the abstract leading ultimately to the acquisition of the power of conceptualization.

The purpose of this chapter is to review what is currently known about the motor abilities of mentally retarded children and to identify some of the factors that seemingly affect their motor proficiency. The chapter includes brief sections on the physical development and primitive responses in the retarded, followed by a rather detailed discussion of age and sex differences in motor abilities and the interrelationships among motor functions in these children. Attention is also given to the problem of motor learning in retarded children, the acquisition of perceptual-motor skills, and the possible impact of perceptual-motor and gross motor training upon their cognitive processes. Recent findings are presented on the physical fitness and working capacity of the retarded. A final section is devoted to a discussion of what the research findings mean in terms of improving educational and recreational programs for the retarded, changes which should provide a better and more meaningful life for this segment of our population.

I. Physical Development of Mentally Retarded Children

One cannot attribute the mentally retarded child's motoric inadequacies solely to mental deficiency, for most of these children have health and developmental problems which are related to their physical, mental, and motor development. Hence, it would seem important to examine what is known about the physical growth of this segment of the population. This would seem particularly appropriate in view of the fact that much of the variance in gross motor performance of normal children and adolescents can be attributed to differences in physical development.

Studies without exception have disclosed that retarded children are on the average both shorter in stature and lighter in weight than children of normal intelligence. While a few of the less retarded children may be taller than average, the vast majority are short for their age. In general, research has shown that the greater the degree of mental retardation the more pronounced is the lag in physical development. For example, Kugel and Mohr (1963) in a summary of early research concluded that a large proportion of intellectually subnormal children were below the seventeenth percentile of normal children in height and weight. It should be pointed out, however, that Rundel and Sylvester (1965) reported that within the group of intellectually subnormal children that were examined by his group there were two populations, one whose growth in height and weight was similar to that of the normal population and another whose growth was pathologically stunted. Roberts and Clayton (1969), in a study of the physical growth of undifferentiated retarded children, confirmed the earlier findings in regard to both the high frequency of retardation in physical growth among the mentally retarded as well as the bimodal distribution of stature described above. In the moderately retarded population approximately 5½ times the expected number were below the third percentile in physical growth, whereas in the profoundly retarded the figure was about 11 times that expected. Among the most severely retarded in physical growth, a large proportion had skeletal ages below the tenth percentile. Thus, according to Roberts and Clayton (1969), one is tempted to propose that some of these children failed to grow normally because of undefined neurophysiological reasons, conceivably the faulty mechanism being the same as that causing the mental retardation. This is borne out by the findings of Pozsonyi and Lobb (1967) and Dutton (1959) who found classes of mentally defective children with no stunting of linear growth or skeletal development. These two studies agreed that linear and skeletal growth of nonpathological (functional and cultural familial) retarded children was not retarded, whereas mongoloid and meta-bolically abnormal children showed severely retarded growth.

While some authorities have proposed that the retardation in physical growth in some cases of mental retardation may be a reflection of inade-quate function of the hypothalamus, which has been shown to regulate the secretion of the growth hormone in animals (Reichlin, 1961; Hinton and Stevenson, 1962), the recent research of Lowrey et al. (1968) has shown that the serum level of the growth hormone in the fasting state is no differ-ent for mentally retarded children of well below average height (below the third percentile of normal children) than for children of normal intelligence or for retarded children of normal stature.

The concensus of the research is that while mental retardation is often ac-companied by failure of physical growth, particularly where the retardation is caused by chromosomal aberration or multiple congenital anomalies (Pryor and Thelander, 1967), the evidence of a common cause that relates the two has not yet been firmly established.

II. Reflex and Reaction Time in Mental Retardation

Some 40 years ago researchers hypothesized that the deficiencies in the intellectual capabilities of the retarded might be reflected in abnormally slow reflex responses. This, however, did not prove to be the case. Research such as that done by Travis and Dorsey (1930) showed that the patellar tendon responses of intellectually superior and mentally defective children did not differ. Similarly, Whitehorn et al. (1930) reported that there was no corre-lation between mental age and tendon reflex time. These findings are reason-able since such responses are localized at the segmental level of the cord and hence require no conscious direction.

The findings from reaction time studies (Ellis and Sloan, 1958; Berkson, 1960; D. Jones and Benton, 1968) however, show that intellectually normal children respond faster than retarded children of the same chronological age. Baumeister and Kellas (1968) demonstrated that high intra-individual variability on serial reaction time tasks is perhaps as characteristic of the be-havior of the retarded as is the depressed level of response. This was evident in the distribution of reaction times of a small sample of normal retarded subjects exposed to 600 trials over a period of 2 days (see Fig. 1). The work of D. Jones and Benton (1968) showed that retarded children's reac-tion times were significantly slower than normal children for both simple and choice reaction times when responding to either auditory or visual stim-uli. Both groups responded in the characteristic way to the two stimuli, responding faster to the auditory than to the visual stimuli, although the dif-

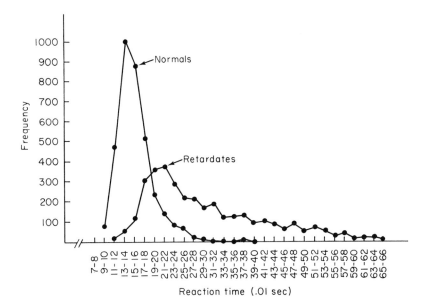

FIG. 1. Frequency polygons of reaction times for 6 normal and 6 retardates. Six-hundred scores are included for each subject. From Baumeister and Kellas (1968).

ferences in response to the two stimuli were less consistent for the choice than for the simple reaction time.

Berkson (1960) pointed out that in the retarded it is likely that the slowness in simple reaction time may be a reflection of difficulty in the initiation of performance of the response rather than any slowness of sensory or central nervous system functions. The more marked slowness and lesser consistency in choice reaction time is, however, indicative of impaired central functions. Here, we are dealing with responses, which, while simple, require conscious attentive effort. Thus, it is not surprising that the retarded do not do as well on this type of task as intellectually normal children and adults.

III. Motor Abilities in the Retarded

The search for a single general factor of motor ability for the normal and mentally retarded segments of the population has been unsuccessful. It is now generally recognized that motor skills are to a considerable extent specific to the task, although there is an increasing body of evidence that certain general abilities underlie the successful performance of many motor skills (Hempel and Fleishman, 1955; Fleishman and Hempel, 1956; Guilford, 1958).

Such general qualities as static strength, dynamic strength, explosive muscular force, agility, balance, coordination, and endurance have emerged from
factor analytic studies as basic components in the broad spectrum of motor
abilities (Fleishman, 1964). These basic qualities account for a large proportion of the variance among individuals in the performance of wide
range of sensorimotor and gross motor tests.

A. The Assessment of Motor Abilities

Physical educators and psychologists over the years have designed and
employed tests which purport to measure the basic components of both fine
and gross motor abilities. One of the earliest and most comprehensive in-
struments of this kind was designed in 1923 by the Russian, Oseretsky, a
test battery comprising a series of fine and gross motor tests normed by age
and sex over the age range 4–16 years and patterned after the Stanford-
Binet Intelligence Scale (Doll, 1946). The test according to Oseretsky assessed six components of motor ability, namely, general static coordination,
dynamic manual coordination, general dynamic coordination, motor speed,
simultaneous voluntary movement, and asynkinesia (lack of precision or su-
perfluous movement). The test enjoyed considerable use in Europe, was
published in English in 1946, but has never been widely used with normal
children in this country.

The Oseretsky tests have in recent years been modified in this country for
use with retarded children. Cassel (1949) reported that his revision of the
Oseretsky tests when given to Vineland subjects differentiated endogenous

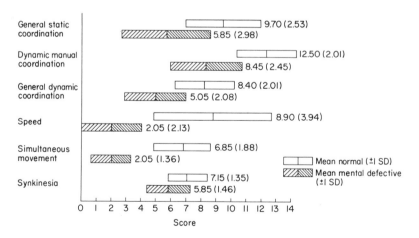

Fig. 2. The performance of normal and Institutionalized mentally defective
children on the Lincoln-Oseretsky scale. Drawn from Sloan (1951).

from exogenous groups of retarded, particularly on tasks requiring the integration of sensory cues and the coordination of simultaneous movements.

The most carefully normed and most widely used version of the Oseretsky scale is the Lincoln-Oseretsky Motor Development Scale adapted by Sloan (1955) for use with mentally retarded children. Sloan's revision eliminated many of the original test items resulting in a scale of 35 items, providing separate age norms for boys and girls ages 6 through 14 years. The norms by sex are not strictly comparable since the scores for the girls' test items were not based on the same standards as those for the boys' items.

Numerous investigations employing the Lincoln-Oseretsky scale have clearly shown that there are substantial differences in motor proficiency between mentally retarded and intellectually normal children. For example, Sloan (1951) reported significant differences in performance on the Lincoln-Oseretsky scale between a group of 20 institutionalized retarded children free of known organic defects (I.Q. 45–70) and 20 normal subjects (I.Q. 90–110) matched by age and sex. Significant differences favoring the intellectually normal subjects were found on all six of the categorized subtests (see Fig. 2).

B. Gross Motor Performance

The assessment of gross motor abilities in normal children has for the most part involved the use of measures of static dynamometric strength and

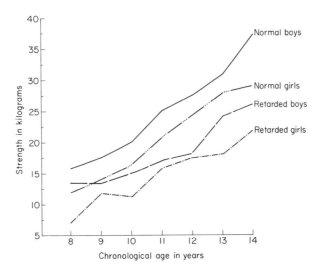

Fig. 3. Mean strength of right grip of normal and mentally retarded boys and girls (Francis and Rarick, 1960). Data on normal boys and girls from Meredith (1935), Metheny (1941), and H. E. Jones (1949).

a battery of track and field tests. The rationale for employing strength tests rests on the assumption that the ability to develop muscular force as reflected by dyamometric strength tests is a good predictor of performance in a variety of gross motor skills. While recent research (Henry *et al.,* 1962) does not fully support this assumption, measurements of dynamometric strength do provide an assessment of an important developmental phenomenon, one which changes systematically with age and maturity (H. E. Jones, 1949), reflecting differences by sex and mode of life.

The rationale for using track- and field-type tests is based upon rather sound logic, for many of the athletic and sports activities of our culture require the use of the basic skills of running, jumping, and throwing in varying degrees. Furthermore, track and field tests when combined with other motor performance measurements cover the broad spectrum of such primary abilities as static and dynamic strength, speed, power, agility, balance, coordination, and endurance. Since measurements of strength and track- and field-type tests have traditionally been used in assessing the motor abilities of normal children, it is not surprising that they have been used extensively in research on mentally retarded boys and girls.

C. STRENGTH

Measured static muscular strength of educable mentally retarded children has been shown to lag well behind that of normal children of the same age and sex (Francis and Rarick, 1959). In both manual and shoulder girdle strength retarded children are from 18 months to 36 months behind the

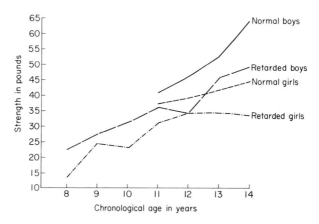

FIG. 4.   Mean shoulder girdle strength of normal and mentally retarded boys and girls: pull (Francis and Rarick, 1960). Data on normal boys and girls from H. E. Jones (1949).

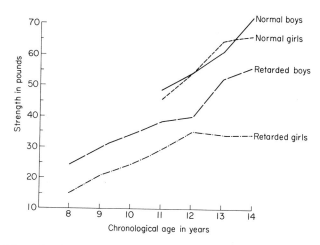

FIG. 5. Mean shoulder girdle strength of normal and mentally retarded boys and girls: thrust (Francis and Rarick, 1960). Data on normal boys and girls from H. E. Jones (1949).

schedule of development of their intellectually normal counterparts (see Figs. 3–5). The age trends and the strength differences by sex are, however, not materially different for the retarded as compared with the normal children. While the basis for the inferior performance of the retarded in tests of strength is not known, it is probably the result of one or both of the following: (1) deficiency in the quantity and quality of the muscle tissue of the retarded children primarily because of the physically inactive life of the retarded or (2) inability or unwillingness of the retarded to mobilize his neuromuscular mechanism for a maximal effort in tests of strength.

D. DYNAMIC STRENGTH, POWER, AND COORDINATION

Many of the physical activities of childhood and adolescence involve in one way or another the fundamental neuromuscular skills of running, jumping, and throwing. Successful execution of these skills requires that the individual develop controlled muscular force, frequently at near maximum levels. Research with normal children has shown that a weighted combination of three measurements, speed of running, the standing broad jump, and the throw for distance in combination with a measure of strength is a good predictor of general motor ability (McCloy, 1946). Recent research studies with mentally retarded children have employed these measurements in one form or another along with tests of balance in assessing the motor performance of the retarded.

In terms of maximum running speed, independent investigations (Francis

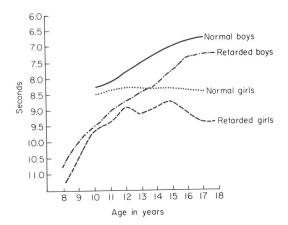

Fig. 6. Differences between normal and mentally retarded boys and girls in mean performance on the 50-yard dash. Drawn from data by Rarick *et al.* (1970).

and Rarick, 1959; Rarick *et al.,* 1970) have shown that educable mentally retarded boys and girls are on the average well below their intellectually normal counterparts at all age levels. The magnitude of the retardation as reported by Francis and Rarick (1959) is substantial, being 4–5 years behind schedule for both sexes at all age levels. The year-by-year changes are,

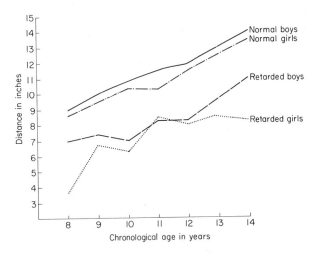

Fig. 7. Mean performance in the vertical jump of normal and mentally retarded boys and girls (Francis and Rarick, 1960). Data on normal boys and girls adapted from Neilson and Cozens (1934).

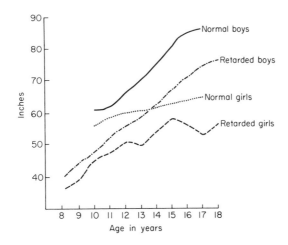

FIG. 8. Differences between normal and mentally retarded boys and girls in mean performance on the standing broad jump. Drawn from data by Rarick *et al.* (1970).

however, similar when comparing the retarded with the normal, so that at 14 years of age the difference between the retarded and the normal is almost identical to that at 8 years of age. Data from a recent national survey (Rarick *et al.*, 1970) of over 4200 educable retarded children provide similar results (Fig. 6).

Performance of retarded boys in the vertical jump as reported by Francis and Rarick (1959) follows very much the same pattern as in the run, the discrepancy being 4–5 years at all age levels (see Fig. 7). Interestingly, the retarded girls at no age level reach the mean performance level of normal 8-year-old girls.

The standing broad jump and the throw for distance both require high levels of neuromuscular coordination in the application of near maximum muscular force. Since these skills represent a somewhat higher order of neuromotor complexity than the run and the verticle jump one would expect even less proficiency in the retarded in these skills. Francis and Rarick (1959) found this, in fact, to be the case. While the discrepancy between the retarded and the normal was greater in the standing broad jump, where at no point did the retarded at any age reach the performance level of 8 year olds of the equivalent sex, the differences in the throw were almost as dramatic. Data obtained from the national survey (Rarick *et al.*, 1970) previously mentioned did not show as extensive a retardation in the standing broad jump as indicated above, although the retarded boys and girls throughout the age range fell one standard deviation below the AAHPER norms on intellectually normal children (see Fig. 8).

On the distance throw, the magnitude of the retardation as reported in the national study was substantially less than that found by Francis and Rarick in 1959, the boys at all ages being one standard deviation below the AAH-PER norms and the retarded girls approximately one-half standard deviation under AAHPER norms. Even so, the retardation in the national survey placed the retarded boys about 2–3 years behind schedule in both events, the girls slightly less (see Fig. 9).

There would seem to be no question that educable retarded children are well below their normal counterparts in measurements of strength, power, and coordination. The extent to which the retardation is a function of intellectual inadequacy or lack of opportunity to participate is not yet clear. It is evident that most mentally retarded children tend to be excluded from the play world of normal children and have few opportunities to participate in a well-planned instructional physical education program. One cannot, however, ignore the intellectual component, for it directly or indirectly must play a role. Even within the educable retarded population itself, differences in the level of motor performance of children of different I.Q. levels have been noted. Widdop (1967), for example, reported statistically significant differences in mean motor performance levels of retarded children in all of the modified AAHPER motor performance test items when comparing groups having a mean I.Q. of 55 with those having an I.Q. of 65. The relatively low intellectual drive and the limited intellectual capacity of the retarded to learn the more complex gross motor skills cannot be ignored in accounting for the inferior performance of these children.

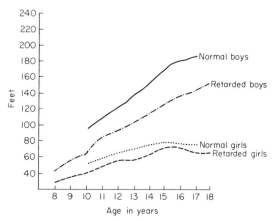

FIG. 9. Differences between normal and mentally retarded boys and girls in mean performance on the distance throw. Drawn from data by Rarick *et al.* (1970).

In 1968, norms by age and sex were published on the AAHPER Physical Fitness test appropriately modified for use with educable mentally retarded (EMR) children. Data for the norms were obtained on a nationwide sample of some 4200 children in the age range 8–18 years inclusive (Rarick *et al.*, 1970). The test battery patterned after the AAHPER Physical Fitness tests (Hunsicker and Reiff, 1965) included seven items, namely: (1) the flexed arm hang, (2) sit-ups, (3) standing broad jump, (4) shuttle run, (5) 50-yard dash, (6) softball throw for distance, and (7) the 300-yard

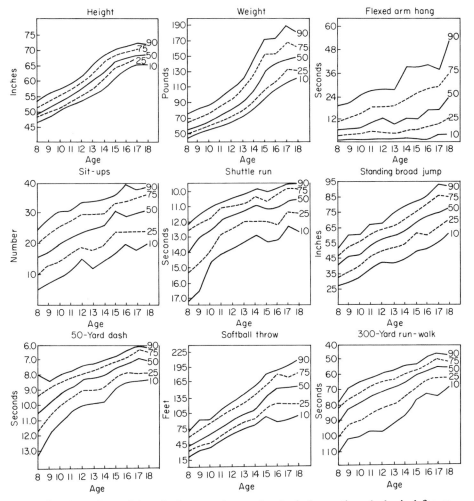

FIG. 10.   Record for plotting age changes in physical growth and physical fitness scores. Percentile zones based on normative data for EMR boys, ages 8–18 years. From Rarick *et al.* (1970) by permission of *Exceptional Children*.

run-walk. The data have been tabled by age and sex in such a way that raw scores can be easily converted to percentile values. An extension of this procedure in graphic form makes it possible to plot the year-by-year performance of an individual child (raw score and percentile values) on each of the seven test items, plus height and weight (see Figs. 10 and 11). Thus, a longitudinal record of the physical growth and motor performance of each child can be kept from age 8 through 18. The value of such a procedure is

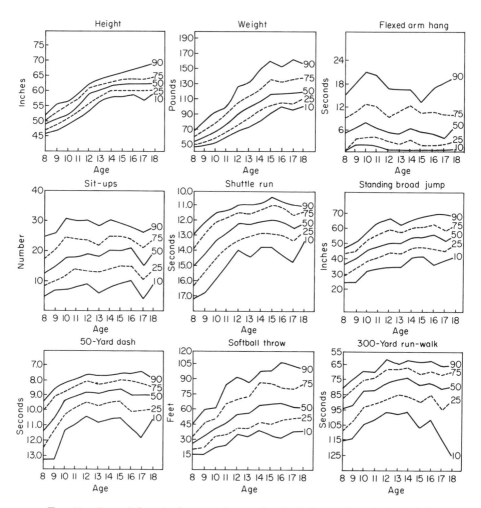

FIG. 11. Record for plotting age changes in physical growth and physical fitness scores. Percentile zones based on normative data for EMR girls, ages 8–18 years. From Rarick *et al.* (1970) by permission of *Exceptional Children*.

threefold: (1) the child's performance on each item can be judged both against his own past record and that of national standards on similar children; (2) by comparing his percentile position from one test item to another information is provided on those abilities in which he needs help; and (3) by comparing his percentile standings on the motor performance events with his percentile standing in height and weight, judgments can be made regarding his motor proficiency as viewed in the light of his physical development.

E. BALANCE AND AGILITY

Complex motor skills, namely, those involved in coordinating and timing the movements of various body parts in either the execution of a single burst of activity involving multiple joint actions or a sequence of bodily movements are difficult for the retarded individual to execute and to learn. This is supported by both research findings and clinical observations of those who work with retarded children and adults. Generally, the more pronounced the mental retardation the poorer the performance in these activities and the more difficult it is to learn these skills. Activities which require rapid recovery of body equilibrium seem to be particularly difficult for the retarded. As mentioned earlier in this section, although retarded boys and girls are well behind schedule in the vertical jump, the extent of the age–sex discrepancy in this event is far less than in the standing broad jump, an activity in which the forward projection of the body's center of gravity and the subsequent recovery of balance during the flight and landing phase of the jump are crucial in its successful execution.

The research by Heath (1953) on rail-walking performance of high grade mentally retarded boys at the Vineland school showed substantial correlations between the rail-walking test and mental age with familial subjects, less substantial with nonfamilials. In Heath's judgment the rail-walking task requires the coordination of a greater number of sensory modalities than is required in simpler motor tasks where the correlations with mental age are low. Interestingly, Heath found that the standing high jump was the only one of the eight motor tests that he used which at all approximated the rail-walking test's correlation with mental age, a finding supportive of the one mentioned in the preceding paragraph.

The squat thrust test, which has been used with retarded children, is a test involving rapid sequential movements of large body parts in which the center of gravity is also sequentially shifted. In this sense it would be considered a reasonably complex task, one in which retarded children would be expected to have some difficulty. Francis and Rarick (1959) reported that at no age level, 8–14 years inclusive, did educable mentally retarded boys or girls reach the mean performance level of 8-year-old boys or girls of normal intelligence. Fait and Kupferer (1956) found that for children in the I.Q. range of 42–87, correlations between I.Q. and the squat thrust were of the order of .49 but were only .19 between I.Q. and the vertical jump. These investigators interpreted the difference in the magnitude of the correlations to the greater complexity of the squat thrust task, particularly in respect to both remembering and executing the somewhat involved movement sequences.

F. Interrelationships among Motor Abilities

While, in general, both fine and gross motor skills tend to be specific, those skills which use similar patterned movements or call upon similar reserves of muscular strength and power, show reasonably high correlations. For example, it has been shown that the intercorrelations among the so-called power events such as the vertical jump and the standing broad jump range in normal boys from .44 to .82 and in girls from .29 to .78 (Espenschade, 1940). Similarly, intercorrelations among measurements of static dynamomentric strength in normal adolescents range from .50 to .90 (H. E. Jones, 1949). Research with retarded children has given intercorrelations among the power events only slightly lower than those reported above, namely, .35–.80 for retarded boys and .20–.60 for the retarded girls (Francis and Rarick, 1959). While the range of intercorrelations among strength measures reported on older mentally retarded boys and girls are similar to the correlations on normal children, there is evidence of considerable specificity in strength at the lower age level among the retarded for both sexes.

From data on balance beam performance and an agility run (Francis and Rarick, 1959; Brown, 1954), it appears that the components of balance and agility as assessed by these events show similar, but low, correlations in both retarded and normal individuals, correlations of the magnitude of .30–.60.

It is evident that there is a great deal of specificity in motor functions in educable retarded children, just as there is in individuals of normal intelligence. Worthy of note, however, is the apparent fact that moderate correlations do exist among those skills which call upon similar abilities the magnitude of the correlations being highly similar in the retarded as compared to the normal.

G. Motor Performance and Intelligence

It is widely recognized that within the normal range of intelligence correlations between motor tests and measurements of intelligence in school age children are positive, but low. Studies with retarded children have almost without exception reported the same. Francis and Rarick (1959) found that among 84 correlations between motor tests and intelligence for each sex, 81 were positive for boys and 56 for girls. There was, however, not a clear-cut nor consistent pattern of intercorrelationships between I.Q. and the motor performance variables for either sex, for variables having reasonable high correlations with intelligence at one age level dropped close to zero at another. The data from a recent national survey (Rarick et al., 1970) of over 4200 educable mentally retarded boys and girls gave similar results. Out of 154 correlations between the motor performance tests and I.Q. 152 were positive, the majority, however, being in the .20's and the .30's.

Other investigators (Heath, 1953; Malpass, 1960; Distefano et al., 1958) have reported somewhat higher correlations than those indicated above. Heath (1953) found correlations of .66 between rail-walking performance and mental age of familial retarded children. Malpass (1960) reported correlations of .440 between the Lincoln-Oseretsky scale and the WISC for public school educable retarded boys and girls and .447 for institutionalized retardates. Correlations ranging from .04 to .58 were found by Distefano et al. (1958) using institutional males and females when employing the Lincoln-Oseretsky and the Vineland rail-walking tests. It should be pointed out that the higher correlations in these studies may, in part, be a reflection of the wide age range used in computing the correlation (5–6 years) and, in part, a function of possible intellectual loadings on the items of the test. The Oseretsky test was constructed as a developmental test and the scoring is based on a sequence of tasks that become progressively more difficult and has been normed with retarded children on a chronological age continuum. Sloan (1955), in discussing the nature of the Oseretsky scale, indicated that some items do appear to have substantial intellectual loadings. Cassel (1949) suggested that in the original test Oseretsky was fully aware of this since the proficiency required to skillfully execute many of the tasks is amplified if an intellectual component is involved.

IV. Motor Ability Structure in the Retarded

It was pointed out earlier that although motor skills tend to be specific in nature, there are certain basic abilities which underlie the execution of specific motor tasks. Factor analysis has been used with considerable success in identifying the basic components of motor performance in normal humans and in determining the proportion of variance in a broad range of motor tasks attributable to these components. Such general factors as strength, speed of movement, agility, balance, coordination, and endurance have been shown to account for individual differences in performance attainments of both adolescents and young adults (Fleishman, 1964; Hempel and Fleishman, 1955; Cumbee, 1954; Rarick, 1937).

Only in recent years have motor performance data from retarded children been subject to factor analysis, although one might logically hypothesize that the factor structure of motor abilities of these children would not be materially different from that of normal children. The studies to date tend to support this hypothesis for the research of Vanderberg (1964) in which a factor analysis of Lincoln-Oseretsky test scores obtained from a mixture of public school retarded and normal school children identified such components as speed of movement, balance, dynamic coordination, and age. Clausen (1966) isolated some 12 factors from a large number of perceptual-motor tasks in educable mentally retarded children of three different age groups (8–10 years, 12–15 years, and 20–24 years). Seven of the 12 factors which were found (general ability, intellectual-perceptual ability, spa-

TABLE I

ROTATED FACTOR LOADING FOR MENTALLY RETARDED BOYS BY AGE[a]

	Ages					
Factor	9		12		14	
A	35-Yard dash	−.95	35-Yard dash	−.97	35-Yard dash	−.89
	Agility run	−.86	Balance	.46	Balance	.73
	Vertical jump	.73			Agility run	−.66
	Balance	.62				
	Throw	.61				
B	Left grip	.92	Left grip	.89	Left grip	.91
	Right grip	.90	Push	.88	Right grip	.88
	Broad jump	.65	Pull	.86	Vertical jump	.73
	Balance	.59	Right grip	.83	Pull	.65
	Pull	.56	Height	.70	Push	.63
	Push	.54	Weight	.70	Broad jump	.55
	I.Q.	.51	Throw	.53		
C	Squat thrust	.75	Squat thrust	.79	Squat thrust	.85
	Throw	.56	I.Q.	.77	I.Q.	.73
			Broad jump	.70		
			Throw	.59		
D	Height	.90			Weight	.95
	Weight	.72			Height	.86
	Pull	.50				
E	Age	.93			Age	.86

[a] Data from Rarick (1968).

TABLE II

<small>ROTATED FACTOR LOADING FOR MENTALLY RETARDED GIRLS BY AGE[a]</small>

Factor		Ages				
		9		**12**		**14**
A	35-Yard dash	−.91	Agility run	−.88	30-Yard dash	−.96
	Squat thrust	.71	30-Yard dash	−.81	Agility run	−.80
	Vertical jump	.57	Age	.76	Right grip	.68
	Broad jump	.51	Throw	.60	Left grip	.51
	Balance	.49	Vertical jump	.56	Balance	.57
B	Right grip	.83	Left grip	.87	I.Q.	.96
	Left grip	.71	Right grip	.86	Left grip	.68
	Age	.68	Balance	.59	Right grip	.51
C	I.Q.	.99	I.Q.	.82	Push	.94
	Agility run	−.83	Squat thrust	.58	Squat thrust	.91
					Vertical jump	.86
					Pull	.69
					Throw	.56
D	Weight	.98	Weight	.82	Weight	.95
	Vertical jump	−.50	Height	.81	Height	.90
	Broad jump	−.60			Broad jump	.65
	Agility run	.61				
E					Age	.98

[a] Data from Rarick (1968).

tially related intellectual ability, auditory acuity, reactive motor speed, and cutaneous space discrimination) were identified in all three groups.

The research of Rarick (1968) in which the motor performance data obtained from three age groups of educable mentally retarded boys and girls (ages 9, 12, and 14 years) was subjected to factor analysis, showed clearly that there is a well-defined factor structure of motor abilities in these children and that the factor structure does not change materially in the age range 9–14 years and differs only slightly between the sexes. Tables I and II give the loadings on each of the five isolated factors for the two sexes by age. The magnitude of the factor loadings provides one with a reasonably logical basis for identifying the general factors isolated by the factor analytic approach. Thus, Factor A might logically be considered to be explosive muscular force for it has high loadings on the dash, agility run, and jumping events, each of which requires the rapid development of muscular force in a brief burst of muscular effort. The high loadings of static strength measurements of Factor B suggest that this is a static strength factor. Factor C is dif-

ficult to define, but on the basis of the high loadings with tests that require
the integration of body parts in complex time-sequenced movements, it
would seem reasonable to consider this as a factor of general muscular coordination. Factors D and E, in view of their factor loadings, are clearly factors of body size and maturity, respectively, although neither has significant
loadings with the 12-year-yld boys, and the maturity factor has no significant loadings with the 9- and 12-year-old girls.

It is interesting that, although the battery of motor performance tests
upon which this factor analysis was based included only twelve motor performance events, the general factors which emerged, namely, explosive muscular force, static strength, and coordination, are general factors which have
often been identified in factor analytic studies of motor abilities of normal
adolescents and young adults.

While there is not yet sufficient evidence to make certain that the structure of motor abilities of retarded children is the same as that for children of
normal intelligence, the work cited here suggests that the differences, if any,
are not great. Of some importance is the probability that the basic components of motor performance which account for a major part of the variance
in motor performance among educable retarded children is highly similar
from one age to another and does not differ appreciably between the sexes.

V. Motor Learning

While it is generally believed that mentally retarded children have more
difficulty in the acquisition of fine and gross motor skills than do intellectually normal children, only limited research has been brought to bear on
the nature or extent of this difficulty. While it is self-evident that there are
intellectual or intellectually dependent factors which are necessary in the
processes of motor learning, we cannot with confidence say what these are.
Most retarded children can learn to perform simple motor skills and many
can with extensive practice master reasonably complex coordinations. The
rate of learning and the complexity of the skill acquired have been traditionally assumed to be dependent to no small extent on the individual's level of
intelligence.

The early research of Holman (1933) in which 18 normal 13 year olds
and 33 mentally defective (I.Q. 58) of the same age were trained on a
ball-and-slot task,* four practice periods (200 trails each) per week over a
period of 4 weeks, showed that although initially the retardates were much
poorer, they improved more rapidly than the normal children, and by the

* Subject is required to tip a small platform upon which a small steel ball rests in
such a way that the ball rolls into a specific slot at the bottom of the box.

end of the fourth week the differences between the groups were negligible. Furthermore, the investigator reported that those who started at the same level as the normal group achieved a higher final score than the normal group and those in the normal group who had an initial score as poor as the retardates finished with a lower final score than the retarded group. The implication here is that with sufficient practice in a rather difficult perceptual-motor task, mentally retarded children can reach the performance level of normal children.

The above findings are supported by the more recent research of Clarke (1962) and S. Gordon (1962) which clearly demonstrated that with practice retardates approximated the performance levels of normal subjects on tasks of manual dexterity. The initial poor performance of the retardates would seem to be a reflection of the novelty of the task and lack of relevant experiences upon which to draw, for the differences in performance between subjects with high and low initial scores diminished materially as a function of practice. While S. Gordon (1962) reported a modest correlation between scores on early trials and I.Q., the relationship between I.Q. and percentage improvement was negligible. In other words, performance changes resulting from practice seem to vary widely among individuals, although as in the case of normal individuals those with the poorest initial scores tend to gain the most. It would seem that there is little evidence to support the hypothesis that the frequently observed low level of motor performance of retardates is indicative of a generalized organismic deficit. While it takes the poor performing retardate a long time to accustom himself to a new task, this in no sense means that he will not with practice and with the proper incentive achieve a reasonable level of proficiency.

Incentive is clearly a factor of importance in the motor learning of the mentally retarded. While S. Gordon et al. (1955) found that the learning of motor tasks is affected differently by varied incentive conditions, the work of Kahn and Burdett (1967) disclosed that tangible awards in the form of candy produced no greater motor learning in adolescent educable retardates than self-competition. It is clear that educable retarded children can learn a wide range of manipulative tasks within reasonable limits of time at a level to be competitive with normal persons.

Only a limited number of research studies have looked at the learning capabilities of retarded children on gross motor tasks. For the most part these studies have indicated that although the performance of retarded children does improve with practice the rate of learning is not comparable to that of children of normal intelligence, is apparently independent of I.Q., but is related to chronological age and the complexity of the task. Kirby (1969) using the stabilometer with educable retarded boys and girls of two age levels

TABLE III

PERCENTAGE GAIN IN PERFORMANCE ON THE STABILOMETER OF MENTALLY RETARDED
AND INTELLECTUALLY NORMAL CHILDREN[a]

Age	Boys		Girls	
	Retarded[b]	Normal[c]	Retarded[b]	Normal[c]
6–7 years	7.4	62	−6.3	60
13–14 years	26.5	58	16.3	58

[a] Based on 12 30-sec trials for the retarded and 10 30-sec trials for the normal children.
[b] Data from Kirby (1969).
[c] Data from Bachman (1961).

(9.5 and 13.5 years, respectively) reported that the older children made statistically significant gains after 12 30-sec trials whereas the younger children showed no significant improvement. The percentage gains by age and sex in comparison to normal children (Bachman, 1961) are given in Table III. Variance analysis of the learning scores clearly demonstrated that age was a significant factor in learning this task, the older children showing substantially greater gains than the younger. While there were no statistically significant differences in learning scores by sex, the learning scores of the boys were on the average higher than those of the girls. The variability of performance within the age and sex groups was substantially greater at the termination of the learning experience than at the start. It was apparent that the children varied widely in their performance levels as a result of the learning experience, those with initially high scores showing reasonably consistent improvement, whereas those who performed poorly at the start showed highly variable performance, many of them performing less well as the experiment continued. In spite of the increased variability in performance with practice the correlations between initial and final scores were moderately high being .73 for the boys and .83 for the girls. Thus, those whose performance score was high initially tended to have high scores at the end of the experiment and vice versa.

Overlearning is generally held to be a significant factor in the retention of motor skills. To test the hypothesis that the mentally retarded respond favorably to overlearning, Chasey (1971) taught two groups of institutionalized mentally retarded children and adults two gross motor tasks (hopping tasks). One group stopped practice at the point at which a learning criterion was reached, the other practiced the skill until it was overlearned. When the subjects were tested 4 weeks after practice was discontinued, the group exposed to the overlearning was significantly the better. Not only was it apparent that overlearning produced superior performances, but those whose initial learning was the best profited the most from the overlearning. In these respects the motor learning of the retarded is similar to that of the intellectually normal.

It is generally recognized that in children of normal intelligence motor learning is unrelated to measures of intelligence. Brace (1948), in his study of 50 mentally retarded girls in the age range 13–18 years with an I.Q. range of 23–82, found that learning scores based on performance changes over 30 trials on each of three athletic-type motor tasks showed negligible correlations with I.Q. when proper adjustments were made for differences in levels of motor ability. In other words, the positive but low correlations between learning scores on these tasks were attributable largely to differences in motor capabilities rather than to differences in intelligence.

VI. Physical Working Capacity

The physical working capacity or the ability of the individual to sustain a high level of energy expenditure for relatively long periods of time is perhaps the best single indicator of a person's organic fitness. The chief factor limiting maximum working capacity is the ability of the organism to supply oxygen to the working muscles. With advancing age and the accompanying increase in body size working capacity increases. Prior to puberty sex differences in maximal aerobic capacity are minimal, but thereafter the male shows a substantial superiority (Astrand, 1956). Individuals in the trained state consistently show higher maximal oxygen intakes than the untrained. Thus, physical condition, age, and sex are factors of considerable significance in accounting for individual differences in working capacity during the growing years.

While considerable research has been done on the physical work capacity of normal children (Adams et al., 1961; Astrand, 1956), it is only very recently that an attempt has been made to study the work capacity of the mentally retarded. The exercise tolerance of four groups of boys and girls (age range 6–15.5 years) of substandard intelligence (mean I.Q. of 90) was assessed by Bar-Or et al. (1970). Maximum oxygen intake and heart rate responses were taken with the subjects walking at a standard rate on a motor driven treadmill of progressively increasing grades. The results showed that the boys had a significantly higher maximal oxygen intake (cc/kg min) than the girls in all but one of the age groups (10–11.9 years). As a group the girls had a significantly higher maximal heart rate than the boys. There was a tendency for the heart rate at each work load to be lower in the older subjects, the oldest boys having significantly lower heart rates at submaximal loads than the youngest boys. The oldest boys and oldest girls showed a significantly lower oxygen intake per unit of body weight at submaximal work loads than did the younger children.

The investigators concluded that this group of marginally mentally retarded boys and girls compared favorably with normal children in respect to their maximal aerobic capacity and their cardiovascular responses in heavy exercise. It should, however, be noted that a sizable proportion of the sample (29 of 159 children) either refused or were unable to perform the treadmill test. The average I.Q.'s of the group placed the children at the low normal intellectual level rather than at the intellectual level usually defined as retarded. Clearly, more research is needed on the exercise tolerance and work capacity of children classified as mentally retarded. On the basis of the retarded physical and motor development of these children one would hypothesize that they would also show a poor capacity for work and low exercise tolerance.

VII. School Physical Education Programs and Changes in Motor Performance

The extent to which motor performance levels of retarded children change after exposure to structured physical activity programs has been the focus of several recent investigations. These studies have generally shown that the motor performance of mentally retarded children improves materially as a result of this experience. For example, Oliver (1958) demonstrated quite clearly that within a period of 10 weeks educationally subnormal adolescent boys engaging in a well-structured physical education program 3 hr daily over a period of 10 weeks made significantly greater gains in measurements of muscular strength, physical fitness, and athletic proficiency than a control group of the same age and intellectual level. Similar results were obtained by Corder (1966) in which significantly higher gain scores on the youth fitness test were obtained by a small experimental group in comparison with an official's group and a control group that were in no way involved in a formal physical education program.

The nature of the physical education program and the incentives seem to be of some significance in eliciting changes in motor performance of educable retarded children involved in school programs of physical education. For example, Rarick and Broadhead (1968) found that children exposed to a basically individualized physical activity program scored higher on a battery of motor performance tests at the conclusion of a 20-week experimental program than children involved in a group-oriented program. Stein (1965) reported that when educable mentally retarded boys (I.Q.'s 59–75) were integrated into physical education classes with intellectually normal boys for a period of one academic year, approximately half of the scores of the retarded boys (grades seven and eight) surpassed the mean of the AAPHER Youth Fitness Test battery. According to the investigator, differences in the motor performance of normal and educable retarded boys are more a function of experience and opportunity than inherent differences.

The possible impact of physical activity programs on the motor proficiency of trainable mentally retarded children has received only limited attention. Nunley (1965) involved a group of 11 trainable mentally retarded children (ages 9–14 years) in a physical activity program extending over a period of 15 months. The program consisted of basic neuromuscular activities and modified exercise patterned after published physical fitness programs for children and youth. The test administered at the end of the 15-month training period (scaled on a pass or fail basis) showed that the group made substantial gains in strength, endurance, and basic locomotor and movement skills. Unfortunately, no controls were involved so the gains attributable to maturation and to training could not be differentiated.

It is evident that improvements in motor performance of mentally retarded children do result from well-designed programs of physical education. The extent, however, of permanency of such gains and the possible effect of different types of reinforcement upon the performance changes has received only limited attention. Solomon and Pangle (1966) in a study of 41 adolescent educable retarded boys found that the gains in motor performance achieved in an 8-week daily physical education program were largely retained 6 weeks after the program was discontinued. Furthermore, they reported that immediate reinforcement (knowledge of results) was more effective in achieving certain aspects of physical fitness than remote or indirect reinforcement. Whether the retention in the gains 6 weeks after the program was discontinued represented a retention of the performance levels established or to possible familiarity with the tests is, of course, not known.

VIII. Effects of Perceptual-Motor Training and Physical Activity Programs upon Intelligence and Academic Performance

In the last decade there has been a wave of enthusiasm for the use of physical activities as a means of improving the intellectual level and the learning capabilities of children with learning disabilities. It is well known that the relationship between performance on motor skill tests and measured intelligence in the general population is too low to be of predictive value. Yet in the infant and very young child motor and perceptual-motor tests constitute the primary means of assessing behavioral development. There are good reasons for the use of such measurements, since at this early stage of development sensorimotor abilities are undergoing well-defined and rapid changes, whereas measurable components of conceptual development have not yet made their appearance. It is through the sensorimotor medium that the infant interacts and adjusts to environmental stimuli. Piaget (1953), in observing the behavioral changes of infancy, concluded that as the child interacts with the forces and objects in his environment and as he coordinates the movements of his body and body parts, mental images are formed which are available for recall to be used as constructs in conceptual and symbolic ways. Thus, the child develops concepts of form, weight, texture, stability, time, and space. The meaning which the child attaches to these phenomena comes primarily through his sensory modalities as he responds actively to his environment.

Kephart (1964), as a result of his clinical observations of children with learning disabilities, has proposed that the learning problems of many of these children are a reflection of faulty early motor development. Building on Piaget's theories, Kephart held that it is through perceptual-motor matching in which sensory data are systematically compared and processed through the motoric system that the perceptual and behavioral world of the child is brought into a meaningful relationship. The motor patterns which Kephart believed are of particular significance and which are often not functioning properly in these children are balance, locomotion (walking, running, and jumping), contact (manipulative skills), and receipt and propulsion of objects. The latter involves the movements of objects in space, retrieval of objects (as in catching), and the propelling of objects as in throwing or striking. It is through this array of movements that the child learns about the relationships in his environment and develops a system which gives meaning to these relationships. While Kephart's theories have not yet been adequately tested experimentally his approach in treating children with learning disabilities has met with considerable success.

The effect of general programs of physical activity or structured perceptual-motor learning experiences upon the intelligence and learning capabilities of mentally retarded and emotionally disturbed children has not yet been demonstrated. One of the most controversial approaches used in the treatment of children with learning disabilities is the therapeutic program designed by Delacato (1959, 1963). The Theory of Neurological Organization, as proposed by Delacato, in essence, holds that neurological development progresses in an orderly anatomical way from the lower centers of the brain to the cortex, culminating in hemispheric dominance, the final step in neurological organization. The development of the individual's motor, visual, auditory, and speech functions are held by Delacato to be intimately related to these anatomical changes. The therapeutic program which he developed around this theory is built upon the premise that many children with developmental and learning problems benefit from activities designed to remedy their state of faulty neurological organization. While many clinicians who have used Delacato's approach testified to its success, only limited study has been made of its effectiveness under experimentally well controlled conditions (Robbins, 1965).

The research of Robbins (1966) indicated that the program of activities proposed by Delacato when used with unselected children (without reading or other disabilities) failed to support the postulated relationship between neurological organization (as assessed by creeping and laterality) and reading achievement. There was no evidence to show that the addition of the experimental program to the ongoing curriculum in any way enhanced reading or laterality development in this sample of second grade children. While this research does not support the theory, it should be kept in mind that Robbins recognized that the original theory was developed with brain injured children and with children who had reading difficulties. While the theory is not limited to this class of children solely, Robbins recognized that the results of his own research might have been different had children with these particular problems been the focus of his study.

Further doubts regarding the effectiveness of the Delacato procedure is evidenced by the research of O'Donnell and Eisenson (1969) in which it was found that children who were classified as "disabled readers" and who participated in 10 min of cross-pattern creeping, 10 min of cross-pattern walking, 2 min of visual pursuit, and 5 min of filtered reading, daily over a period of 20 weeks, performed no better on either the Gray Oral Reading Test or the Stanford Diagnostic Reading Test than a similar group receiving the limited Delacato training or a control group given selected physical education activities for the same period of time.

The effect of generalized physical education programs on the intellectual and scholastic ability of educable mentally retarded children is far from clear. Oliver (1958) reported statistically significant gains in I.Q. (Termen, Merrill) of a small group of adolescent educationally subnormal boys, who engaged in a 10-week 3-hr daily program of physical education in comparison to a control group. The mean gain in I.Q. of the experimental group, while statistically significant, was only four points greater than that of the control group. Hence the meaningfulness of this change is open to question. The big change in the experimental group during the course of the experiment was in the realm of personal–social adjustment, an enhanced feeling of self-confidence resulting from the successful achievement. This change was apparent in their attitude to their school work and in their other relationships in the school setting. Doubt about the improvements in measured intelligence coming from physical activity programs is reflected by the research of Solomon and Pangle (1966) who reported no change in I.Q. as a result of a 7-week physical education program and the work of Corder (1966) who reported that the gains in I.Q. which were achieved from the physical activity program were no greater than those made by an official's group included to control for the Hawthorne effect.

The effects of structured and unstructured programs of physical activity on symbol and form perception of children with learning disabilities is a present controversy. Cratty and Martin (1970) reported that young Negro and Mexican-American children within the "central city" of Los Angeles who were involved in a "Learning Games" program for a period of 18 weeks learned and were able to write the letters of the alphabet better than a control group and as well as a group given special tutorial help. In terms of serial memory, the children in the Learning Games group performed better than the other groups at the end of the training period. The investigators concluded that a program of Learning Games can be of benefit in certain academic operations if properly administered and if given to the proper children. On the other hand, Chasey and Wyrick (1970) found that a developmental program of gross motor activity had no significant effect on the ability of educable mentally retarded children to perceive and copy geometric forms. It should be kept in mind that the two studies are in no sense parallel since the one used subjects from the inner city who had learning problems, whereas the other involved institutionalized retarded children.

There would not seem to be any logical reason to expect great changes in intellectual functioning to result from special exercises or from specially designed physical activity programs. Changes in performance on standardized intelligence tests or measures of scholastic achievement might conceivably occur because of a shift in the child's attitudinal structure, his concept of self, or his outlook on life resulting from satisfying experiences in physical activity programs.

IX. Summary and Implications

The findings of the research cited in this chapter clearly show that mentally retarded children lag well behind children of normal intelligence in the development of both fine and gross motor skills. What is not so evident is the cause of the deficit. One might ask if the retardation in motor development is primarily a function of intellectual inadequacy or a reflection of environmental circumstances. Research findings indicate that both nature and nurture play a role. These children can, when given help, acquire reasonable levels of skill, although the time required for learning is related to the extent of intellectual impairment. While the schools provide considerable instruction in manual skills, opportunities afforded to the retarded for the development of gross motor skills under competent supervision is generally lacking. For example, a recent national survey (Rarick *et al.,* 1967) disclosed that half of the educable mentally retarded children in our public schools had no organized classwork in physical education and only 25% received 60 min or more of instruction each week. The research cited in this chapter points out that retarded children make substantial gains in motor proficiency in relatively short periods of time if they have competent instruction.

Evidence regarding the impact of perceptual motor training on other aspects of behavior is not entirely clear. As pointed out earlier, there is some clinical data but little, if any, experimental evidence to indicate that conceptual responses of the retarded are positively affected by perceptual motor training. In instances where such changes have been noted, these probably reflect attitudinal changes toward the intellectually oriented task rather than alterations in intellectual functioning.

It is generally recognized that mentally retarded children, because of repeated failure, develop a defeatist attitude, and this expectation of failure is reflected in the way in which they apply themselves. Wider use of physical activities with these children may help reverse this situation. Since motor skills are not highly loaded with intellectual components, the chance for success here is more favorable than in other phases of the school program. Favorable changes in academic performance of retarded children have been attributed by some to attitudinal changes coming from success in physical activity programs. At present the evidence on this is far from conclusive.

Mentally retarded children are usually characterized by substandard levels of physical vitality. This is supported by general observation and by independent investigations. As pointed out in this chapter the muscular strength and aerobic work capacity of these children has been shown to be well below that of normal children of similar age. It is apparent that most retarded children do not lead a sufficiently active life to maintain a reasonable level of fitness nor do they receive sufficient physical activity to insure optimum physical growth. While the extent to which the retardation in physical growth of these children can be attributed to inadequate levels of physical activity is not known it is generally believed that physical activity is supportive of physical growth.

The evidence is clear that our schools have not given adequate attention to the physical activity needs of the mentally retarded. This is, in part, because of a lack of appreciation of the importance of physical activity for them. It is also partly because of apathy among physical educators who have been reluctant to serve this segment of the population. Be that as it may, the mentally handicapped need, and are entitled to, good instructional programs of physical education. The need is perhaps greater for these children than for their intellectually normal counterparts, for the chances of profiting from free play experiences on the playground is for the retarded highly unlikely.

References

Adams, F. H., Leonard, M. L., and Hisazumi, M. (1961). *Pediatrics* **28,** 55.

Astrand, P. O. (1956). *Physiol. Rev.* **36,** 307.

Bachman, J. C. (1961). *Res. Quart.* **32,** 123.

Bar-Or, O., Skinner, J., Bergstein, V., Haas, J., Shearburn, C., and Buskirk, E. (1970). *In* "Factors Related to the Speech-Hearing of Children of Below Normal Intelligence" (B. M. Siegenthaler, ed.), Proj. No. 8–0426. U.S. Dept. of Health, Education and Welfare, Office of Education, Bureau of Education for the Handicapped, Washington, D.C.

Baumeister, A. A., and Kellas, G. (1968). *Amer. J. Ment. Defic.* **72,** 715.

Berkson, G. (1960). *J. Ment. Defic. Res.* **4,** Part 1, 51.

Brace, D. K. (1948). *Res. Quart.* **19,** 269.

Brown, H. S. (1954). *Res. Quart.* **25,** 8.

Cassel, R. H. (1949). *Train. Sch. Bull.* **46,** 11.

Chasey, W. C. (1971). *Res. Quart.* **42,** 145.

Chasey, W. C., and Wyrick, W. (1970). *Res. Quart.* **41,** 345.

Clarke, A. B. (1962). *Proc. London Conf. Sci. Study Ment. Defic. 1960,* Vol. 1, p. 89.

Clausen, J. (1966). "Ability Structure and Subgroups in Mental Retardation." Macmillan, New York.

Corder, W. O. (1966). *Except. Child.* **32,** 357.

Cratty, B. J., and Martin, M. M. (1970). "The Effects of a Program of Learning Games Upon Selected Academic Abilities in Children With Learning Difficulties," Program Grant U.S.O.E. (0–0142710) (032). University of California, Los Angeles.

Cumbee, F. Z. (1954). *Res. Quart.* **25,** 412.

Delacato, C. H. (1959). "The Treatment and Prevention of Reading Problems." Springfield, Illinois.

Delacato, C. H. (1963). "The Diagnosis and Treatment of Speech and Reading Problems." Springfield, Illinois.

Distefano, M. K., Ellis, N. R., and Sloan, W. (1958). *Percept. Mot. Skills* **8,** 231.

Doll, E. A. (1946). "The Oseretsky Tests of Motor Proficiency: A Translation From The Portuguese Translation." Educational Test Bureau, Minneapolis, Minnesota.

Dutton, G. (1959). *Arch. Dis. Childhood* **34,** 331.

Ellis, N. R., and Sloan, W. (1958). *Amer. J. Ment. Defic.* **63,** 304.

Espenschade, A. S. (1940). "Motor Performance in Adolescence." Univ. of California Press, Berkeley.

Fait, H. F., and Kupferer, H. J. (1956). *Amer. J. Ment. Defic.* **69**, 729–32.

Fleishman, E. A. (1964). "The Structure and Measurement of Physical Fitness." Prentice-Hall, Englewood Cliffs, New Jersey.

Fleishman, E. A., and Hempel, W. E. (1956). *J. Appl. Psychol.* **40**, 96.

Francis, R. J. and Rarick, G. L. (1959). *Amer. J. Ment. Defic.* **63**, 292.

Francis, R. J., and Rarick, G. L. (1960) "Motor Characteristics of the Mentally Retarded," Coop. Res. Monogr. No. 1. U.S. Dept. of Health, Education, and Welfare, Washington, D.C.

Gordon, S., O' Conner, N., and Tizard, J. (1955). *Amer. J. Ment. Defic.* **60**, 371.

Gordon, S. (1962). *Proc. London Conf. Sci. Study Ment. Defic., 1960,* Vol. 2 p. 587.

Guilford, J. P. (1958). *Amer. J. Psychol.* **71**, 164.

Heath, S. R. (1953). *Train. Sch. Bull.* **50**, 110–127.

Hempel, W. E., and Fleishman, E. A. (1955). *J. Appl. Psychol.* **39**, 12.

Henry, F. M., Lotter, W. S., and Smith, L. E. (1962). *Res. Quart.* **33**, 70.

Hinton, G. G., and Stevenson, J. A. F. (1962). *Can. J. Biochem.* **40**, 1239.

Holman, P. (1933). *Brit. J. Psychiat.* **23**, 279.

Hunsicker, P. A., and Reiff, G. G. (1965) "A Survey and Comparison of Youth Fitness, 1958–1965." Coop. Res. Proj. No. 2418. University of Michigan, Ann Arbor, Michigan.

Jones, D., and Benton, A. L. (1968). *Amer. J. Ment. Defic.* **73**, 143.

Jones, H. E. (1949). "Motor Performance and Growth." Univ. of California Press, Berkeley.

Kahn, H., and Burdett, A. D. (1967). *Amer. J. Ment. Defic.* **72**, 422.

Kephart, N. C. (1964). *Except. Child.* **31**, 201.

Kirby, J. (1969). Unpublished Master's Thesis, University of California, Berkeley.

Kugel, B. R., and Mohr. J. (1963). *Amer. J. Ment. Defic.* **68**, 41.

Lowrey, G. H., Bacon, G. E., Fisher, S., and Knoller, H. (1968). *Amer. J. Ment. Defic.* **73**, 474.

McCloy, C. H. (1946). "Tests and Measurements in Health and Physical Education." Crofts, New York.

Malpass, L. F. (1959). "Responses of Retarded and Normal Children to Selected Clinical Measures," Sect. I. Southern Illinois Univ. Press, Carbondale.

Malpass, L. F. (1960). *Amer. J. Ment. Defic.* **64**, 1012.

Meredith, H. V. (1935). *Univ. Iowa Stud. Child Welfare* **11**, No. 3, 1–128

Metheny, E. (1941). *Res. Quart.* **12**, 115.

Neilson, N. P., and Cozens, F. W. (1934) "Achievement Scales in Physical Education Activities for Boys and Girls in Elementary and Junior High Schools." Barnes, New York.

Nunley, R. L. (1965). *J. Amer. Phys. Ther. Ass.* **45**, 946.

O'Donnell, P. A., and Eisenson, J. (1969). *J. Learn. Disabil.* **2**, 10.

Oliver, J. N. (1958). *Brit. J. Educ. Psychol.* **28**, 155.

Piaget, J. (1953). "The Origins of Intelligence in the Child" (M. Cook, transl.). Routledge, London.

Pozsonyi, J., and Lobb, H. (1967). *J. Pediat.* **71**, 865.

Pryor, H. B., and Thelander, H. E. (1967). *Clin. Pediat.* **6**, 501.

Rarick, G. L. (1937). *Res. Quart.* **8**, 89.

Rarick, G. L. (1968). *In* "Expanding Concepts in Mental Retardation" (G. A. Jervis, ed.), pp. 238–246. Thomas, Springfield, Illinois.

Rarick, G. L., and Broadhead, G. D. (1968). "The Effects of Individualized Versus Group Oriented Physical Education Programs on Selected Parameters of the Development of Educable Mentally Retarded and Minimally Brain Injured Children." Univ. of Wisconsin Press, Madison.

Rarick, G. L., Widdop, J. H., and Broadhead, G. D. (1967). "The Motor Performance and Physical Fitness of Educable Mentally Retarded Children." Dept. Phys. Educ., University of Wisconsin, Madison.

Rarick, G. L., Widdop, J. H., and Broadhead, G. D. (1970). *Except. Child.* **36,** 509.

Reichlin, S. (1961). *Endocrinology* **69,** 227.

Robbins, M. P. (1965). Unpublished Doctoral Dissertation, University of Chicago, Chicago, Illinois.

Robbins, M. P. (1966). *Except. Child.* **32,** 517.

Roberts, G. E., and Clayton, E. C. (1969). *Develop. Med. Child. Neurol.* **11,** 584.

Rundel, A. T., and Sylvester, P. E. (1965). *Amer. J. Ment. Defic.* **69,** 635.

Sloan, W. (1951). *Amer. J. Ment. Defic.* **55,** 394.

Sloan, W. (1955). *Genet. Psychol. Monogr.* **51,** 183.

Solomon, A. H., and Pangle, R. (1966). "IMRID Behavioral Science," Monogr. No. 4. George Peabody College for Teachers, Nashville, Tennessee.

Stein, J. U. (1965). *Rehab. Lit. (Spec. Rept.).* **26,** 205.

Travis, L. E., and Dorsey, J. M. (1930). *J. Exp. Psychol.* **13,** 370.

Tredgold, A. F., Tredgold, R. F., and Soddy, K. (1956). "A Textbook of Mental Deficiency," 9th ed. Williams & Wilkins, Baltimore, Maryland.

Vanderberg, S. G. (1964). *Percept. Mot. Skills* **19,** 23.

Whitehorn, J. C., Lundholm, H., and Gardner, G. E. (1930). *J. Exp. Psychol.* **13,** 293.

Widdop, J. H. (1967). Unpublished Doctoral Dissertation, University of Wisconsin, Madison.

CHAPTER

11

Play, Games, and Sport in the Psychosocial Development of Children and Youth

John W. Loy and Alan G. Ingham

I. The Socialization Process

Albeit brief, this first part of the chapter attempts to (a) define the term "socialization," (b) set forth the fundamental consequences of socialization, (c) outline the major modes and basic mechanisms of socialization, (d) describe the primary agencies of socialization, and (e) summarily characterize the socialization process.

A. SOCIALIZATION DEFINED

The concept of socialization is manifestly broad and complex and thus not readily definable. As Clausen (1968) has written:

Socialization may be viewed from the perspective of the individual or from that of a collectivity (be it the larger society or a constituent group having a distinct subculture). Further, individual development may be viewed generically within a given society or it may be viewed in terms of the experiences and influences that lead to significant differences among persons (both social types and unique personalities).

Human beings learn to be social beings. It appears that little of the social behavior of the human being can be traced to genetic or hereditary sources. Nurture outweighs nature in its contribution to social development. The individual has to learn to participate effectively in social groups. Thus, Elkin (1960) defined socialization ". . . as the process by which someone learns the ways of a given society or social group so that he can function within it." According to Becker (1962) socialization is an interactional process because the human being is self-reflexive and symbolizes and must not only learn to place himself in a social group but must also learn the social definitions of behavior which enable him to confront a multitude of individuals without creating anxiety by an inappropriate presentation of self.

In short, then, socialization is a process that involves interaction and learning. On the one hand, it centers attention on the adaptation of individuals to their social situations; on the other hand, it centers attention on how individuals develop social identities as a result of their participation in various social situations. Taking these several themes into account, *socialization* may be defined for present purposes as an interactional process whereby a person acquires a social identity, learns appropriate role behavior, and in general conforms to expectations held by members of the social systems to which he belongs or aspires to belong.

The human being affiliates and reaffiliates with social groups throughout life. "Life (may) be viewed as continuous socialization, a series of careers, in which old identities are sacrificed as new identities are appropriated, in which old relations are left behind as new relations are 'joined'" (Stone, 1962). Each reaffiliation requires a redefinition of the situation. However, one should not assume that affiliation with one group necessarily denies affiliation with another. Each individual occupies many places or positions in many groups. This matrix of statuses with their contingent roles demands that the individual be capable of transforming his identities. In any given day the typical individual experiences many transformations; he may change from boss to employee, from father to son, from salesman to customer, etc. In sum, socialization is a process of preparing the social actor to perform appropriately in all of his social settings.

B. CONSEQUENCES OF SOCIALIZATION

Caplow (1964) stated . . . "In every situation in which a member or an aspirant is to be transformed into a successful incumbent, there are at least four requirements that must be met. The candidate must acquire: (a) a new self-image, (b) new involvements, (c) new values, and (d) new accomplishments." These requirements outlined by Caplow may be viewed as the major consequences of socialization.

1. A New Self-Image

"Identities are socially bestowed. They must also be socially sustained, and fairly steadily so" (Berger, 1963). As the individual takes a position (status) in a group, he learns to define himself in response to the expectations which the group has of a person occupying that status. That is, the person attempts to introject a group-defined identity into his own identity. The more congruent the projected self is with the group-defined self, the more social sustenance (reinforcement) one expects. Once having established one's claim to an identity, one works to preserve it, and if possible to enhance it. Therefore, it is suggested that the more established the claim to an identity becomes through adequate performance, the more an individual can feel secure in assuming that identity.

2. New Involvements

When an individual joins a new group he encounters new people. He finds that he must modify his performance, as a social actor, to the interests of the new audience which in all cases demands that the individual be that which he portrays (Berger, 1963; Goffman, 1959). That is, his new affiliates demand sincerity in the acting out of a social role. "When an actor takes on an established social role, usually he finds that a particular front has already been established for it" (Goffman, 1959). The performance required of a role in a group has probably been group-defined before an individual joins the group. Not only has the role been group-defined but in a task-oriented group, such as a football team, the reciprocal behaviors of other roles have probably also been defined. Socialization may be considered successful when a candidate for the role has learned to present an identity and fulfill the behavioral obligations to the satisfaction of the group. To borrow a concept from Hollander (1958), the individual attempts to secure, by conformity to group expectations, a good credit rating. Thus credit ". . .represents an accumulation of positively-disposed impressions residing in the perceptions of relevant-others. . ." In the initial phases of membership in a new group, the individual seeks to establish an identity through conduct which is acceptable to his new affiliates. New involvements are anxiety producing until the social actor has sufficient credits for his performance to be taken for granted.

3. New Values

Through accepting and internalizing the communicated values and norms of the group which is acting as a reference, the social actor is able to ". . .visualize his proposed line of action from this generalized standpoint, anticipate the reactions of others, inhibit undesirable impulses, and thus guide his conduct" (Shibutani, 1961).

Socializing a neophyte into a group involves the lengthy process of imbuing the neophyte with the necessary ideology to enable him to justify to non-group members what is done by the group. Thus, the professionally oriented physical education student is indoctrinated through such courses as "Principles of Physical Education" which supposedly prepare him to cognitively handle those outsiders who question his existence. This sharing of a perspective with a reference group is, in Mead's (1964) terms, taking the attitudes of "generalized others."

4. New Accomplishments

Berger (1963) suggested that ". . .identity comes with conduct and conduct occurs in response to a specific social situation." When we assume a status (position) it is often necessary to work at occupying that status. When we assume a status within a group we not only work to maintain our own identity but also to preserve the performance of the whole group. As Goffman (1959) suggested: ". . .it would seem that while a team-performance is in progress, any member of the team has the power to give the show away or to disrupt it by inappropriate conduct." In some cases this simply means an observance of professional etiquette, in others, it requires that a whole series of complex skills must be learned to maintain the group performance at an adequate level.

C. Mechanisms of Socialization

Every neophyte in any given social system is exposed to one or more principal modes of socialization. These fundamental means of socialization include training, schooling, apprenticeship, mortification, trial-and-error, assimilation, and anticipatory socialization (Caplow, 1964). Common to each of these major modes of socialization are patterned forms of social learning, control, and influence.

New members of a group are asked to learn the prevalent definitions or meanings upon which the social reality of the organization is based. The new members learn to act out the identities which are designated to them. They learn the roles that are contigent with the assumption of such identities. And, in general, they learn the culture of their new social system. However, socialization is not merely a learning process. "The individual not only learns objectivated meanings but identifies with them and is shaped by them. He draws them into himself and makes them his meanings" (Berger, 1969). In sum, all modes of socialization imply attitudinal and behavioral changes resulting from certain basic mechanisms of social influence.

Processes of social influence have long been of central concern for behavioral scientists, and many theories and models have been set forth explaining the nature of social influence. One such theory is that proposed by Kelman (1961).

Kelman's theory suggests the circumstances under which attitude change will or will not occur and identifies the conditions leading to either temporary or permanent attitude change. Kelman distinguished three primary processes of social influence which he called (1) compliance, (2) identification, and (3) internalization.

1. Compliance

"Compliance can be said to occur when an individual accepts influence from another person or from a group because he hopes to achieve a favorable reaction. . ." (Kelman, 1961). That is to say, a person learns the appropriate responses to situations as defined by others in order to obtain reinforcement or avoid punishment. Initially, "the child's dependency and needs for affection established during early infancy provide a positive basis for maternal authority and for the child's compliance with it" (Deutsch and Krauss, 1965). The individual responds to influence from his significant others and eventually learns that groups also have authority over him and can reward or punish him similarly. Compliant behavior leads to reinforcement (the presentation of a reward or the removal of noxious stimuli). Noncompliant behavior leads to punishment. This orientation to the study of learning has been labeled "operant conditioning." "In operant conditioning we 'strengthen' an operant in the sense of making a response more probable or, in actual fact, more frequent" (Skinner, 1968). A response must be emitted before it can be reinforced. How this response is elicited is not clear. One possible method is direct tuition, as tuition engenders the intentional guidance of behavior and the manipulation of reinforcement (Secord and Backman, 1964). However, direct tuition presupposes that an individual knows the contingencies between the behavior and the cue-producing word or symbol (Miller and Dollard, 1941).

Compliant behavior is largely a function of the patterns of reward presented to an individual. By patterns we refer to the schedules of reinforcement. Under laboratory conditions, reward schedules can be manipulated. Some schedules (e.g., continuous) facilitate the rapid acquisition of responses. Other schedules (e.g., intermittent) while not resulting in such rapid acquisition render the response, once acquired, more resistant to extinction (Bandura and Walters, 1963).

Intermittent reinforcement can be presented in a variety of ways. One main mode of experimental reinforcement is a "fixed-ratio schedule" where only every other, or every nth response of the subject is reinforced. Another main mode of experimental reinforcement is a "fixed-interval schedule" where selected time intervals intervene between presentations of rewards. Bandura and Walters (1963) observed that:

> Examples of fixed-ratio schedules of reinforcement in everyday life, particularly in child-training procedures are difficult to find. On the other hand, in most modern social systems the socializing agents, who are dispensers of reinforcers, have to organize their lives on the basis of the time schedules of others. Consequently, in most families some responses of the children are reinforced on a relatively unchanging fixed-interval schedule. Feeding, the availability of the father or school-age siblings for social interaction, and, in general, events associated with household and family routines may serve as reinforcers, positive or negative, that are dispensed at relatively fixed intervals.

In terms of compliant behavior it must be recognized that an individual may upon occasion learn that when one particular preferred response is reinforced, other responses contiguous in time and space may also be simultaneously reinforced. Effective social learning thus requires sharp discrimination and adequate generalization of learned patterns of responses. Generalization enhances the efficiency of the socialization process in that social situations which are defined similarly can be handled using the same set of responses. However, an individual sometimes learns responses other than those intended by his tutor. This may be the result of overgeneralizing or generalizing on the basis of cues not relevant to the situation in which he is placed. Fine discrimination among stimuli must often be exercised in the selection of a response which is appropriate to the situation as it is socially defined. For example, a young child must learn that physical aggression is particularly appropriate in some situations but most inappropriate in other social settings.

2. Identification

"Identification can be said to occur when an individual adopts behavior derived from another person or group because this behavior is associated with a satisfying self-defining relationship to this person or group" (Kelman, 1961). Identification is one explanation of imitative behavior. An individual attempts to establish himself in an identity by imitating persons who already possess that identity. Secord and Backman (1964) presented seven principles which account for the choice of a model. Suffice it here to merely generalize these seven principles into reinforcement terminology. Models are chosen for imitation mainly because they have coercive power (i.e., they can reward or punish) or because they are capable of obtaining rewards or approval which the individual also desires but does not have access to (the identifier is vicariously reinforced).

Two other theoretical orientations which have utilized the mechanism of identification are "role theory" and "reference group" theory.

a. Role Theory. Role theory conceives of society as a stage and views the individual as a social actor. Accordingly, the task of socialization is to insure adequacy of performance from each and every actor. From this dramaturgical perspective role theory considers the ways in which an individual presents himself and his performance to others. In short, it is concerned with the management of impressions and the sustaining of identities (Goffman, 1959).

The individual learns to share definitions of the situations in which he places himself. Through interaction he attempts to perfect his performance in the identities he assumes. "The term role is usually applied to situations in which the prescriptions for interaction (i.e., the script for the performance) are culturally defined. . ." (Deutsch and Krauss, 1965). "A role, then, may be defined as a typified response to a typified expectation" (Berger, 1963). Conventional ethics demand that the prescribed role and the subjective role be in symmetry. That is, the individual should render a sincere playing of the role.

As a child we imitate our significant others (those with whom we identify) who aid us in self definition. In taking the role of others the child begins to develop a self-system, that is, he begins to view himself as a social object. Through imitative interaction and performance he learns the prescribed behavior for the statuses he occupies and he realizes that the prescribed behavior patterns are bolstered by social norms. He learns that only a range of behavior is acceptable to his significant and generalized others.

b. Reference Group Theory. The major problem for reference group theory is centered on the process whereby ". . .a person orients himself to groups and to other individuals and uses them as significant frames of reference for his own behavior, attitudes or feelings" (Deutsch and Krauss, 1965). Merton (1963) distinguished several types of reference groups. One of his major distinctions was between positive and negative reference groups. The former's standards are valued and adopted. The latter's standards are rejected. "Reference-group theory indicates that social affiliation or disaffiliation normally carries with it specific cognitive commitments" (Berger, 1963).

In brief, the degree of influence a reference group exerts in the formation of attitudes among its members is dependent on the degree to which any individual member identifies with the group. It should be recognized that by means of "anticipatory socialization" nonmembership groups may also serve as important reference groups and thus substantially influence attitude development.

> For example, the son of an unskilled worker who aspires to middle-class status will tend to accept middle-class values and attitudes. His middle-class outlook will embrace such diverse objects as sexual practices and political issues. Upward mobile, he will reject the values and attitudes of his lower-class family (Krech *et al,* 1962).

As can be surmised from the preceding discussion, the effect of groups on social learning and attitude development is often indirect and typically complex.

3. Internalization

Internalization can be described as the appropriation of social reality and its transformation ". . .from structures of the objective world into structures of the subjective consciousness" (Berger, 1969). That is, society has outposts stationed in our heads (this metaphor is attributed to Kempton, 1970) which enable us to regulate our own behavior. Obviously, the external society reified in our own consciousness is rarely in a state of symmetry. Individuals modify collective definitions in the light of their idiosyncratic definitions of social reality. However, the more successful the socialization process is, the more symmetrical they become (Berger, 1969). The internalization of social norms and values enables us to confront our actions and reflectively appraise them. This appraisal may result in feelings of guilt (self-punishment) or esteem (self-reinforcement). Although many theories share the concept of internalization, we would like to focus very briefly on the psychoanalytic orientation.

Although biologistic and behavioristically interpretable, Freudian theory attempts to reinstate the organism as an intervening entity between the stimulus and the response. In fact, Freud attempts to move from stimulus–response reaction to ego-controlled reaction (Becker, 1962). The ego operates on a reality principle, it either represses or defers gratification of id impulses. By reality we refer to the social world. The ego tempers the demands of the id in terms of social reality. Both the ego and superego are generated by interaction. The superego *is* the introjected cultural value system. Introjection refers to the internalization of standards. Hence, through the introjection of cultural values and norms, the individual anticipates the reactions of others to his behavior and appraises his own behavior in the light of these introjections. Psychoanalytic theory entertains the possibility of an internal dialectic between the individual's impulses and society reified in consciousness. It might be assumed that Freud's concern with this internal dialectic is manifested in his discussions of cathexis and anticathexis, defense mechanisms (e.g., repression and fixation), and the origins of neurotic and moral anxiety.

D. AGENCIES OF SOCIALIZATION

At the moment of birth a child obtains membership in a number of social systems, including his family, community, and society. These social systems and all others in which he may acquire membership at a later date constitute agencies of socialization. Initially, of course, socialization is limited to the micro-social system of the family, composed of parents, siblings, and perhaps immediate relatives. Members of the family constitute the first "significant others" for the child and initially define the world for him. As the child's social circle gradually extends he comes under the influence of other social systems which come to constitute his "generalized others" (Mead, 1964). Although any given individual is influenced by a multitude of social systems during the course of his life cycle, the principal agents of socialization for children and youth are held to be the family, the peer group, the school, and the community.

1. The Family

In industrial nations the family's all-encompassing role as a socializing agent has been diminished. Other societal agencies have taken over many of its functions. However, the family still retains its importance in primary socialization in that the parents (significant others) provide the child with his first exposure to rules and role behaviors. Initially, the child is totally dependent upon the family and is subordinated by it (Parsons and Bales, 1955). In these initial stages of life (the oral stage—Freud), a child is dependent upon its mother for the relief of oral stress. Food and suckling reduce this drive and they become associated with love and approval which are generalized reinforcers. Parental authority is based upon this dependency relationship (Hall, 1954). The child complies in order to gain approval and to avoid punishment (Miller and Dollard, 1941). As the symbolizing capacity of the child increases, shaping of behavior through the reinforcement of approximate imitations (Skinner, 1968) is supplemented by direct tuition. The child can be taught values and sentiments. The family

can now interpret the wider community to the child and transmit
". . .segments of the wider culture to the child, the particular segments
being dependent upon (the family's) social positions in the community"
(Elkin, 1960).

2. The Peer Group

The importance of the peer group as an agent of socialization in present-
day society is increasing. During the latency period (Freud), the child extends his social circle beyond that of the family. The generations gradually
move apart so that the peer group replaces the parents as a source of information on contemporary know-how (Broom and Selznick, 1963). In what
Riesman called "the other-directed society" the peer group becomes a measure of all things (Riesman et al., 1953). The child channels his competitive
drives into a constant searching for peer group approval.

Socialization in the peer group is not formalized. Membership consists of
knowing the norms and values of the in-crowd and being able to perform
adequately in those tasks or performances which involve the group. Peer
groups are egalitarian in that there is no formal authority structure (Broom
and Selznick, 1963; Chinoy, 1963). However, it is the task of the older
members of the group to socialize the neophytes. The neophyte identifies
with the group values in order to gain approval and maintain his relationship with the group. Because an individual chooses his affiliates in an act of
self-definition, it is likely that a certain uniformity among group members
prevails (Berger, 1963; Jones and Gerard, 1967). Uniformity and cohesive-
ness are maintained not only by sanctioning deviates but also by group idio-
syncrasies such as "slanguage," dress codes, and the group's vantage points
in the universe of social relationships.

3. The School

The industrialization of societies has necessitated the development of a
formal (organized) system of education (Toby, 1964; Worsley et al.,
1970). Historically traced, the family structure has slowly evolved from an
extended system to a nuclear system and increasing geographic mobility has
loosened kinship ties. As a consequence of these developments, the formal
education of maturing individuals is a deliberate attempt to provide the
knowledge and skills required by the technical and social roles of complex
modern society (Worsley et al., 1970). The child spends most of each day
within the confines of the school. The school ". . .is the theater in which
much of the drama of the child's life is played" (Jersild, 1968). In dealing
with the adult significant others (teachers) who stand in loco parentis, the
child is provided with one more setting in which he must accommodate him-

self to adult authority. These adults have the same mechanisms of socialization available to them as have the child's parents. They can reward or punish. They provide models for imitation and suggest values for the child to internalize (Elkin, 1960). The teacher has available the generalized reinforcers of approval and praise or the formalized rewards of grades and trophies. Besides accommodating himself to adult authority, the child has to assimilate the experience presented to him in the school. Jersild (1968) suggested that "the learner perceives, interprets, accepts, resists, or rejects what he meets at school in the light of the social system he has within him." This statement recognizes the possibility of a potential conflict between the home and the school or the peer group and the school. What self-system does the child bring with him to school and will the social origins from which this self has emanated conflict with the middle-class orientations of the teachers within the school? Will the activities valued by the peer group coincide with the valued activities of the school (Coleman, 1961)?

4. The Community

Generally, the word "community" has been used as an ideal-typical concept which orientates us to a social setting usually defined by territorial boundaries, a specific subculture, a degree of autonomy, and a fairly homogeneous populace (Frankenberg, 1957; Minar and Greer, 1969; Sanders, 1958; Young and Willmott, 1967).

The child because he is placed in a community ". . .forms an appreciation of the manner in which various categories of people are evaluated and incorporates customary patterns into his way of approaching the world" (Shibutani, 1961). The child responds to both the recurrences in social behavior (customs) and to the value orientations which define the meaning of life. He responds to the institutions of community life which have been "established by some common will" (MacIver, 1917). Socialization attempts to fit man to his community. In essence, the socially produced personality is asked to endure the society and continually sacrifice some personal wishes to preserve it. The child is asked to believe the subcultural ideologies of his community as disbelief destroys the perceived unanimity upon which the community publicly operates.

E. SOCIALIZATION AND SOCIETY

In the preceding sections, the modes, mechanisms, and agencies of socialization have been presented, albeit in a somewhat mechanistic fashion. These sections have only alluded to the societal purpose of socialization. In the final part of Section I some reasons why the socialization process is necessary for society are presented.

Society is the most encompassing, comprehensive, and consequential group to which an individual belongs. All other groups may be viewed as integrated subsystems of society. Thus, the purpose of socialization can be defined in general terms as the process by which an individual ". . .learns the ways of a given society so that he can function within it" (Elkin, 1960). In order for socialization to occur, society must be conceived of as a reality, an object in itself (Durkheim, 1950). Reified in this way it can proceed to exert influence over the individual. However, the ". . .individual is not molded as a passive inert thing. Rather he is formed in the course of a protracted conversation in which he is a participant. That is, the social world is not passively absorbed by the individual, but actively appropriated by him" (Berger, 1969). The socialization process conceived of in this way stresses the ongoing dialectic which exists between man and his society. Socialization is one side of this dialectic where the social reality attempts to impose itself on the individual so that he conforms to, and believes in, the behavioral dictates necessary to ensure stability in the society. Stability refers to a minimum consensus of beliefs in the "goodness" of the social reality, as it is presently constructed, so that it can be perpetuated. Hence, if socialization were successful in all cases, symmetry (Berger, 1969) would exist between the values of society and the values of the individual and social stability would be ensured. If socialization were successful and an introjected society were all that there was to the self ". . .the account of the relationship between man and society would be an extreme and one-sided one, leaving no room for creativity and reconstructive activity; the self would merely reflect the social structure, but would be nothing beyond that reflection" (Morris —introduction to Mead, 1970). In the mechanistic presentation which preceded, we ignored this essential criticism which makes the plea for a dialectical conception of socialization. Out of a need for brevity and convenience, rather than our background assumptions concerning the nature of man and society, the discussion was confined to a one-sided approach in that a picture was presented whereby man needs only to be capable of learning and internalizing to fit into a socially ordered reality.

Socialization is obviously not as successful as society might wish it to be. Social change does occur. Creativity is evident. To speak of a dialectic is to admit to a process whereby the individual not only learns established meanings but also contributes to those meanings. Although a minimum of symmetry seems to exist (that is, there is sufficiency of social order), complete symmetry is probably not humanly possible. This is to admit that the individual is capable of redefining situations using rules as guides but not mandates. It suggests, as was alluded to in the discussion of Freud, that a second dialectic can occur within the individual's own consciousness. Hence, confrontation can occur between the internalized social system and the self-system.

In the discussion of modes, mechanisms, and agencies, both levels of the dialectic were somewhat ignored. Such an approach rightfully should have led to criticisms that the discussion suffered from an oversocialized conception of man (Wrong, 1961). The discussion of agencies and mechanisms implied that human conduct is totally shaped by common norms or institutionalized patterns of interaction (Wrong, 1961). For instance, under Section I, C, 1, "Compliance," the basic tenets of operant conditioning ignore man's ability to interpret and define the situations in which he places himself. Rather the discussion concentrates with ratlike experimental simplicity on shaping the cognitions of man through reinforcement and punishment. Also, internalization should not be merely viewed as habit formation (Wrong, 1961). If internalization were merely habit formation, then it would deny the dialectic between the self-system and social system which occurs in consciousness.

The discussion which now follows is essentially aimed at indicating the role of play, games, and sports in the socialization process. It is assumed that the agencies of socialization see ludic activities* as a means of portraying the value system of society to a child so that through participation he can learn it. It is also necessary to make the assumption that games provide definitions of meaningful interaction which, if participation is encouraged in them, have some transformable values to other situations defined as meaningful to society.

II. Play, Games, and Sport in Socialization

As previously defined, "socialization is an interactional process whereby a person acquires a social identity, learns appropriate role behavior, and in general conforms to expectations held by members of the social systems to which he belongs or aspires to belong." Thus, every social system can be conceived of as an agent of socialization modifying individuals' behavior through various processes of social influence.

The purpose of this part of the chapter is to show, suggest, and speculate as to how play, games, and sport serve the socialization process within selected micro- and macro-social systems. Specifically, Section II,A focuses on the role of play during early childhood socialization in the family. Section II,B examines the importance of games within peer groups for preadolescent

---

* The phrase "ludic activities" is used throughout this chapter to represent all activities generally described as play, games, or sports.

development and analyzes the function of sport for adolescent development in school settings. Section II,C centers attention on the complex nature of the socialization process and illustrates its complexity by describing the dialectic between individuals and society in terms of game involvement. Finally, Section II,D deals with the utility of simulated agonistic activities* in meeting explicit socialization objectives of particular physical education programs.

A. THE FAMILY

The first social system and thus the initial agent of socialization to which the newborn infant is exposed is that of the family. It is in the context of the family that an individual's fundamental concept of self is largely molded. In this section an attempt is made to show what function play fulfills in this particular process of socialization.

1. Development of Self

Mead (1934), in his eminent work titled *Mind, Self, and Society,* described in detail how an individual obtains a full development of self. In his theoretical treatment of the genesis of self he discussed two general sets of background factors and two general stages in the full development of self. The first set of background factors concerns the conversation of gestures between animals, and the acquisition of language and consequent communication among humans. The second set of background factors "is represented in the activities of play and the game." Mead (1934) depicted the two general stages of the development of self as follows:

> At the first of these stages, the individual's self is constituted simply by an organization of the particular attitudes of other individuals toward himself and toward one another in the specific social acts in which he participates with them. But at the second stage in the full development of the individual's self, that self is constituted not only by an organization of these particular attitudes, but also by an organization of the social attitudes of the generalized other or the social group as a whole to which he belongs.

Play primarily relates to the first stage of development and games to the second stage of development in the genesis of self-conceptualized by Mead. In play, the child takes on and acts out roles which exist in his immediate, but larger, social world. By acting out such roles he organizes particular attitudes about them. Moreover, the child in the course of role playing becomes cognitively capable of "standing outside himself" and formulating a reflected

* The phrase "agonistic activities" is used in this chapter to denote those ludic activities which are essentially competitive in nature, e.g., games and sports.

view of himself as a social object separate from but related to others. In contrast, a child in a game situation must be prepared to take on the role of every player in the game. He must sense what all other players are going to do in order to make his own particular plays. Mead (1934) summarily stated that "The game is then an illustration of the situation out of which an organized personality arises. In so far as the child does take the attitude of the other and allows that attitude of the other to determine the thing that he is going to do with reference to a common end, he is becoming an organic member of society."

2. Stages of Early Childhood Play

The influential role of play with respect to personality development in particular and the socialization process in general can first be seen in infancy, and its effects are well illustrated throughout early childhood. Erikson (1963) recognized three stages of infantile play which he labeled "autocosmic play," "microcosmic play," and "macrocosmic play." During the period of autocosmic play the child centers attention on his own body and play ". . .consists at first in the exploration by repetition of sensual perceptions, of kinesthetic sensations, of vocalizations, etc." Play in the microsphere is confined to "the small world of manageable toys." "Finally, at nursery school age playfulness reaches into the macrosphere, the world shared with others."

3. Autocosmic Play

Several recent studies by Call (1964, 1965, 1968, 1970; Call and Marschak, 1966; Work and Call, 1965) show the importance of autocosmic play in early ego development. Call (1968) observed that "The capacity for social play in infancy is developed only in the presence of a reciprocating party . . . impressions (ideas and feelings) about the self are influenced from the beginning by the mother's contribution to play in infancy and the capacity to play is crucially dependent on being played with." He further noted that "In the infant's play there is obvious utilization of congenital ego equipment. There is a testing of this equipment, there is the exploration of environmental responses, a survey of environmental responses, and under special conditions a selection of certain reciprocal experiences which become institutionalized as a game."

An example of a type of play experience which is an early precursor to full game behavior is the playful interaction within the dyadic social system of mother and child in feeding situations. For instance: "Near the end of the feeding the mother often tests his hunger by withdrawing the nipple to see if he will go after it again. Or the infant may let go and turn away from the nipple then grab it again and suck or chew lightly without getting milk" (Call, 1970). Similar kinds of play experiences include various forms of lap and finger play in infancy (Call, 1968). These types of play experiences provide the foundation for the construction and conduct of relatively complex games between the child and either parent and or sibling within the wider social system of the whole family. In turn, such games assist the child in his struggle for self-identity and aid him in his resolution of inner conflicts arising from basic anxieties. The traditional game of "peek-a-boo," for example, ". . .gives the infant practice at separating himself from someone else, then experiencing delightful reunion" (Call, 1970). The game of peek-a-boo can be considered a way for the infant to master anxiety when the mother is absent, i.e., mastery of object loss (Call and Marschak, 1966). The major social function of low order games in the sphere of auto-cosmic play is perhaps revealed in Erikson's theory (1963) that ". . .child's play is the infantile form of the human ability to deal with experience by creating model situations to master reality by experiment and planning."

4. Microcosmic Play

Social scientists have virtually ignored the microsphere of play and its realm of dolls and toys. This state of affairs is both surprising and dismaying in view of the socially significant nature of microcosmic play. The underlying social import of play in the microsphere is clearly pointed out in Ball's (1967) pronouncement that:

> Almost from the moment of birth and for many years thereafter, children are in frequent and intense contact with toys in great variety, of diverse type, complexity, and composition. These toys are, from a very early age, an important part of the child's experiently perceived reality, operating in several related ways over and above their more manifest recreational purpose. For example, they function as socializing mechanisms, as educational devices, and as scaled down versions of the realities of the larger adult-dominated social world.

Ball further suggested that toys serve the socialization process in two major ways. First, toys function as role rehearsal vehicles for the practice of role-associated activities. Second, toys act as role models by representing diverse social categories. This second function is exemplified in the following quotation from Sessoms (1969):

> The small girl playing with her doll, acting as the mother of Raggedy Ann, provides a case in point. Her caring for the doll may be viewed as preparation for the traditional role of mother-homemaker. She performs the tasks as she

has observed her mother perform them. The values and concepts of mother-hood are being internalized through the play process just as the boy participating in baseball is learning to conform to the rules of the group and assumes the role assigned to him by that body.

In addition to providing a means of role rehearsal and the learning of important cognitive styles of behavior, toys ". . .also serve at the same time as surrogates for human companionship, as important targets for affective expression" (Ball, 1967). Herein one calls immediately to mind the "security blanket" sported by Linus in *Peanuts,* the popular cartoon series drawn by Charles M. Schulz. Notwithstanding this humorous example, one should not lightly dismiss the importance of early affective attachments to playthings. For as Call (1970) showed in his citation of a psychiatric case study:

> Infantile attachments can remain strong and influential even into adult life. A 40-year-old woman talked in analysis about her "wag-a-dollie," a little cloth doll she associated with her father. While she was 21 and away at college, her mother burned wag-a-dollie. When she learned of this she became confused and depressed, even dropped out of college for awhile. She tried a series of substitute transitional objects, and even called her doctor "Doll" for a time. It was possible for her, by tracing the origins of wag-a-dollie, to recapture the feelings of her earliest attachment to her father, which helped her to understand the nature of the difficulties she had been having both in her marriage and in the raising of her children.

In sum, the microsphere of play provides a vast universe of symbolic discourse which serves the socialization process in many diverse ways both manifest and latent, conscious and subconscious, temporarily and permanently.

5. Macrocosmic Play

When a child reaches the state of macrocosmic play he enters into a play world shared by others and becomes involved in a world characterized by social drama. As Stone (1965) pointed out, drama is ". . . basic in the child's development of a conception of self as an object related to, yet different from, other objects. Drama is a vehicle for the development of identity." Stone (1962, 1965) suggested that there are two basic modes of socialization associated with drama which he termed "anticipatory socialization" and "fantastic socialization." In the first case, the child plays roles that he will likely perform in later life; for example, the child may play at being father, salesman, customer, etc. In the second case, the child plays roles which he is highly unlikely to perform in later life; for example, a child may pretend to be a king, a cowboy, or a space cadet. Although anticipatory drama appears to hold greater import for later life experiences, fantastic drama is not without functional importance since it ". . .often serves to maintain and keep viable the past of society—its myths, legends, villains and heroes" (Stone, 1965).

A third form of early childhood play associated with drama is what Stone (1965) referred to as "childish tests of poise." Stone observed that:

> It is not enough only to establish an identity for one's self it must be established for others at the same time. Identities are *announced* by those who appropriate them and *placed* by others. Identities must always be validated in this manner to have reality in social interaction.

It is suggested that childhood play related to tests of poise affords a suitable medium whereby a child may learn in a rudimentary manner the arts and techniques of "impression management" so crucial to role performance in adult careers. Goffman (1959), in his major work *The Presentation of Self in Everyday Life,* discussed in depth the arts of impression management. His dramaturgical perspective posits that each individual in the social intercourse of everyday life presents himself and his performances to others and attempts to influence the impressions others form of him by employing various techniques which help sustain his performance in a manner not unlike the actor's portrayal of a particular character before an audience. Goffman discussed the arts of impression management in terms of personal performances, team performances, and regions of behavior. He used the term "performance" to refer to all the activity of an individual which occurs during a period marked by his continuous presence before a particular set of observers and which has some influence on the observers. The two major elements of a "performance" are what Goffman referred to as "the setting" and the "personal front." The setting consists of the physical environment in which the performance takes place and includes any expressive equipment required for the performance. Personal front may include such things as "insignia of office or rank; clothing; sex, age, and racial characteristics; size and looks; posture; speech patterns; facial expressions; bodily gestures; and the like." Goffman dichotomized the stimuli which make up the personal front into "appearance" and "manner." Manner refers "to those stimuli which function at the time to warn us of the interaction role the performer will expect to play in the oncoming situation." Appearance refers to "those stimuli which function at the time to tell us of the performer's social statuses."

The relationships between appearance, play, and self have been treated at length by Stone (1965). He showed how "child's play demands costume and body control, and is facilitated by props and equipment (toys) appropriate to the drama." Stone (1962) especially emphasized the importance of clothes in the meaning of appearance. He spoke of "investure" during the preplay period, "dressing out" in the play period, and "dressing in" as related to the game stage of early socialization. An example of investure is a mother dressing her son in blue clothes and her daughter in pink in order to denote masculinity and femininity, respectively. Dressing out involves "costume" and is associated with acting out a social role in another's clothing (e.g., playing housewife in one of mother's old dresses or shoes). Dressing in requires a uniform and is related to a "real" identity (e.g., a boy wearing a Little League baseball uniform).

By way of summary, it may suffice to state that macrocosmic play serves the socialization process by the fact that "much of the drama of childhood replicates the interaction of the larger society in which it occurs" (Stone, 1965).

B. The Peer Group and the School

Following early childhood socialization experiences within the family, the peer group and the school are the next significant social systems to impinge upon the child. Although these two social structures are analytically distinguishable, they are often, if not usually, empirically intertwined in American society. Therefore, these two agencies of socialization are treated together in this section.

1. Types of Peer Groups

The earliest and most informal peer grouping to emerge is the *play group*. Children become involved in such groups as early as the third or fourth year of life and continue to interact in play groups during the early school years. Two chief characteristics distinguish play groups from other kinds of peer groupings: First, "The choice of playmates is relatively restricted, in kind and number"; second, the play group is the child's *"first* introduction to a group which assesses him as a child from a child's point of view, and teaches him the rules of behavior from the same point of view" (Bossard and Boll, 1966).

Later age peer groups which the child usually enters between the ages of 8 and 12 are more formalized than play groups. They have a more complex social structure, a greater degree of permanency, and more rigid requirements for membership. According to Bossard and Boll (1966) there are two major types of later age peer groups, the *clique* and the *gang*. "A clique may be defined as a small, intimate social participation group consisting of persons of the same social status and in agreement concerning the exclusion of others from the group." A gang is a more formal group than a clique, less exclusive, more permanent in nature, and usually has an identifying subculture marked by nicknames, symbols, slogans, passwords, particular dress styles, etc.

While several kinds of peer groups can be differentiated for special purposes of analysis, all forms of peer groups share three characteristics in common (Havighurst and Neugarten, 1957). First, unlike the adult world where a child occupies a subordinate status position, in his peer world he holds an equal status with others. Second, the interpersonal relationships among peers in any given group tend to have a transitory quality. While peer relationships and friendships are often intense in nature, they are also often of short duration. This is especially true in childhood and preadolescence. Third, the influence of the peer group increases as children grow older.

The peer group as a socializing agent fulfills many functions. But perhaps the three primary socializing functions performed by the peer group are teaching the culture, teaching new social roles, and teaching social mobility (Havighurst and Neugarten, 1957).

2. Teaching the Culture

While each particular peer group has its own special subculture, every peer group nevertheless tends to reflect in important ways the culture of the society at large. Perhaps the most critical cultural content imparted by the peer group are codes of moral conduct. "A child learns through his peers the prevailing standards of adult morality—fair play, cooperation, honesty, responsibility—that, while they may at first be child-like versions, become adult-like with increasing age" (Havighurst and Neugarten, 1957).

Undoubtedly, the most famous description of the role of play in the moral development of the child is contained in Piaget's (1965) classic work, *The Moral Judgment of the Child*. In the first part of his book, Piaget described how children gradually develop a mature understanding of the rules of a game. His account is based upon a series of observational studies of children engaged in the game of marbles which Piaget characterized as a microcosmic moral system.

Piaget focused his attention on two particular phenomena associated with game rules: first, the *practice* of rules; second, the *consciousness* of rules. With respect to the practice of rules, Piaget (1965) distinguished three kinds of behavioral patterns which appear in successive stages; namely, motor behavior, egocentric behavior, and cooperative behavior. Corresponding to these three types of behavior are, according to Piaget, three kinds of rules. "There is the *motor rule,* due to preverbal motor intelligence and relatively independent of any social contact; the *coercive rule* due to unilateral respect; and the *rational rule* due to mutual respect." As succinctly summarized by Berlyne (1969): "Competitive games thus exemplify for Piaget the advance from a view of morality characterized by 'heteronomy' and 'moral realism,' based on one-sided respect for persons in authority and belief in immutable moral laws comparable to the laws of nature, toward a mature 'autonomous' moral sense, rooted in mutual respect among equals, and capacity for cooperation."

Although recent research reveals that moral judgments made by children do not always conform to Piaget's analysis (Berkowitz, 1964; Hoffman, 1970; Kohlberg, 1963b), his investigations nevertheless offer significant insights into the role of games for "internalization" of moral values. Moreover, recent empirical findings strongly support the important implication of Piaget's work that play patterns and attitudes can be used as indices of socialization (Herron and Sutton-Smith, 1971; Seagoe, 1969; Webb, 1969).

An excellent empirical illustration of play as an index of socialization as well as an example of how cultural content is transmitted through play is Webb's (1969) study of the professionalization of attitudes toward play among adolescents. Webb observed that "The transition from 'child's play' to games, and then to sport, involves increasing complexity and rationalization of the activities and increasing professionalization of attitudes." By professionalization Webb meant ". . . the substitution of 'skill' for 'fairness' as the paramount factor in play activity, and the increasing importance of victory."

The findings of Webb's investigation are based on questionnaire responses from random samples of students enrolled in the public and parochial schools of Battle Creek, Michigan in 1967, stratified by grades (three, six, eight, ten, and twelve). At each grade level he asked students to rank in order of importance what they thought was personally most important in playing a game: to play it as well as you can, to beat your opponent, or to play the game fairly. Results of his study show the diminishing importance of the fairness factor and the increasing importance of the "success" factor (i.e., beat one's opponent) as age increases.

Webb (1969) drew a number of interesting parallels between the business and economic institutions of Western society, emphasizing the work-oriented values of equity, skill, and success, and the world of sport, emphasizing the play-oriented values of fairness, skill, and victory. He suggested that sport is a reflection of society in that ". . . it provides in fact in one institution what is essentially ideology in another." In conclusion, Webb (1969) commented:

> Clearly, if nothing else, this investigation demonstrates that participation in the play world is substantially influential in producing that final result, the urban-

industrial man. Although it is true that play attitudes, as demonstrated, are extensively influenced by other factors, it is the final isomorphism of the play arena to the economic structure, and the fact of participation in it, at a time when participation in other areas is virtually nonexistent, that makes that participation the significant factor it now appears to be. Thus to continue the sophomoric and even moronic insistence on play's contribution to the development of such "sweetheart" characteristics as steadfastness, honor, generosity, courage, tolerance, and the rest of the Horatio Alger contingent, is to ignore its structural and value similarities to the economic structure dominating our institutional network, and the substantial contribution that participation in the play arena thus makes to committed and effective participation in that wider system.

In summary, the work of Piaget and Webb indicate how play, games, and sport in peer group and school settings "fit the child to his society" through relatively indirect and nonobtrusive means of socialization.

3. Teaching New Social Roles

Associated with the transmission of the general culture by the peer group is the teaching of new social roles. The context of the peer group provides opportunities for youth to occupy a variety of statuses and social positions and offers a suitable situation for children to try out various behavioral styles. Since sports and games are an almost universal feature of peer groups, it can be assumed that they play an important part in the teaching of new social roles. That such is the case is shown by Helanko (1958, 1963, 1969) in his several studies of the developmental pattern of sports participation among Scandinavian children.

Helanko spoke of the "expansion of socialization" and described how children first learn to interact in pairs, then in larger primary groups, followed by small secondary collectivities, and finally by still larger social systems. He showed how children via sport involvement at different age levels are exposed to new social norms and roles as a result of interacting with peers in progressively more complex social systems.

In his examination of the interrelationships between the peer groups, sport, and socialization, Helanko described the socialization process in terms of three stages which he refers to as the pre-gang period, the gang period, and the post-gang period. These stages of socialization are outlined in Table I and contrasted with Piaget's stages of moral development. Sports participation within the gang serves two basic functions according to Helanko. First, sport involvement serves as a source of pleasure. Second, sports participation acts as a means of status definition. As Helanko (1963) observed: "Sports and the gang together constitute the social milieu in which, for the first time in his life, the boy is called upon to create a social position for himself among his equals."

TABLE I

STAGES OF SOCIALIZATION[a]

| Ages (years) | Stages in learning how to play marbles, according to Piaget | | Stages in the process of socialization among boys, according to Helanko |
	Practice of rules	Consciousness of rules	
1			
	Stage I	Stage I	Yard-aggregation-stage
2	Motor behavior of	No comprehension	Egocentricity
	an individualistic	of rules	Primitive pairs appear
3	nature		No actual groups are formed
			No sports
4			
	Stage II		
5	Egocentricity		
		Stage II	Play-gang stage
6		Rules are regarded	Primitive groups are formed
		as "sacred" and	Weak solidarity
7		absolute	Primitive sports
8			
	Stage III		First period of gang age
9	Cooperation		Solid groups
			Strong in-group feeling
10			Centripetal interaction
			Clubs are formed within the
11			gang
			Sports (ballgames) appear
12	Stage IV	Stage III	Second period of gang age.
	Codification of rules	Rules are regarded	At the outset of this
13		as relative	period solid groups still
			appear
14			At the termination of this
			stage interaction becomes
15	Termination of in-		centrifugal
	terest in marbles		Gangs and gang clubs begin
16			to dissolve
			Interest in sports continues
17			but becomes more individ-
			ualistic
18			The gang age terminates
			Individuals who have left
			their gangs form pairs
			which aggregate

[a] Adapted from Helanko (1963).

Helanko and other investigators who have attempted to analyze the relations between sport and socialization in the peer group usually have not been very specific about what particular kinds of social roles are learned in the course of play with peers. It is evident, however, that one of the most significant social roles largely learned in the peer group is an individual's sex role. Although gender is socially assumed at birth, and while sex role learning is an important part of the familial socialization process, peer groups extend and elaborate earlier sex role acquisition in profound ways. As expressed by Havighurst and Neugarten (1957): ". . . the peer group is a powerful agency in molding the behavior of males and females in accordance with current American versions of manhood and womanhood." But, as Tryon (1944) took care to point out in her discussion of adolescent peer culture: "It is a long, complex, and often confusing learning-task to achieve manhood or womanhood in our society with the skills and behaviors, the attitudes and values appropriate to the role which a given individual must take. For the most part boys and girls work at these tasks in a stumbling, groping fashion, blindly reaching for the next step without much or any adult assistance."

Tryon's comments concerning the difficulty which adolescents experience in achieving an adequate concept of masculinity or femininity while made in 1944 nevertheless seem particularly relevant today as the mass media's exclamatory treatment of such topics as "unisex," "women's liberation," and the "gay liberation front" has brought once again the whole matter of sexual identity to the fore. Winick (1968), for example, treated at length the process of desexualization in American life in his recent book titled *The New People*. He suggested that current modifications of sex roles may be "intimately related to our society's ability to survive." Winick further forecasted that "Archeologists of the future may regard a radical dislocation of sexual identity as the single most important event of our time."

Regretfully, "sex-role behavior is one of the least explored areas of personality formation and development" (Brown, 1956). Thus, while it is commonly assumed that boys and girls learn appropriate sex role behavior as a by-product of playing games in their peer groups (Brown, 1958; Gump and Sutton-Smith, 1955; Rosenberg and Sutton-Smith, 1960; Rosenberg and Sutton-Smith, 1964), relatively few empirical efforts have been made to determine the degree of learning and even fewer investigations have been made to determine the specific means whereby such learning takes place. Therefore, in view of the present state of research regarding sex-role socialization, one can at best only examine revealed sex differences in play activities and speculatively infer what socialization processes are associated with them.

One of the more extensive examinations of sex differences in play choices is Sutton-Smith's and Rosenberg's work (1961) titled "Sixty Years of Historical Change in the Game Preferences of American Children." Their investigation is based on the comparative analysis of similar studies of game preferences of children conducted in 1896, 1898, 1921, and 1959. One of the most marked findings of their investigation is the fact that the game preferences of girls have become increasingly like those of boys since the turn of the century. They noted that this finding was not unexpected "in light of the well known changes in woman's role in American culture during this period of time." They were surprised, however, to find that "boy's play roles have become increasingly circumscribed" and that girls currently engage in a wider variety of play activities than boys. They stated that "there is little doubt that boys have been steadily lowering their preference for games that have had anything to do with girl's play"; but they noted that this finding may be interpreted in various ways. On the one hand, "it contributes to clear-cut role definition of appropriate boys' behavior and perhaps facilitates the development of those boys who have particular skills required by the games that are in demand. On the other hand, it must as surely penalize those many other boys who find that there is a discrepancy between their own abilities and those required in the play roles of their own age sex category." In short, the findings of Sutton-Smith and Rosenberg implicitly support the results of other studies (Cratty, 1967) which indicate that there is less social stigma attached to the display of masculine behavior by girls than to the show of feminine behavior by boys.

Several social observers have commented on the fact that many boys have difficulty in achieving appropriate sex-role identification since they often lack exposure to male role models (Parsons, 1942, 1947). And a number of articles and books both popular and scholarly have expressed concern about the perceived decline of masculinity in American society (Winick, 1968). An extended treatment of these issues is given by Sexton (1969) in a book titled *The Feminized Male,* and subtitled *Classrooms, White Collars and the Decline of Manliness.*

Sexton (1969) persuasively argued that our school systems are effectively emasculating adolescent males. She pointed out that 68% of the teachers in our public schools are women and contends that women teachers set the standards for adult behavior, favoring those who most conform to their own behavior norms. She further observed that:

> The feminized school simply bores many boys; but it pulls some in one of two opposite directions. If the boy absorbs school values, he may become feminized himself. If he resists, he is pushed toward school failure and rebellion. Increasingly, boys are drawn to female norms.

Sexton presented a good deal of empirical evidence to buttress her case. Primary data in her investigation consist of school records and interview and questionnaire responses of 1000 ninth grade boys and girls in a community which she refers to as Urbantown. Secondary data in her study consist of records of academic achievement and deviant behavior for all of Urbantown's 12,000 students. In addition, her data for ninth grade boys in Urbantown are compared with data from a national survey of ninth graders. Sexton focused her basic analyses on achievement and masculinity. High achievers were operationally defined as students with all A's on their report cards, middle achievers were those with mostly B's and C's, while low achievers were those with mostly D's and F's. Masculinity was assessed by means of Gough's masculinity scale from the California Personality Inventory.

Comparing her measures of masculinity and academic achievement, Sexton (1969) found that (1) "The less masculine boys had better marks in most school subjects"; (2) "Only in physical education and science did boys with middle masculinity scores tend to get the best marks"; and, (3) "the most masculine boys usually received their worst grades in English."

Having determined a relationship between masculinity and academic achievement, Sexton (1969) attempted to assess the relationship of sports to scholastic success and masculinity. In general she found that:

> In all sports except tennis, bowling, and volleyball (a popular sport with girls), the interest of the most masculine boys exceeds that of the least masculine boys. The pattern of interest tends to follow that of high- and low-achieving boys. The interests of low achievers are most like those of the highly masculine, and most unlike the interests of girls.

In particular, Sexton discovered that high achievers prefer sports stressing individual performance and involving minimal aggression and body contact, whereas low achievers prefer team sports of a physically demanding nature. Exceptions to the latter finding are the individualistic sports of boxing and pool which Sexton found to be very popular among low-achieving, masculine boys and disliked by high-achieving, less masculine boys. Finally, with respect to sport involvement Sexton reported that:

> Low achievers were more likely to say they had been on a school athletic team, and that they regarded it as important to be among the first in sports (52 percent of low and 43 percent of high achievers felt it was *very important* to be first in sports). Highly significant differences were found in the time boys spend playing sports. Low achievers spend far more time on sports, both on weekend and weekdays (50 percent of low, 38 percent of middle, and 15 percent of high achievers spend *three hours or more* each weekday on sports).

Girls spend very little time on sports. About one in three said she never or rarely played sports on weekdays, and about half spend less than a half-hour. In their aloofness from sports and their general use of leisure, high-achieving boys are more like girls than like other boys.

Sexton went on to outline the pros and cons of the place of sports in the schools. She forcefully concluded that "If sports are replaced by greater stress on the sedentary and passive academic work, the masculine quality of our males, already weakened, may collapse. Other masculine activities may be found, but they may not be able to carry such a heavy load."

4. Teaching Social Mobility

Related to the teaching of new social roles is the teaching of social mobility by the peer group. As Havighurst and Neugarten (1957) have suggested: "A lower-class boy or girl who, through an organized youth group or through the school, becomes friendly with middle-class boys and girls learns from them new ways of behaving. He may be encouraged to acquire the values and goals of his new friends, and this may eventuate in his moving up from the social position of his family." The importance of teaching social mobility is underscored by the fact that in our high schools "27 percent of boys who are in the *highest ability quartile* but the *lowest socioeconomic quartile* never go to college"; and, "in the *top ability half,* 82 percent of the poorest boys and 20 percent of the richest" do not go on to any kind of higher education (Sexton, 1969).

Few efforts have been made to determine the degree to which lower-class students informally interact with middle-class students in school and peer group settings, and even fewer investigations have been conducted to determine the social consequences of such interaction. On the basis of research to date one is led to conclude that boys and girls of diverse social strata interact to a very limited degree in our senior high schools. For example, Hollingshead (1949) in his study of *Elmtown's Youth* examined the clique relations of adolescent boys and girls enrolled in Elmtown High School. He stratified the students according to the socioeconomic background of their families into five prestige classes. Hollingshead reported that:

Analysis of the 1,258 clique ties when the factor of school status is ignored reveals that approximately 3 out of 5 are between boys or girls of the same prestige class, 2 out of 5 are between adolescents who belong to adjacent classes, and 1 out of 25 involves persons who belong to classes twice removed from one another.

Findings such as those of Hollingshead implicitly indicate that interscholastic sports may play a role in teaching social mobility since one might expect greater interaction to occur between lower-class and middle-class students in sport situations than in other types of peer settings in the school community. Data in support of this assumption are presented in a study by Schafer and Armer (1968) wherein they analyzed a number of relationships between athletic participation and academic achievement. Their study is based upon data obtained from the school records of 585 boys in two Midwestern high schools. They classified 164 (28%) of the boys in the sample as athletes, and they matched each athlete with a nonathlete in terms of intelligence test scores, occupational status of fathers, type of high school curriculum, and grade point averages (G.P.A.s) for the final semester of junior high school. Schafer and Armer (1968) found that:

> More than half of the athletes in each category exceed their matches, and the average G.P.A.s of athletes is always higher than that of their matched non-athletes. Moreover, on father's occupation and curriculum, the gap is greater between athletes and their matches in the *lower* categories than in the higher. For example, greater percentages of blue-collar athletes than white-collar athletes exceed their matches in G.P.A.s (63.0 percent versus 53.7 percent). An even greater spread separates non-college-preparatory athletes from college-preparatory athletes (69.0 percent versus 53.7 percent). In short, the boys who would usually have the most trouble in school are precisely the ones who seem to benefit most from taking part in sports.

It is, of course, a yet-unanswered question as to why sports participation apparently has its greatest positive effect on the academic performance of blue-collar athletes. Schafer and Armer stated that "A plausible interpretation of this finding is that, compared to nonathletes with the same characteristics, blue-collar and non-college-bound athletes are more likely to associate and identify with white-collar and college-bound members of the school's leading crowd."

Another type of educational achievement examined by Schafer and Armer is graduation from high school. Their findings indicate that participation in interscholastic sports has a "holding influence" on students. They report that "whereas 9.2 percent of the matched nonathletes dropped out of school before graduating, less than one-fourth as many (2.0%) of the athletes failed to finish." Since social mobility is in part a function of high school graduation, it can be assumed that sport serves this process to some degree.

Although there are not as yet any systematic studies of the effects of athletic participation and upward mobility, several recent investigations report interesting results regarding the relationships between sports participation, educational expectations, and educational achievement (Bend, 1968; Rehberg and Schafer, 1968; Rehberg *et al.* 1970; Snyder, 1969; Spreitzer and

Pugh, 1971). For example, Rehberg and Schafer discovered a positive correlation between athletic particpation and educational expectations in their analysis of data obtained from 785 senior males in six urban Pennsylvania high schools in the spring of 1965. Moreover, they found that the relationship was strongest toward a college education. For instance, while only 36% of working class nonathletes stated that they expected to complete at least 4 years of college, 55% of the working class athletes reported that they planned to finish at least 4 years of college.

Longitudinal analyses of the relationship between perceived educational expectations and actual educational achievement (Bend, 1968; Snyder, 1969) imply that there may be long-term positive effects accruing from interscholastic athletic participation. Snyder made a 5-year follow-up study of the 1962 high school graduating class of the only high school in a Midwestern community of 38,000. Results of his study show that high school social participation (including sports participation) is positively correlated with both high school and post-high school educational achievement. In addition, findings of his investigation reveal a positive association between high school social participation and occupational status 5 years after graduation.

Snyder's findings at the local level are supported by the results reported by Bend (1968) at the national level. On the basis of data drawn from the Project Talent data bank, Bend made an extensive examination of the social correlates and possible consequences of interscholastic athletic involvement. The sample for his study consisted of approximately 14,000 senior males who completed questionnaires for Project Talent in 1960 and who returned follow-up questionnaires mailed to them in 1965. For purposes of analysis Bend designated four types of athletic groups, ranging from nonathletes to superior athletes. A superior athlete was defined as a student who was a member of more than four teams and who received more than two athletic awards during his high school years. In addition, each subject was classified into high or low endowment categories on the basis of a nine-item socioeconomic environment index. Among low-endowed subjects Bend found that superior athletes in comparison with nonathletes held greater educational, financial, and occupational expectations in 1960 and had achieved greater educational and occupational status by 1965. As an example, although only 6.2% of the nonathletes in the low endowment category attended college as full time degree students, 17% of the superior athletes in the lower socioeconomic group attended college as full time students.

As a final illustration of the longitudinal analysis of the relationship between athletic participation and academic variables related to upward mobility, reference is made to the very recent study of Rehberg et al. (1970). Their investigation is based on data collected from a sample of 1170 high school males surveyed during both their freshman and sophomore years in seven public and parochial, urban, and suburban schools in southern New York.

> In comparisons with non-athletes, the varsity and junior varsity participants were found: (1) to have higher educational aspirations and expectations, both as freshmen and as sophomores, (2) to be more likely to increase their educational goals from two years of college or less to four years of college or more during the one-year interim, (3) to spend as much time on homework, (4) to value academic competence, (5) to receive more advice from teachers and guidance counselors to enroll in a four-year college, and (6) to be as conforming to the official normative structure of the school.

Rehberg and associates stated in conclusion:

> . . . our data do indicate a difference between athletes and non-athletes with respect to scholastic pursuits and the difference more often than not is a positive one favoring those who participate in interscholastic sports. Furthermore, our data lend themselves to the interpretation that at least a portion of this difference is a function of certain achievement-relevant socialization experiences encountered by the athlete but not by the non-athlete. It is toward the identification and measurement of these socialization experiences and their separation from associated selection variables that further research should be dedicated.

C. SOCIALIZATION AND SOCIETY

By centering our attention on only the family, peer group, and school we have no doubt inadvertently given the reader a rather simplistic overview of the socialization process since there are a number of other socializing agencies in society which impinge in important ways upon the family, peer group, and school. Within a given community, for example, there are such additional agencies as ethnic groups, neighborhoods, religious associations, social classes, and various forms of the mass media. The complexity of the socialization process is further compounded at the societal level by significant cross-cultural differences. Regretfully, space limitations preclude a detailed discussion of the relationships between agonistic activities and socialization processes associated with subcultural variations at the community level and cross-cultural variations at the societal level of analysis. Thus, it may suffice to present in conclusion a speculative summary of the dialectic between individuals and society in terms of game involvement.

The fundamental problem of social order is the Hobbesian puzzle of how every individual can intensely pursue his own self-interests without society having an internal war of all against all. Games of children and youth well illustrate the Hobbesian puzzle and afford many key insights into the nature of social order in society. "In particular, they demonstrate the process by which institutions with their elements of cooperation and morality and their concepts of right and justice can emerge from the actions of an originally unorganized aggregation of individuals each selfishly seeking to maximize his own personal satisfactions" (Lenski, 1966).

On the one hand, by providing a source of personal pleasure, games serve to express the needs of the people who choose to play them. On the other hand, games serve as models of cultural activities. Games are an important subset of a larger class of expressive models, including art, dance, drama, folktales, and music.

> They provide a way of teaching people in a society (particularly the young) some ways of getting important things done; they also provide a kind of therapy, in that individuals who are in conflict about getting their actual cultural work done can live for a time in an easier fantasy world of expressive models and evade the world of those things which are modelled until their feeling of conflict passes (Lambert and Lambert, 1964).

The findings of the several studies conducted by Roberts and Sutton-Smith (1959, 1962, 1966) offer a number of indications as to how games model the maintenance problems of societies, provide significant socialization situations, and act as a means of assuaging intrapersonal conflicts. In an early investigation, Roberts et al. (1959) established a threefold classification of games based on the major modes of game outcome. Their triparite game typology consists of (1) games of physical skill, (2) games of strategy, and (3) games of chance. Having established these basic game categories, Roberts et al. then determined the distribution of game types in 50 tribal societies and made an exploratory examination of the relations between the predominant game form of given societies and certain cultural characteristics. They discovered that (1) games of strategy are related to cultural complexity as assessed by degree of political integration and social stratification, (2) games of chance are associated with religious beliefs and matters of the supernatural, while (3) games of physical skill may be related to environmental conditions.

In a second study, Roberts and Sutton-Smith (1962) extended the previous cross-cultural analysis into the area of childhood socialization. Specifically, they studied the relationships between game types and child-training variables at two levels of generality. First, at the intercultural level, the investigators examined data from the Cross-Cultural Survey Files and Human Relations Area Files at Yale University for 56 tribal societies. Second, at the intracultural level, the investigators compared the game preference of 1900 elementary school boy and girls in 12 Midwestern communities. Analysis of the cross-cultural data showed that games of strategy are related to obedience training, games of chance are related to responsibility training, and games of physical skill are related to achievement training. Subsystem validation of these cross-cultural findings with data obtained from Midwestern children confirmed the researchers predictions that (1) girls with their higher training in obedience should show a greater preference for games of strategy than boys, (2) girls with their higher training in responsibility should show a greater preference for games of chance than boys, and (3) boys with their higher training in achievement should show a greater preference for games of physical skill than girls. Roberts and Sutton-Smith (1962) interpreted these findings in terms of a *conflict-enculturation model* which implies:

> (1) that there is an over-all process of cultural patterning whereby society induces conflict in children through its child-training processes; (2) that society seeks through appropriate arrays and varieties of ludic models to provide an assuagement of these conflicts by an adequate representation of their emotional and cognitive polarities in ludic structure; and (3) that through these models society tries to provide a form of buffered learning through which the child can make enculturative step-by-step progress toward adult behavior.

The causal assumptions of Roberts and Sutton-Smith's conflict-enculturation model are schematically outlined in Fig. 1. Roberts and Sutton-Smith noted that the assumptions and hypotheses underlying their model have not as yet been fully validated. However, they observed that their model does provide

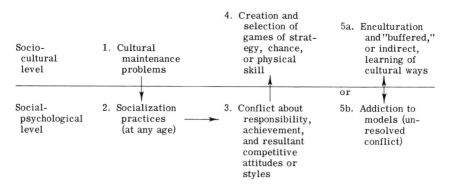

Fɪɢ. 1. Conflict-enculturation model. Reprinted from Lambert and Lambert (1964).

an explicit theoretical foundation for future research. And in a third study with American adults they replicated their research as outlined with children (Sutton-Smith *et al.*, 1963). On the basis of a comparative analysis of three state and national survey polls made in 1940 and 1948, they reported that (1) games of strategy are preferred by higher status groups as compared with lower status groups, and by females as compared with males; (2) games of chance are associated with lower status rather than high status categories, and with women in contrast to men; and, (3) games of physical skill are preeminent among upper as compared with lower status groups, and are held in greater preference by men than women (Sutton-Smith *et al.*, 1963).

Finally, in yet a fourth study, Roberts and Sutton-Smith (1966) examined the cross-cultural correlates of games of chance with particular reference to their conflict-enculturation model. They presented and discussed a wide variety of findings, but stated by way of conclusion that:

> It is our provisional formulation, then, that games of chance are linked with antecedent conflict, powerlessness in the presence of uncertainty, the possibility of both favorable and unfavorable outcomes within the area of uncertainty, and certain compatible projective beliefs, particularly in the area of religion. The motivations produced in this situation are assuaged by play with uncertainty models, and the resulting learning may give individuals and groups strength to endure bad times in the hope of brighter futures.

In sum, Roberts and Sutton-Smith's conflict-enculturation model of game involvement affords many exciting insights into the relationships between dominant game categories in a culture or subculture and selected socialization practices. The model as well provides a theoretical integration of diverse findings from several studies and gives explicit leads and directions for future research.

However, notwithstanding the several studies made by Roberts and Sutton-Smith, no definitive experiments have been conducted to date to test critically the hypotheses underlying the conflict-enculturation model. Perhaps the least investigated aspect of the Roberts and Sutton-Smith model is the notion that individuals, who are unable to resolve strong psychological conflicts through buffered learning in the normal course of game involvement, become addicted to certain games and thus remain obsessively involved in game play. Although research is limited, some empirical support can be found for Roberts and Sutton-Smith's assumptions regarding game addiction. For example, ". . . preliminary work points up the probability that college students who are addicted to such games of strategy as poker, even to the point of playing the game as much as 40 hours a week, are laboring under a particularly strong set of conflicts about their positions in the social system" (Lambert and Lambert, 1964). More in the realm of sport is Nicholi's (1970) work on the "motorcycle syndrome" which well illustrates the relationships between inner psychological conflicts and involvement in a motor sport. On the basis of in-depth psychiatric study of accident-prone student motorcyclists, Nicholi outlined the essential characteristics of a motorcycle syndrome as follows:

1. Unusual preoccupation with the motorcycle.
2. A history of accident-proneness extending to early childhood.
3. Persistent fear of bodily injury.
4. A distant, conflict-ridden relationship with the father and a strong identification with the mother.
5. Extreme passivity and inability to compete.
6. A defective self-image.
7. Poor impulse control.
8. Fear of and counterphobic involvement with aggressive girls.
9. Impotence and intense homosexual concerns.

Nicholi described how the motorcycle functions as a powerful emotional prosthesis and illustrated his description with the following colorful report of a patient:

> You sit as tightly on it as you can . . . and all of a sudden it responds to you. It's a throaty, gusty kind of sound. To go from 30 to 70 miles per hour sends a quiver through you . . . most people treat a motorcycle as an animal . . . they're almost human you know . . . it's a thrill, a joyous thing, like suddenly being free . . . the noise is all you hear . . . there is a strength and power in it. It's masculine and makes me feel strong. I approach a girl on a cycle and I feel confident. Things open up and I am much more at ease.

In summarizing his analysis of the motorcycle syndrome Nicholi stated that:

> Suffering a serious ego defect that stems from a distant and difficult father-son relationship and results in a tenuous masculine identification, these patients experience the motorcycle as an integral part of their body image. The vehicle serves to confirm and delineate a shadowy inner definition of masculinity and to reinforce a fragile ego. Used adaptively to provide a sense of pleasure, virility, power, and freedom, the machine's primary use is defensive: The patient substitutes a sense of "getting somewhere" for substantive effort.

Thus, the results reported by Nicholi reveal how persons may become addicted to "expressive models," and his findings indicate how involvement with such models be they games or sporting machines may provide a kind of therapy. Moreover, Nicholi's examination of student motorcyclists graphically portrays the dialectic between the individual's search for self-identity and society's needs and pressures for adequate achievement and masculine performance on the part of its male members.

D. GAMES IN PHYSICAL EDUCATION

Numerous examples have been given throughout the chapter showing how games afford a significant social context for the "indirect" and "buffered" learning of important cultural content and positively sanctioned patterns of social behavior. The playing of games is, of course, an ancient and widespread form of learning and socialization. Yet it is only recently that games have received serious study with regard to their scientific and instrumental value. The recent attention given to games is, however, rather phenomenal. Presently, a number of academic disciplines offer courses on "game theory," and so-called serious games are being developed and utilized in several areas of education, industry, and the military. Thus we have educational games, management games, and war games which simulate important "real-life" activities within their respective institutional sectors (Abt, 1970; Avedon and Sutton-Smith, 1971; Boocock and Schild, 1968; D. L. Miller, 1970; Robinson, 1966; *Simulation and Games,* 1970–1971).

In view of the heightened interest in games by diverse scholars, it is a pathetic paradox that those professionals who traditionally have been most involved with games appear to have the least understanding of their nature and social import. Physical educators have professed for over a century that play, games, and sport provide a unique and particularly appropriate medium for meeting the main objectives espoused by our educational institutions. Similarly, they have made strong claims concerning the importance of participation in selected gross motor activities for the emotional, intellectual, physical, and social development of children and youth. Physical educators, however, have made few attempts to ascertain the nature of games, and they have done very little to confirm their assumption that participation in certain physical activities either directly or indirectly enhances learning in general or socialization in particular. Moreover, physical educators have virtually ignored the implications of the educational usage of games in other academic areas, business, government, and the military.

It seems apparent that physical educators have taken hold of selected recreative games for ostensibly educational purposes and have naively taken for granted that students somehow achieve a substantial degree of personal growth as a result of undergoing significant learning and socialization experiences assumed to be inherently associated with their participation in such games. It is suggested that if physical educators seriously wish to pursue educational objectives (especially those related to social development) through physical activity, then they must attempt to operationalize their particular educational aims and explicitly design, develop, and conduct innovative games and sports which are likely to aid the student in attaining these specific goals.

Needless to say, the design, development, and conduct of special games and sports related to particular educational objectives are a major undertaking since it requires an understanding of both the structure of games and the nature of the socialization process. More specifically, the physical educator must be cognizant of (1) the intrinsic virtues of games as teaching devices, (2) the developmental syntax of games, (3) the primary stages of moral development, and (4) the basic mechanisms of social influence. By way of concluding this epilogue the nature of these four factors are briefly described.

1. Characteristics of Academic Games

In commenting on the substantive consequences of the research on which his book *The Adolescent Society* is based, Coleman (1967a) wrote the following:

> . . . perhaps the major problem that must be faced in socialization of neophytes into responsible members of the system is how and when to give over autonomy and responsibility to the neophyte. It is this problem, I propose, that is the most difficult one educational institutions of industrial societies have to solve today.

> This has led in a direction in which I would never have anticipated going: toward experimentation in schools. The research in which I am presently engaged involves the construction of social and economic games to be used in schools. For a variety of reasons, games appear to me to be extremely interesting socializing devices: the reward structure they furnish, the self-governing system they establish, the autonomy they provide without dire consequences, the lack of necessity for outside judges.

In a later article titled "Learning through Games," Coleman (1967b) extended and elaborated upon this list of intrinsic virtues of games. His work and that of others suggest that games make good teaching devices because they possess the following characteristics.

a. Active Learning. Unlike most classroom situations wherein the student is largely a passive participant, games involve nearly total involvement on the part of students. As Coleman (1967b) has stated the case: "(an) important asset of simulation games is that they constitute an approach to learning that starts from fundamentally different premises than does the usual approach to learning in schools. The first premise is that persons do not learn by being taught; they learn by experiencing the consequences of their actions."

b. Rewards. In game situations a student receives immediate reinforcement from his peers for performance of successful acts. Moreover, the reward structure of games focuses on group achievement as well as individual achievement. Thus, unlike many classroom situations wherein the success of one student implies the failure of another, in game situations individual success often results in team success.

c. Attention and Motivation. It is often difficult in a classroom to motivate students and to hold their attention for relatively long spans of time. Games, however, seem to have an intrinsic attention-focusing quality. "The depth of involvement in a game, whether it is basketball, *Life Career,* or bridge, is often so great that the players are totally absorbed in this artificial world" (Coleman, 1967b). This intensive involvement is no doubt a function of immediate pleasure and perceived utility in gaining future pleasure.

d. Teacher's Role. Classrooms tend to be teacher-dominated. During the course of games "the teacher's role reverts to a more natural one of helper and coach." Thus, games are uniquely student-oriented.

e. Discipline. Closely associated with the factor of teacher's role is the matter of discipline. In game situations the teacher's role of judge and jury is greatly diminished. Furthermore, one seldom encounters discipline problems as peer sanctions are so strong. Young players do not permit game disruptions by spoilsports and can usually effectively cope with cheaters.

f. Self-Instruction and Testing. It is often difficult to adopt classroom materials to the ability levels of all students in a given course. Games may be developed at almost any level of complexity, and they more readily allow both high and low achievers to participate in the same social context. Games encompass a wide range of skills, and their rules and structure can be easily modified. In short, ". . . possibilities for creativity are opened up that the classroom situation often inhibits" (Coleman, 1967b).

g. Modeling. Because children have a certain familiarity with games in general, because games typically entail no life or death consequences, and because games present a unique time perspective in terms of such variables as frequency, duration, and intensity, games are consequently a highly suitable medium for conveying abstract ideas and concepts. "When a game situation simulates aspects of student's present or future life, the student begins to see how his future depends very directly upon present actions, and thus gives meaning to these actions" (Coleman, 1967b).

h. Self-Development. Games require communication and role playing abilities; they stress rational decision-making and the understanding of cause-and-effect relationships; and they reward impulse control and self-restraint; consequently, games provide a medium for self-development. As Coleman (1967b) has written:

> A special value of academic simulation games appears to be the capacity to develop in the player a sense that he can effect his own future. A massive study conducted by the U.S. Office of Education shows that one attribute strongly related to performance on standard achievement tests is a child's belief that his future depends on his own efforts rather than on a capricious environment. Many disadvantaged pupils appear to lack this belief.

In summary, it is clearly evident that the special pedagogical properties of simulated games make them useful devices for learning and socialization. Although simulated games are admittedly not an educational panacea, they nevertheless appear to be particularly suited for teaching the various contemporary categories of disadvantaged youth. Most surprisingly, few, if any, efforts have been made to determine the relative merits of educational games involving large amounts of physical activity with educational games involving only a minimal degree of gross motor activity. Moreover, few physical educators have used learning games as opposed to recreative games in their adaptive and special physical education programs. The recent work of Cratty (Cratty and Martin, 1970; Cratty, 1971) is an important exception to the preceding statement. His work well illustrates the nature and use of "active learning games" by physical educators to enhance the academic abilities of children handicapped by various learning disorders.

2. The Structure of Games

The discussion of the preceding section implies that an analysis of simulated games may give important insights into the nature of games and sport in general. In this section the other side of the case is taken up. Specifically, it is suggested that an analysis of general games and sport should be a prerequisite for the design and development of particular learning games. In addition, it is argued that if better inferences are to be made regarding the socializing functions of games and sport then greater knowledge is required concerning the basic structural dimensions of agonistic activities at different age levels.

While several attempts have been made to outline the general structure of sports and games (Avedon and Sutton-Smith, 1971; Loy and Kenyon, 1969), it is difficult to discover developmental accounts of the structural dimensions of games. One of the few behavioral scientists to have studied play, games, and sport from both a developmental and a structural perspective is Brian Sutton-Smith (Herron and Sutton-Smith, 1971).

TABLE II

DEVELOPMENTAL SEQUENCE OF APPROACH–AVOIDANCE GAMES[a]

Level	Sample game	Actors	Act	Space	Time
I 5–6 years	(a) Hide-and-seek (b) Tag	Central person of high power	Hide-and-seek Escape and chase	Hideways	Episodic
II 7–8 years	Release	Central person of lower power	Capture and rescue	Hideways and prisoner's base	Cumulative
III 9–10 years	Red rover	Powers change between central and others	Capture	Two home bases	Climax
IV 11–12 years	Prisoner's base	Diffuse teams	Capture-rescue Chase-escape	Home and Prisoner's bases	Team-cumulative

[a] Adapted from Sutton-Smith (1971).

An illustration of Sutton-Smith's developmental approach to play, games, and sport is given in a recent paper presented at the *Second World Symposium on the History of Sport and Physical Education* (Sutton-Smith, 1971). In this paper he examined the developmental sequence of "approach-and-avoidance" games played by children between the ages of 5 and 12. He discussed "the particular spatial and temporal relationships in games at different age levels, the approach and avoidant actions that are special to particular categories of games, and the relationships between the players as actors and counter-actors." Sutton-Smith's analysis is schematically outlined in Table II. With respect to the material given in the table, Sutton-Smith stated the following:

> What is occurring across these four levels is a testing of powers first against "magical" IT figures, and finally against other players of relatively the same skill. The actions in this sequence are those of chasing, escaping, capturing and rescuing with the final game of prisoner's base containing both sets of elements. There is a new form of spatial and temporal arrangement at each level. We know that these arrangements of space and time correspond to parallel forms of cognitive organization in children of these age levels. But we may

assume that when presented in these exciting forms, the spatial and temporal qualities take on a vividness which they may not have when presented more impersonally.

Professor Sutton-Smith observed that "similar levels can be illustrated for other types of games"; and he noted that "When game progress is viewed in this developmental fashion it is difficult to resist the view that important qualitative properties in the understanding of social relations, social actions, space and time are being learned by the children that proceed through the series."

3. Stages of Moral Development and Processes of Social Influence

Since physical educators have been particularly concerned with the "character development" aspects of games and sports, it is both surprising and dismaying that they have not given greater attention to recent research in the behavioral and social sciences regarding moral development. It seems sensible to assume that unless physical education teachers acquire an adequate knowledge of the developmental nature of moral behavior, they will have difficulty in either assuring or assessing the moral values and conduct associated with the play and games of children and youth. In addition to a knowledge of the levels of moral development, physical educators must attain an understanding of the processes of social influence underlying the stages of moral behavior.

Kohlberg (1963a,b, 1964, 1966, 1968) has perhaps made the most extensive contemporary analysis of moral development. On the basis of an examination of Piaget's work and as a result of data collected in several longitudinal and cross-cultural studies, Kohlberg has established a highly comprehensive outline of the evolution of moral behavior. He distinguished three levels of development, divided into six stages as follows (Kohlberg, 1968; Maccoby, 1968):

A. Pre-conventional Level
   1. Punishment and obedience orientation (obey rules to avoid punishment)
   2. Naive instrumental hedonism orientation (conform to obtain rewards, have favors returned)
B. Conventional Level
   3. Good-boy, good-girl orientation (conform to obtain approval from others)
   4. Authority orientation (conform to avoid censure by legitimate authority and the resulting guilt)
C. Post-conventional Level
   5. Social contract orientation (conform to maintain the respect of the impartial spectator judging in terms of community welfare)
   6. Individual conscience orientation (conform to avoid self-condemnation)

Maccoby suggested that Kohlberg's three levels represent different degrees of internalization of moral values, and her treatment of his typology reveals what appears to be a close correspondence between Kohlberg's three levels of moral development and Kelman's description of three primary processes of social influence. For example, Maccoby stated that at the preconventional level, "standards of judgment are external to the child, and the motivation for conforming to the standards is also external in the sense that the child is governed by external rewards and punishments" (1968). Kelman (1961), in a parallel manner, stated that in the case of *compliance* an individual is ". . . interested in attaining certain specific rewards or in avoiding certain specific punishments that the influencing agent controls." At the conventional level the individual accepts the rules and norms from identified-with authority figures (Maccoby, 1968). Similarly, *"identification* can be said to occur when an individual adopts behavior derived from another person or a group because this behavior is associated with a satisfying self-defining relationship to this person or group" (Kelman, 1961). Finally, at the postconventional level, "the standards as well as the motive to conform have become inner; they are felt as emanating from the self, and no longer depend upon the support of external authority" (Maccoby, 1968). And as Kelman (1961) concluded: "Finally, *internalization* can be said to occur when an individual accepts influence because the induced behavior is congruent with his value system."

Regretfully, studies have not been conducted to date which attempt to relate Kohlberg's typology of moral development and Kelman's typology of processes of social influence to attitudinal and behavioral changes correlated with play and games. Their conceptual frameworks though provide the needed theoretical foundations for conducting critical experiments to test the claims made by professional physical educators regarding the importance of sport involvement for the social development of youth. Does moral conduct in sport situations, for example, represent compliance, identification, or internalization? Does athletic participation further or retard mature moral development?

Although the functional effects of play, games, and sport upon the socialization process have been stressed throughout the chapter, it must be recognized that involvement in agonistic activities may also have dysfunctional consequences for socialization. Richardson (1962), for example, reported rather disheartening results in his study of ethical conduct in sport situations. He ascertained the beliefs of 233 senior male physical education majors in 15 institutions regarding sportsmanship in sport situations by means of Haskins and Hartman's Action-Choice Tests for Competitive Sport Situations. Richardson found that (1) "non-letter winners indicated a higher degree of sportsmanship than did letter winners," (2) "those students receiving no athletic grants scored much higher than respondents receiving athletic grants," and (3) "football players ranked below all other sports in test scores."

Richardson's results and recent autobiographical accounts of "dropouts" from professional sports (Meggysey, 1970; Sauer and Scott, 1971) indicate that the ethical nature of the competitive world of sport is one of conventional morality. That is to say, socialization into or via sport may be largely a matter of compliance and identification rather than internalization. For instance, Sauer (Sauer and Scott, 1971), the recent outstanding wide receiver for the New York Jets, in speaking about his retirement from professional football, stated:

> I think that . . . most anybody who plays football on the college and professional levels is doing something he's done ever since he was preadolescent . . . Given all the rewards that an athlete can get, it's alluring to a kid to want to be a good athlete. It's a means of identity, and as long as you stay in organized athletics, your identity is based on something you wanted to be back when you were a very young person . . . The bad thing about football is that it keeps you in an adolescent stage, and you are kept there by the same people who are telling you that it is teaching you to be a self-disciplined, mature and responsible person. But if you were self-disciplined and responsible, they wouldn't have to treat you like a child.

In conclusion, socialization via play, games, and sport is a complex process having both manifest and latent functions, and involving functional and dysfunctional, intended and unintended consequences. Since research on the topic is limited, one must regard with caution many present empirical findings and most tentative theoretical interpretations of these findings.

References

Abt, C. C. (1970). "Serious Games." Viking Press, New York.
Avedon, E. M., and Sutton-Smith, B., eds. (1971). "The Study of Games." Wiley, New York.
Ball, D. W. (1967). *Sociol. Quart.* **8**, 447–458.
Bandura, A., and Walters, R. H. (1963). "Social Learning and Personality Development." Holt, New York.
Becker, E. (1962). *Amer. J. Sociol.* **67**, 494–501.
Bend, E. (1968). "The Impact of Athletic Participation on Academic and Career Aspiration and Achievement." National Football Foundation, New Brunswick, New Jersey.
Berger, P. L. (1963). "Invitation to Sociology: A Humanistic Perspective." Doubleday, New York (Anchor Edition).

Berger, P. L. (1969). "The Sacred Canopy: Elements of a Sociological Theory of Religion." Doubleday, New York (Anchor Edition).

Berkowitz, L. (1964). "The Development of Motives and Values in the Child." Basic Books, New York.

Berlyne, D. E. (1969). In "The Handbook of Social Psychology." (G. Lindzey and E. Aronson, eds.), 2nd ed., Vol. 3, Chapter 27, pp. 795–852. Addison-Wesley, Reading, Massachusetts.

Boocock, S. and Schild, E. O., eds. (1968). "Simulation Games in Learning." Sage Publ., Beverley Hills, California.

Bossard, J. H. S., and Boll, E. S. (1966). "The Sociology of Child Development," 4th ed. Harper, New York.

Broom, L. and Selznick, P. (1963). "Sociology," 3rd ed. Harper, New York.

Brown, D. G. (1956). Psychol. Monogr. 70, No. 14, 1–19 (Whole No. 421).

Brown, D. G. (1958). Psychol. Bull. 55, 232–242.

Call, J. D. (1964). Int. J. Psycho-Anal. 45, 286–294.

Call, J. D. (1965). Sobrefiro Cuaderno Psicoanal. 1, 237–256.

Call, J. D. (1968). Int. J. Psycho-Anal. 49, 375–378.

Call, J. D. (1970). Psychol. Today 3, No. 8, 34–37 and 54.

Call, J. D., and Marschak, M. (1966). J. Amer. Acad. Child Psychiat. 5, No. 2, 193–210.

Caplow, T. (1964). "Principles of Organization." Harcourt, New York.

Chinoy, E. (1963). "Society," 4th ed. Random House, New York.

Clausen, J. A. (1968). In "Socialization and Society" (J. A. Clausen, ed.), Chapter 1, pp. 1–17. Little, Brown, Boston, Massachusetts.

Coleman, J. S. (1961). "The Adolescent Society." Free Press, Glencoe, Illinois.

Coleman, J. S. (1967a). In "Sociologists at Work" (P. E. Hammond, ed.), Chapter 8, pp. 213–243. Doubleday, New York.

Coleman, J. S. (1967b). NEA J. pp. 69–70 (January).

Cratty, B. J. (1967). "Social Dimensions of Physical Activity." Prentice-Hall, Englewood Cliffs, New Jersey.

Cratty, B. J. (1971). "Active Learning." Prentice-Hall, Englewood Cliffs, New Jersey.

Cratty, B. J., and Sister Martin, M. M. (1970). "The Effects of a Program of Learning Games upon selected Academic Abilities in Children with Learning Difficulties." [Evaluation of a Program Grant awarded by the U.S. Office of Education, (0–0142710) (032).]

Deutsch, M. and Krauss, R. M. (1965). "Theories in Social Psychology." Basic Books, New York.

Durkheim, E. (1950). "The Rules of Sociological Method." Free Press, New York.

Elkin, F. (1960). "The Child and Society." Random House, New York.

Erikson, E. H. (1963). "Childhood and Society." 2nd rev. ed. Norton, New York.

Frankenberg, R. (1957). "Village on the Border." Cohen & West, London.

Goffman, E. (1959). "The Presentation of Self in Everyday Life." Doubleday, New York.

Gump, P. V., and Sutton-Smith, B. (1955). Group 17, 3–8.

Hall, C. S. (1954). "A Primer of Freudian Psychology." New American Library, Inc., New York (Mentor Edition).

Havighurst, R. J., and Neugarten, B. L. (1957). "Society and Education." Allyn & Bacon, Boston, Massachusetts.

Helanko, R. (1958). "Theoretical Aspects of Play and Socialization." Turan Yliopiston Kustantama, Turku, Finland.

Helanko, R. (1963). *In* "Personality and Social Systems" (N. J. Smelser and W. T. Smelser, eds.), pp. 238–247. Wiley, New York [article reprinted from *Acta Sociol.* **2**, 229–240 (1957)].

Helanko, R. (1969). *Int. Rev. Sport Sociol.* **4**, 177–187.

Herron, R. E., and Sutton-Smith, B., eds. (1971). "Child's Play." Wiley, New York.

Hoffman, M. L. (1970). *In* "Carmichael's Manual of Child Psychology" (P. H. Mussen, ed.), 3rd ed., Vol. 2, Chapter 23, pp. 261–359. Wiley, New York.

Hollander, E. P. (1958). *Psychol. Rev.* **65**, 117–127.

Hollingshead, A. B. (1949). "Elmtown's Youth." Wiley, New York.

Jersild, A. T. (1968). "In Search of Self," 6th ed. Teachers' College Press, Columbia University, New York.

Jones, E. E., and Gerard, H. B. (1967). "Foundations of Social Psychology." Wiley, New York.

Kelman, H. C. (1961). *Pub. Opinion Quart.* **25**, 57–78.

Kempton, S. (1970). *In* "Sociology 101 Study Guide" (P. Barber and T. O. Wilkinson, eds.), pp. 39–52. Univ. of Massachusetts, Amherst, Massachusetts.

Kohlberg, L. (1963a). *Vita Humana* **6**, 11–33.

Kohlberg, L. (1963b). *In* "Child Psychology— 62nd Yearbook of the National Society for Education" (H. W. Stevenson, ed.), Chapter 7, pp. 277–332. Univ. of Chicago Press, Chicago, Illinois.

Kohlberg. L. (1964). *In* "Review of Child Development Research" (M. L. Hoffman, ed.), Vol. 1, pp. 383–431. Russell Sage Foundation, New York.

Kohlberg, L. (1966). *Sch. Rev.* **74**, 1–30.

Kohlberg, L. (1968). *Psychol. Today* **2**, No. 4, 25–30.

Krech, D., Crutchfield, R. S., and Ballachey, E. L. (1962). "Individual in Society." McGraw-Hill, New York.

Lambert, W. W., and Lambert, W. E. (1964). "Social Psychology." Prentice-Hall, Englewood Cliffs, New Jersey.

Lenski, G. (1966). "Power and Privilege." McGraw-Hill, New York.

Loy, J. W., and Kenyon, G. S., eds. (1969). "Sport, Culture and Society." Macmillan, New York.

Maccoby, E. E. (1968). *In* "Socialization and Society" (J. A. Clausen, ed.), Chapter 6, pp. 227–269. Little, Brown, Boston, Massachusetts.

MacIver, R. M. (1917). "Community: A Sociological Study." Macmillan, New York.

Mead, G. H. (1934). "Mind, Self and Society" (C. W. Morris, ed.). Univ. of Chicago Press, Chicago, Illinois.

Mead, G. H. (1964). "George Herbert Mead on Social Psychology" (A. Strauss, ed.). Univ. of Chicago Press, Chicago, Illinois (Phoenix Edition).

Mead, G. H. (1970). "Mind, Self and Society" (C. W. Morris, ed.), 7th ed. Univ. of Chicago Press, Chicago, Illinois.

Meggysey, Dave (1970). "Out of Their League." Ramparts Press, Berkeley, California.

Merton, R. K. (1963). "Social Theory and Social Structure," 8th ed. Free Press, Glencoe, Illinois.

Miller, D. L. (1970). "Gods and Games—Toward a Theology of Play." World Publ. New York.

Miller, N. E., and Dollard, J. (1941). "Social Learning and Imitation." Yale Univ. Press, New Haven, Connecticut.

Minar, D. W., and Greer, S. (1969). "The Concept of Community." Aldine, Chicago, Illinois.

Nicholi, A. M., II. (1970). *Amer. J. Psychiat.* **126**, 1588–1595.

Parsons, T. (1942). *Ameri. Sociol. Rev.* **7**, 604–616.

Parsons, T. (1947). *Psychiatry* **10**, 167–181.

Parsons, T. and Bales, R. F. (1955). "Family, Socialization, and Interaction Process." Free Press, Glencoe, Illinois.

Piaget, J. (1965). "The Moral Judgment of the Child" (transl. by M. Gabain). Free Press, New York.

Rehberg, R. A., and Schafer, W. E. (1968). *Amer. J. Sociol.* **73**, 732–740.

Rehberg, R. A., Charner, I., and Harris, J. (1970). Unpublished paper, Dept. Sociol., SUNY at Binghampton, New York.

Richardson, D. E. (1962). *Proc. Nat. Col. Phys. Educ. Ass. Men, 1962* **66**, 98–104.

Riesman, D., Denny, R., and Glazer, N. (1953). "The Lonely Crowd." Doubleday, New York (Abridged Anchor Edition).

Roberts, J. M., and Sutton-Smith, B. (1962). *Ethnology* **1**, 166–185.

Roberts, J. M., and Sutton-Smith, B. (1966). *Behav. Sci. Notes* **3**, 131–144.

Roberts, J. M., Arth, M. J., and Bush, R. R. (1959). *Ameri. Anthropol.* **61**, 597–605.

Robinson, J. A. (1966). *In* "The New Media and Education" (P. H. Rossi and B. J. Biddle, eds.), Chapter 3, pp. 93–135. Doubleday, New York.

Rosenberg, B. G., and Sutton-Smith, B. (1960). *J. Genet. Psychol.* **96**, 165–170.

Rosenberg, B. G., and Sutton-Smith, B. (1964). *J. Genet. Psychol.* **104**, 259–264.

Sanders, I. T. (1958). "The Community: An Introduction to a Social System." Ronald Press, New York.

Sauer, G., and Scott, J. (1971). Interview by Jack Scott with George Sauer on the Reasons for Sauer's Retirement from Professional Football while at the Height of his Career. Printed by Dept. of Physical Education, California State College at Haywood

Schafer, W. E., and Armer, J. M. (1968). *Trans-Action* **6**, 21–26 and 61–62.

Seagoe, May V. (1969). *Proc. 77th Annu. Conv. Amer. Psychol. Ass.* Vol. 4, Part 2, pp. 683–684.

Secord, P. F., and Backman, C. W. (1964). "Social Psychology." McGraw-Hill, New York.

Sessoms, H. D. (1969). *In* "Recreation—Issues and Perspectives" (H. Brantley and H. D. Sessoms, eds.), pp. 58–68. Wing Publ., Columbia South Carolina.

Sexton, P. C. (1969). "The Feminized Male." Vintage Books, New York.

Sexton, P. C. (1970). *Psychol. Today* **3**, No. 8, 23–29 and 66–67.

Shibutani, T. (1961). "Society and Personality." Prentice-Hall, Englewood Cliffs, New Jersey.

"Simulation and Games" (1970–1971). A Quarterly International Journal of Theory, Design, and Research. Sage Publ., Beverly Hills, California.

Skinner, B. F. (1968). "Science and Human Behavior," 6th ed. Free Press, New York.

Snyder, E. E. (1969). *Sociol. Educ.* **42**, 261–270.

Spreitzer, E. A., and Pugh, M. D. (1971). *Pap., Ohio Valley Sociol. Soc., 33rd Annu. Meet., 1971.*

Stone, G. P. (1962). *In* "Human Behavior and Social Processes" (A. M. Rose, ed.), Chapter 5, pp. 86–118. Houghton, Boston, Massachusetts.

Stone, G. P. (1965). *Quest* **4**, 23–31.

Sutton-Smith, B. (1971). *Pap., 2nd World Symp. Hist. Sport Phys. Educ.,* Banff, Alberta, Canada.

Sutton-Smith, B. and Gump, P. V. (1955). *Recreation* **45**, 172–174.

Sutton-Smith, B., Roberts, J. M., and Kozelka, R. M. (1963). *J. Soc. Psychol.* **60**, 15–30.

Sutton-Smith, B., and Rosenberg, B. G. (1961). *J. Amer. Folklore* **74**, 17–46.

Toby, J. (1964). "Contemporary Society." Wiley, New York.

Tryon, C. M. (1944). *In* "The 43rd Yearbook of the National Society for the Study of Education" (N. B. Henry, ed.), Part 1, Chapter 12, pp. 217–239. Univ. of Chicago Press, Chicago, Illinois.

Webb, H. (1969). *In* "Aspects of Contemporary Sport Sociology" (G. S. Kenyon, ed.), Chapter 8, pp. 161–187. Athletic Institute, Chicago, Illinois.

Winick, C. (1968). "The New People." Pegasus, New York.

Work, H. H., and Call, J. D. (1965). "A Guide to Preventive Child Psychiatry: The Art of Parenthood." McGraw-Hill, New York.

Worsley, P., Fitzhenry, R., Mitchell, J. C., Morgan, D. H. J., Pons, V., Roberts, B., Sharrock, W. W., and Robbin, W. (1970). "Introducing Sociology." Penguin Books, Middlesex, England.

Wrong, D. H. (1961). *Amer. Sociol. Rev.* **26**, 183–193.

Young, M. and Willmott, P. (1967). "Family and Kinship in East London," 5th ed. Penguin Books, Middlesex, England.

Becoming Involved in Physical Activity and Sport: A Process of Socialization

Gerald S. Kenyon and Barry D. McPherson

For well over 50 years, psychologists, anthropologists, and sociologists have been addressing themselves to the social process whereby persons learn to become participants in their society—the process of *socialization*. Most of their efforts have focused upon the socialization of the child, and as a result the consequences of the process which have interested many investigators have been relatively broad behavioral dispositions. Recently, however, there has been a trend toward the study of socialization as it occurs not only during childhood but also throughout the life cycle. Moreover, attention is being directed to more limited aspects of the phenomenon such as occupational socialization and political socialization; consequently, researchers are able to concentrate upon a narrower set of behaviors and dispositions, i.e., those associated with more definitive social roles. Not only have these newer approaches made research on socialization more manageable, but also they have facilitated a more efficacious study of the consequences of socialization.

This chapter is an attempt to build further upon the study of socialization in a focused sense—in this case, *socialization into roles associated with institutionalized physical activity or sport.* Our approach will, in many respects, present the other side of the coin. While Loy and Ingham elsewhere in this volume (Chapter 11) have dealt primarily with socialization via play, games, and sport, most of our efforts will center upon socialization into sport. Section I consists of an attempt to provide a conceptual frame of reference, that is, our view of what it is to be socialized into sport involvement. Section II describes some findings related to the socialization process drawing upon data acquired from a variety of sport role players. Section III concludes the chapter with a brief consideration of future research needs by presenting a set of propositions based upon one approach to modeling the socialization process.

I. Sport Socialization: A Conceptual Frame of Reference

Given, then, our concern for socialization into rather than via sport, the scope of our topic becomes more manageable. However, considerable diversity remains. Although the learning of sport roles* is likely to occur in about the same way as the learning of nonsport roles, the complexity of the process in sport contexts is often greater since sport is engaged in throughout the life cycle, and in a wide variety of interrelated forms. Such complexity becomes obvious when the various dimensions of the sport socialization process are enumerated:

1. Socialization of the elite performer
2. Socialization of the sub-elite performer
3. Socialization of the sport consumer
4. Socialization of the sport producer
5. Socialization of the sport leader
6. Sport resocialization
    Sport to sport (i.e., from one sport to another)
    Sport to nonsport (i.e., the withdrawal from sport to learn nonsport roles)
    Role to role, within sport (i.e., changing roles within a given sport)
7. Sport socialization and social change
8. Socialization via sport

* The reader will note the frequent allusion to role theory in this section. We attempt to justify this in Section I,A,1.

A. EXPLAINING THE SOCIALIZATION PROCESS: A SOCIAL ROLE-SOCIAL SYSTEM APPROACH

A summary of the several approaches to socialization has been presented by Loy and Ingham (Chapter 11). For us, however, we note that among the several theoretical orientations used to study the process including psychoanalysis, psychoanalytically oriented social anthropology, the normative-maturational approach, the developmental-cognitive approach, the genetic and constitutional approach, and the various learning theory approaches, the latter, and particularly the social learning orientations, have become the most popular and most productive in both theory and empirical findings (Bandura and Walters, 1963; Brim and Wheeler, 1966; Clausen, 1968). Thus, given the nature of sport, and given that the major characteristics associated with sport roles are probably acquired after early childhood, a social learning approach, utilizing both psychological and sociological variables, would seem to be most fruitful for the study of sport socialization. However, there are variations within this approach. For example, sport role learning can be studied using a frame of reference the three main elements of the socialization process (see Fig. 1), namely, *significant others** (socializing agents) who exert influence within social situations (socializing agencies) upon role learners (actors or role aspirants) who are characterized by a wide variety of relevant personal attributes. Some findings using this model are presented in Section II. There is an inherent weakness in such a general approach however, namely, the difficulty of specifying a manageable number of variables from a great many plausible ones, that is, the "specification problem" (Heise, 1969).

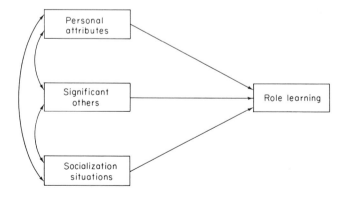

FIG. 1. The three elements of the socialization process.

* Those persons who exercise major influence over the attitudes and behavior of individuals (after Woelfel and Haller, 1971).

A more definitive rationale than the foregoing for the selection of variables is the "social role–social system" approach (Sewell, 1963). Role learning is accounted for by exposure of the role aspirant, who is already characterized by a set of physical and psychological traits, to a variety of stimuli and reinforcements provided by significant others, who act within one or more norm-encumbered social systems. Thus, a careful analysis is needed of the nature of both social roles and social systems. The social role, and more particularly its characteristics, which becomes the dependent variable in any explanatory system, is treated first.

1. Sport Roles: The Dependent Variable Problem

When it comes to specifying the consequences of sport socialization, after only a little reflection one quickly realizes that potentially there are many. Even though the discussion herein is restricted to socialization into sport involvement, and thus eschews considering socialization via sport (whereby the enactment of sport roles become independent variables for such dependent variables as social integration, community identification, social mobility, and social control), the complexity of the dependent variable remains great. To cope with this problem we have found the dramaturgical approach of "role theory" to be helpful.

As one becomes involved in sport he can be characterized as playing one of several roles. A sport role, like any role, implies that the role incumbent possesses knowledges, skills, and dispositions characterizing the role in question. Obviously, a clear definition of these dimensions is required for each sport role. More difficult, however, is the identification and classification of the many sport roles. While some role players actually participate in the contest (primary involvement), others "consume" sport (secondary involvement) either directly (as spectators) or indirectly (as consumers of the mass media). Still others produce sport through the enactment of leadership, organizational, and entrepreneurial roles (A general classification of sport roles is given in Fig. 2.). Moreover, sports themselves differ greatly in kind (e.g., exercise-oriented activity, gamelike activity, expressive activity), social environment [from involvement in the presence of many others to involvement in the absence of all others; or involvement oriented to varying combinations of reference groups such as Kemper's (1968) "normative," "role model," and "audience" groups], and complexity (from unstructured to highly structured).

Clearly, each sport role is conceptually complex. Nevertheless, to be able to specify the characteristics of a particular role is an essential prerequisite to the study of the learning of that role and thus deserves far greater attention than heretofore received.

Mode	Primary	Secondary				
		Consumer		Producer		
		Direct	Indirect	Leader	Arbitrator	Entrepreneur
Role	Contestant	Spectator	Viewer	Instructor	Member of –Sports governing body	Manufacturer
	Athlete		Listener	Coach	–Rules committee	Promoter
			Reader	Manager	Referee	Wholesaler
	Player			Team leader	Umpire Scorekeeper Other officials	Retailer

FIG. 2. Some social roles associated with primary and secondary modes of sport involvement.

2. Social Systems and Socialization: The General Case

Having paid some attention to the role aspects of the social role–social system approach, we turn now to a brief consideration of how certain institutions or social systems such as the family, the school, the church, and the peer group might contribute to role learning. With respect to such systems then, it is argued that given the nature, complexity, and pervasiveness of sport roles, each relevant social system should be treated separately as a potential role learning situation. This takes advantage of considerable knowledge already available concerning the structure and function of institutionalized social systems. Moreover, such a procedure recognizes that factors accounting for socialization are likely to be more salient within, rather than between, systems (Sewell *et al.*, 1969).

If the social system is made central in characterizing the socialization process, it is suggested that a general model of this process can be constructed consisting of a simple causal chain. Thus, in the proposed model: Given a degree of role aptitude (cognitive and motoric) the role aspirant is variously influenced within each of the social situations in which he inevitably finds himself, with the net effect being the acquisition of a propensity for learning the role in question. Furthermore, this motivates him to rehearse the role, which in turn leads to the learning of the role. In propositional form:

1. The greater the role aptitude, the greater the system-induced propensity for role learning.
2. The greater the system-induced propensity for role learning, the greater the role rehearsal, and vice versa.
3. The greater the propensity for role learning, the greater the role learning, i.e., degree of socialization.

These propositions are illustrated in Fig. 3. Central to the second stage of this model, obviously, are the socializing processes or mechanisms which occur within each relevant system. These are given more attention below. However useful the system context may be, it is recognized that ultimately it may be possible to strip away the institutional barriers and utilize only those few variables that reflect the sum total of influence within a system. For example, the Blau and Duncan (1967) model of the occupational attainment process relies on only two status variables (father's education and occupational attainment) and two behavioral variables (individual's education level and prestige of first job).

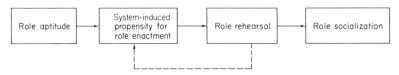

FIG. 3.   Postulated social role–social system model of socialization process.

Sewell *et al.* (1969, 1970), in explaining occupational attainment, have shown that it is possible, and indeed useful, to characterize significant others' influence as a single variable. At the moment, then, it is suggested that the best strategy for identifying the influential variables is through a careful analysis of each relevant social system as illustrated in Fig. 4. Although one might desire parsimony, the selection of a small number of variables is premature with respect to sport socialization since too little is known at this time.

Given the present state of knowledge, then, it would seem that the best research strategy would be somewhat akin to McPhee's "modular" concept of model building (1963), and in particular, Blalock's "block-recursive system" (1969).

3. Relevant Factors in Relevant Social Systems

The general model proposed above would likely account for socialization into most social roles and not just those associated with sport. What is needed, then, is a careful consideration of the sport-relevant factors in those so-

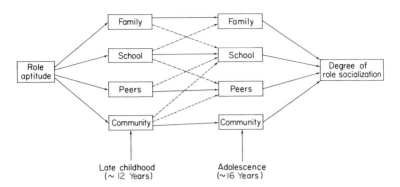

FIG. 4. Postulated social role–social system two-stage block-recursive model for sport socialization.

cial systems likely to be important in inducing a propensity for sport involvement.* An alternative would be to begin with a highly theoretical analysis of the nature of social systems (e.g., Parsons, 1961) and try to show how sport may be functional or dysfunctional for a particular system and the consequent implications for the socialization process.† However, insofar as our objective is not to explain the nature of social systems per se, but rather to account for the processes precipitating sport socialization, it is suggested that only those factors likely to be useful in explaining the socialization process be considered. Therefore, it is proposed that the following elements, drawn largely from Smelser (1962), serve as a frame of reference when considering system-level factors:

Values
Norms
Sanctions
Situational facilities

Thus, for a given social system, both in general and with regard to sport, the social learning that occurs, whether in the form of imitation, identification, or various forms of reinforcement (Bandura and Walters, 1963), would be governed by each of the foregoing elements, separately and in combination.

* It is proposed that the construct "propensity for sport involvement" (PSI) be considered as an intervening variable in the basic model. Each institutionalized social system considered would be assumed to contribute to, or detract from, the development of PSI.

† For an example of this approach and a discussion of its strengths and weaknesses see Lüschen (1969).

4. Explaining Sport Socialization at the System Level

If explanation is taken as theory and if theory is considered as a set of logically interrelated propositions (Zetterberg, 1966), then explanation of socialization should be in propositional form. By way of illustrating this point, a small number of propositions will be presented that reflect sport socialization as a consequence of a social role–social system approach. In each case the dependent variable (intervening variable in the larger system) will be propensity for sport involvement.

First, consider the family as an institutionalized social system. Although Reiss (1966) suggested that the transmission of behavioral patterns through the family may not be as great as in other institutions, in sport socialization it appears that the family is of some importance (Kenyon, 1970b) and therefore should not be excluded from the general model. For example, in the context of values as a social system element, achievement is often associated with both family socialization and sport role enactment. Thus, an analysis of achievement training may lead to a better understanding of the family's contribution to sport socialization, particularly for those sport roles where achievement and role enactment are closely related. Considering only a few variables (Rehberg et al., 1970; Rosen, 1956; Turner, 1970; Zigler and Child, 1969), we might have, in propositional form, something as follows:

1. The greater the independence training, the greater the achievement motivation.*
2. The more democratic the parents, the greater the achievement motivation.
3. The greater the father's entrepreneurial role behavior associated with his occupational status, the higher the son's achievement motivation.
4. The greater the achievement motivation, the greater the propensity for sport involvement.
5. Therefore, the greater the independence training, the more democratic the parents, and the greater the father's entrepreneurial behavior, the greater the propensity for sport involvement

Again, using an element common to all social systems, namely, situational facilities, but again within the context of the family, the following propositions are a few of the several that might help to explain sport socialization (Spady, 1970):

1. The higher the family socioeconomic status, the greater the social participation.

* This proposition should not be taken as axiomatic in view of the recent work of Rehberg et al. (1970).

2. The higher the family socioeconomic status, the greater the propensity for involvement.
3. Therefore, the greater the social participation, the greater the propensity for sport involvement.

Still in the context of situational facilities, but more particularly with regard to role models (Brim, 1958; Stone, 1969), we have:

1. Sport involvement reflects masculinity (at least in North America).
2. Siblings frequently serve as powerful role models.
3. Children take on the personality traits of siblings of the opposite sex.
4. Therefore, as the number of male siblings increases, and the number of female siblings decreases, the propensity for sport involvement increases.

Shifting to the elements, norms and sanctions, the following propositions are presented, again in the context of the family (Campbell, 1969; Mc-Candless, 1969), and again with achievement motivation considered a factor:

1. The less autocratic, the less authoritarian (particularly the father) and the more permissive the parents, the greater the likelihood of their being used as role models.
2. The greater the warmth and nurture provided by the parents, the greater the likelihood of their being used as role models.
3. The less autocratic, the less authoritarian, the more permissive, and the greater the warmth and nurture provided by the parents, the greater the offspring's need to achieve.
4. The greater the need to achieve, the greater the propensity for sport involvement.
5. Thus, given the capability of modeling sport roles, as parents become less autocratic and authoritarian, and more permissive, nurturant, and warm, the propensity for sport involvement increases.

Although the foregoing propositions do not exhaust the factors accounting for family-induced propensity for sport involvement, they do suggest that it is possible to consider sport socialization at the social system level. Moreover, from the point of manageability, they also illustrate the desirability of a social system approach.

Other institutionalized social systems could be analyzed in a similar way, including those reflecting various educational, political, economic, military, and religious institutions. Moreover, if the adolescent peer group is taken as a social system, drawing upon the work of Helanko (1957), Homans (1950), and Spady (1970) it might be possible to partially account for peer influence with the following propositions:

1. The greater the aggregation, the greater the interaction.
2. The greater the interaction, the greater the mutual liking.
3. The greater the mutual liking, the more effective the peer influence (sport provides the means for achieving peer group goals—e.g., "status definition"—Helanko, 1957).
4. The greater the involvement in youth aggregates (groups and clubs), the greater the propensity for sport involvement.

When the contribution to sport socialization by each of several institutions is more completely analyzed, it is likely that some propositions would be system-specific while others would emerge as applicable in all systems. For example, from existing knowledge (LeVine, 1969; Webb, 1969) the following propositions, although somewhat general, would likely apply to all or most social systems having any bearing on sport socialization:

1. The greater the similarity between the values associated with sport and those associated with the institution, the greater the system-induced propensity for sport involvement.
2. The more deliberate the socialization effort, the greater the effect.
3. The more the role learner is aware of the socialization process, the greater the socialization.
4. The greater the number of positive sanctions (and the earlier they are applied) and the fewer the negative sanctions, the greater the system-induced propensity for sport involvement.
5a. The greater the situational facilities (number and kind), the greater the chances for success.
5b. The earlier success is experienced, the greater the propensity for sport involvement.
5c. As the degree of success increases, propensity for sport involvement increases.

5. National Differences in the Socialization Process

In assuming a social role–social system orientation to socialization, it is obvious that national differences exist in the nature of both sport roles (kind, milieu, and complexity) and relevant social system elements (values, norms, sanctions, and situational facilities). Consequently, cross-national research would provide a medium for testing the generality of any model of the sport role learning process. By way of example, consider the following propositions:

1. The earlier the success, the greater the propensity for sport involvement.
2. The more diverse the opportunity structure, the greater the chance of early success.

3. The more diverse the opportunity structure, the greater the propensity for sport involvement.

With wide between-nation differences in opportunity structure and public policy (e.g., achievement-oriented vs. participation-oriented, or, as in Moore and Anderson's terms (1969), "performance" societies vs. "learning" societies) it can be expected that national variations will exist. Again, however, the significance of these for explaining sport socialization can only be ascertained through careful cross-national investigations.

B. SUMMARY

In view of the nature of sport and those involved in it, it has been suggested that a social learning approach would be the most efficacious paradigm in the study of sport socialization. Upon acknowledging the complexity of the process, it was argued that a social role–social system model would be best suited to studying the nature of sport role learning. On the assumption that socialization is situation-dependent, and by analyzing each institutionalized social system separately, the process becomes more manageable. Moreover, advantage can be taken of the considerable knowledge already available related to the structure and function of specific systems.

Thus, it is suggested that the degree to which actors are socialized into sport roles is dependent upon the propensity for sport involvement which has been generated by each system. In operational terms, specific system-related propositions need to be deduced relating relevant factors to system-specific propensities. The dependent variable—degree of socialization into a sport role—also requires careful delineation before data can be collected. As will be seen in the next section, the degree to which the foregoing conceptual considerations have been translated into empirical findings is small at this time.

II. Sport Socialization: Some Findings

Until recently, few sociologists of sport have concentrated their research efforts on problems associated specifically with socialization into sport roles. Some of the findings of studies that have been undertaken however are reported here, using a classification of roles approximating that given in Fig. 2. Thus, results are presented under the headings *Socialization into Roles Associated with Primary Involvement* (Section II, A) and, *Socialization into Roles Associated with Secondary Involvement* (Section II, B).

At this point in time (1971), a small number of sport roles have been studied from a socialization perspective. Unfortunately, the empirical work to date has not mirrored the theoretical efforts presented in Section I (i.e., the social role–social system model). Therefore, the findings are discussed in the context of the "elements of socialization" model (personal attributes, significant others, and socialization situations).

A. SOCIALIZATION INTO ROLES ASSOCIATED WITH PRIMARY INVOLVEMENT

Researchers studying the socialization process as it is applied to primary involvement, for the most part have been interested in the "elite" performer. Although the term "elite" has many connotations, athletes who have reached a level of competition at or near a national standard are arbitrarily labeled elite performers. Data from two (but not entirely mutually exclusive) populations are discussed: first, Olympic aspirants and, second, college athletes. The section on primary involvement is concluded with a brief review of other studies.

1. Olympic Aspirants

In a project coordinated at the University of Wisconsin (Kenyon, 1968), data were acquired from two sets of Olympic aspirants: track and field athletes* and gymnasts.†

In the first study 113 athletes, who were competing for a position on the 1968 United States Olympic Track and Field Team, completed a questionnaire designed to gather data concerning the influence of personal attributes, significant others, and the social situation on the learning of a sport role. The sample consisted of 96 white and 17 black athletes. Although only 30% of these were eventually chosen for the national team, performances were remarkably homogeneous.

For most, involvement in sport per se began early in life with 96% indicating that they participated in football, basketball, or baseball in elementary school. This early general involvement in a number of sports, plus the fact that over 65% were "winners" the first time they competed in a sport event, suggests that a high level of sport aptitude was present at an early age. However, ability does not totally account for their involvement and later success in the role of track and field athlete. For example, 50% of the subjects did not participate or compete in track and field as such until after they entered high school. Therefore, the learning of the role must have been situationally influenced by significant others who taught and reinforced the

---

* The data were collected at the South Lake Tahoe, California altitude training camp through the cooperation of Dr. Jack Daniels (now at the University of Texas) who served as research physiologist for the United States Olympic Committee.

† See F. A. Roethlisberger, "Socialization of the Elite Gymnast," unpublished Master of Science thesis, University of Wisconsin, 1970.

specific role behaviors within specific social settings. For example, over 75% of the respondents indicated that their interest in the activity was first aroused at school by either watching others compete, talking to a peer or teacher about the activity, or receiving instruction in a physical education class. In addition, they reported that they attended a school where 88% of the students and 83% of the teachers valued track and field and considered it to be an important extracurricular activity for students. Similar findings were reported for the years they attended college.* Thus, a social situation which values a pattern of behavior, and provides opportunities for the learning of the behavior, determines which roles will be learned.

The influence of significant others in generating an interest in a sport role appears to be not only important but also sport specific (Table I). For example, when asked who was most responsible for arousing an interest in track and field, 30% indicated their peers, 25% indicated their teachers or coaches, and 21% felt that members of the family were responsible. However, when asked a similar question concerning interest in the traditional team sports of baseball, basketball, and football the results were: peers, 44%; family, 33%; and teachers or coaches, 16%. Whereas peers are important contributors to arousing interest in most sports, teachers and coaches appear to be more influential than peers in stimulating an interest in track and field. In contrast, the family is more influential than school faculty in generating interest in the traditional spectator sports.

TABLE I

RELATIVE CONTRIBUTION OF CLASSES OF SIGNIFICANT OTHERS IN GENERATING AN INTEREST IN SPORT AMONG ELITE ATHLETES

Significant other	% Responsible for first interest in	
	Track and field	Traditional Spectator sports
Peer group	30	44
Teachers or coaches	25	16
Family	21	33

In addition to generating interest in a role, significant others also reinforce the enactment of the role and thereby facilitate the learning of appropriate role behaviors. Figure 5 indicates the percentage of significant others who offered positive sanctions for competing in track and field at each of the four stages in the athlete's competitive career. It should be noted that

* Data were obtained on another situational variable, namely, birth order. Fifty percent of the respondents reported that they were first born.

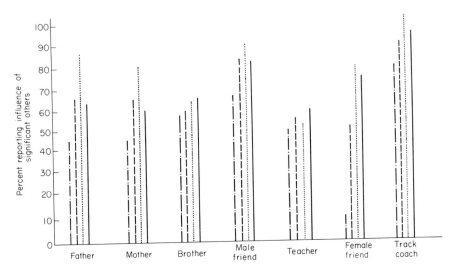

Fɪɢ. 5. Relative proportion of significant others providing positive sanctions at four stages of the life cycle [Olympic track and field aspirants ($n = 113$)]. (———) on the job, (. . .) during college, (- - -) during high school, and (— ● —) before high school.

male peers and track coaches were the most influential at all four stages. It is also interesting to note that while the athlete is in school or college, the number of significant others who offer encouragement increases in a linear fashion. This appears to coincide with the increasing amount of success experienced by the athletes; however, a cause–effect relationship is not implied, nor can it be substantiated with the present data. This latter finding does suggest, however, that the influence of significant others may be differential in a temporal sense, that is, it may increase or decrease over time.

In addition to significant others, reinforcement for continued involvement, and therefore further learning, is derived from other sources. For example, publicity from the mass media and personal satisfaction derived from successful competition (i.e., winning or improving personal times or distances) were found to be important factors in maintaining an interest in track and field throughout the athlete's competitive career. In general, then, it would appear that the elite track and field athlete receives encouragement and reinforcement from many sources, several of which act simultaneously.

The results presented to date have been based upon the data of black and white athletes combined. However, it has been suggested that minority group members may be differentially socialized into sport roles (McPherson, 1971). Thus, when the data from black and white athletes were separated, differences in the socialization process appear. Considering the social situation first, and recognizing the limitations of extrapolations based on a minimal sample of 17 subjects, it was found that compared to white athletes, black athletes were members of larger families (4.5 children compared to 2.4 for white families), were from a lower socioeconomic background, and were raised in large urban centers to a greater extent (56% vs. 29%). Moreover, they became involved in track at an earlier age, developed their first interest in the sport in the neighborhood and home, rather than in the school as the white athletes did, and perceived track to be more highly valued by students at school and by members of the community.

An examination of the influence of significant others suggests that the black athletes (a) *before high school* received more encouragement and reinforcement from the family than from others, received more positive sanctions from the mother than the father, and considered their peers to be more influential than a teacher or coach; while (b) *in high school* they received the most reinforcement from track coaches, peers, and their mother, were discouraged (negative sanction) from competing by "girl friends" whereas the white athletes reported that their girl friends were for the most part indifferent, and considered the opportunity to earn a college scholarship, the admiration of male peers (status), and publicity for success as positive reinforcement essential to maintaining their interest in track; and while (c) *in college* they received the greatest amount of reinforcement from peers, track coaches, and the father. This late appearance of the father as a source of reinforcement suggests that perhaps the fathers' desire is to identify with a success model. Thus, it appears that different role models serve as significant others for blacks, that the role models differ at each stage of the athlete's career, and that economic rewards related to track (scholarships) are more important to the black athlete.

Finally, when comparing the differences in the personal attributes of the two groups, it was noted that the black athletes were more successful in making a team on their first attempt; experienced success (won a race or event) at an earlier age, especially in the track events; appeared willing to sacrifice more in the realm of cultural pursuits and part-time jobs in order to compete in sports (were more successful in making the Olympic team); had less need for social support; were less religious; and had similar levels of aspiration as to their anticipated future socioeconomic level.

In view of the foregoing, and within the limitations of the data, it appears that there is at least minimal support for the hypothesis that the process accounting for sport socialization may differ from one subgroup to another.

In a second study (Roethlisberger, 1970) of United States Olympic Team aspirants, all the "all-around" gymnasts (competitors in six events), who had scored 104.00 points or more in National meets in 1968, completed a questionnaire similar to that used in the track and field study. Of the 16 aspirants, 8 were eventually selected for the Olympic team. The data were treated for both the combined group ($n = 16$) and the two separate groups, classified according to whether or not they were chosen for the Olympic team. Similar to the track and field study, a "nonsystem" model (see Fig. 1) was employed, which considered personal attributes, socializing situations, and significant others.

It was found that with respect to their personal attributes, taken collectively, gymnasts apparently were skillful in a number of sports, having participated in several individual sports before they specialized in gymnastics. In terms of social values, they were not disposed toward social activism such as boycotting or demonstrating at the Olympic games; they had high educational and occupational aspirations; they were not highly committed to religion; and they considered themselves moderate to liberal in terms of their value orientations. A comparison of the chosen and unchosen gymnasts on such personal values revealed that the chosen somewhat more vigorously defended the civil rights of minorities, adhered less to conventional religious thought and attended religious services less often.

An analysis of the early socialization setting indicated that compared with other institutions, the educational system was primarily responsible for socializing an individual into the elite gymnast role; that the elite gymnasts were raised in an upper-middle or upper-class environment; that unlike the track athletes the mass media was of little importance in reinforcing the respondent's interest in gymnastics; that they learned the role in an environment that did not consider gymnastics to be important; and that initial primary involvement usually occurred between the sixth and tenth grade. A comparison of the chosen versus the unchosen indicated that the former were both interested in, and competed in, gymnastics at an earlier age. Those who were selected for the national team came from smaller families and had more older male siblings and fewer female siblings than those who were not selected.

Finally, an analysis of the role of significant others suggested that fathers and coaches were the most influential significant others, followed in order of importance by peers and brothers. Mothers, sisters, and other relatives were not particularly influential. At all stages of the gymnast's competitive career, success was perceived to be the most important factor in maintaining interest in the sport. A comparison of the chosen and unchosen indicated that the fathers and older brothers were even more important as significant others for the chosen gymnasts.

In comparing the data from Olympic aspirants in two sports, it would appear that the school is the most important socializing milieu and the peer group and coaches the most influential significant others.

2. College Athletes

In view of the fact that most Olympic aspirants in the United States are college students or college graduates, it is likely that the overall socialization experiences of elite college athletes and Olympic aspirants are similar in many respects. However, as suggested above, sport differences can be expected. Therefore, the results of two studies which investigated socialization into elite college athlete roles are noted here.

A study of Canadian college ice hockey *(n* = 52) and tennis players *(n* = 19) focused upon the psychosocial factors accounting for college athletes becoming involved in sport (McPherson, 1968). A number of similarities and differences were discovered in the socialization experiences of the two groups. For example, it was found that 96% of the respondents were interested in sport by the age of 10, and that 63% were involved as consumers prior to their participation in sport. Again, this suggests that socialization into sport roles begins early in life. However, specialization in one sport may not occur until later. Whereas hockey players competed on organized teams before 9 years of age, competition in tennis was not attempted until a mean age of 12.7 years. Similar to the Olympic aspirants there was a trend toward specialization in one or two sports as they progressed through high school, with competition at college being restricted to the one sport.

In terms of the social situation, and in particular, class background and residence location, the tennis players were raised in a middle or upper-middle social class milieu and all lived in urban or suburban areas. The hockey players came from a middle-class or lower-middle-class milieu, with 30% having attended a rural or small community school.

The influence of significant others tended to follow a similar pattern for both groups, with the single exception being that mothers were more influential for tennis players than for hockey players. It was noted that interest in sport was initially aroused within the family and mainly by the father. However, during the high school years familial influence decreased and any interest in a new activity was aroused mainly by peers, coaches, and physical education teachers. Again, there appears to be a temporal factor whereby influence is differential over time. In addition, it appears that the initial stimulus to become interested in sport is received from involved peers and more so from a home environment which considers sport to be an important facet of life.* Among the Canadian athletes, peers were second in impor-

* This finding was also suggested by Pudelkiewicz (1970) who noted that a positive evaluation of sport by Polish parents gives rise to sport interests among their children.

tance during the elementary and high school days but became less important at the university. Similarly, teachers and coaches were important significant others early in life and then declined in importance. A final significant other was the professional athlete—almost all respondents reported that they had an "idol." In view of the sample being Canadian, it is not surprising that most of the idols listed by both hockey and tennis players were outstanding professional hockey players. However, as the tennis respondents reached college age, a highly ranked tennis player tended to replace the hockey idol. It was interesting to note that for hockey there was a positive relationship, which increased with age, between the position played by the idol and that played by the respondent.

In a related study (Kenyon and Grogg, 1969) in the United States, 87 athletes in eight sports at the University of Wisconsin were interviewed to determine the factors influencing their socialization into the role of elite college athlete.* Since many of the results are similar to that of the study previously reported, only unique differences will be presented here. The social situation in which interest in a specific sport is first generated varies by sport. For example, interest in baseball was initiated about equally in the home and the school; for fencing and crew, almost entirely in the school; for football, in the home and neighborhood; for hockey, in the home and neighborhood; for swimming, in the home and club or recreational agency; for tennis, in the school and club or recreational agency; and, for track, in the home and school. Similarly, the means by which the respondents first became interested varied somewhat by sport. For example, baseball, football, hockey, and tennis players, together with swimmers and track and field athletes, reported that actual participation was the most important influencing factor in developing a serious commitment to their sport. For fencers and participants in crew, their initial interest was stimulated by personal conversation with those already committed to the sport in question. Exposure to television, reading, or attending contests as a spectator did not contribute significantly to arousing interest in competing. Interestingly, a number of athletes attributed their initial interest not to any of the foregoing factors but rather to a perceived self-motivation. Finally, the data suggest "opportunity set" differences. For example, the athlete's place of residence during high school varied among sports (Table II). For instance, none of the hockey players, swimmers, or tennis players lived in the open country or on a farm, indicating, as one might expect, that certain facilities and a motivational climate generally necessary for learning and perfecting a specific sport role are not readily available in rural areas.

* These data were collected in 1969 by members of a graduate class in the sociology of sport at the University of Wisconsin and analyzed by Mr. Tom Grogg. Their assistance is gratefully acknowledged.

TABLE II

PLACE OF RESIDENCE DURING HIGH SCHOOL AS A FUNCTION OF SPORT
(IN PERCENT OF SUBJECTS)

Sport	Large city	Suburb of large city	Middle-sized city or small town	Open country (not a farm)	Farm	n
Baseball	25.0	33.3	33.3	0.0	8.3	12
Crew	7.7	23.1	53.8	7.7	7.7	13
Fencing	37.5	25.0	12.5	12.5	12.5	8
Football	40.0	0.0	30.0	10.0	20.0	10
Hockey	36.4	0.0	63.6	0.0	0.0	11
Swimming	30.8	46.2	23.1	0.0	0.0	13
Tennis	22.2	66.7	11.1	0.0	0.0	9
Track	18.2	9.1	54.5	9.1	9.1	11
$\overline{X}$	26.4	25.3	36.8	4.6	6.8	87

With regard to encouragement provided by significant others, Table III shows the relative influence of several socializing agents and also suggests that the total amount of encouragement increases over time. However, as shown in Table IV, the amount of reinforcement received by significant others during the critical pre-high school period varies by sport. Taken collec-

TABLE III

MEAN RATING OF SOCIALIZING AGENTS ON ENCOURAGEMENT TO PARTICIPATE BEFORE
HIGH SCHOOL, DURING HIGH SCHOOL, AND DURING COLLEGE (SCALE VALUES 1–5)

Socializing agents	Before High school		High school		College	
	n	$\overline{X}$	n	$\overline{X}$	n	$\overline{X}$
Father	57	4.05	69	4.26	85	4.24
Mother	57	3.96	70	4.07	87	4.14
Brother	49	3.92	63	4.06	76	4.10
Sister	44	3.39	55	3.53	68	3.74
Other relative	46	3.59	58	3.91	72	3.90
Male friend	57	4.28	70	4.38	87	4.20
Female friend	50	3.32	69	3.77	84	3.88
Classroom teacher	55	3.40	69	3.75	71	3.41
School coach	49	4.08	69	4.65	85	4.60
School counselor	39	3.28	64	3.59	60	3.33
Nonschool personnel	40	4.05	47	4.15	51	3.84
Other	26	3.81	32	4.00	44	3.86

TABLE IV

MEAN RATING OF SOCIALIZING AGENTS ON ENCOURAGEMENT TO PARTICIPATE BEFORE
HIGH SCHOOL AS A FUNCTION OF SPORT (SCALE VALUES 1–5)

Socializing agents	Baseball		Football		Hockey		Swimming		Tennis		Track	
	n	$\overline{X}$	n	$\overline{X}$	n	$\overline{X}$	n	$\overline{X}$	n	$\overline{X}$	n	$\overline{X}$
Father	12	4.08	9	3.33	11	4.36	12	4.42	5	4.20	8	3.75
Mother	12	4.17	9	3.11	11	3.91	12	4.42	5	4.00	8	4.00
Brother	12	3.92	6	4.00	9	4.22	9	4.00	5	3.80	8	3.50
Sister	8	3.38	9	3.44	7	3.43	9	3.56	4	3.00	7	3.28
Other relative	10	3.80	9	3.78	7	3.57	8	3.12	4	3.50	8	3.62
Male friend	12	4.25	9	4.44	11	4.55	12	4.17	5	3.60	8	4.38
Female friend	10	3.20	7	3.57	9	3.44	12	3.17	4	3.00	8	3.50
Classroom teacher	11	3.73	9	3.56	11	3.45	12	3.00	4	3.00	8	3.50
School coach	8	4.62	8	4.12	8	3.75	12	3.50	5	3.60	8	5.00
School counselor	8	3.50	4	3.50	8	3.38	9	3.00	3	3.00	7	3.28
Nonschool personnel	9	4.33	4	3.00	11	4.64	7	4.14	3	4.00	6	3.17
Other	6	4.00	4	3.50	5	4.00	6	4.33	1	3.00	4	3.00

tively, these data reveal that college athletes seem to receive encouragement from many agents and that it would be unusual to find socialization occurring when only a single agent functions.

3. Other Studies

In a study of the culture of young ice hockey players (Vaz, 1970), it was found that certain criteria were essential for initiation into the role of professional hockey player. For example, Vaz reported that aggressive fighting behavior is normative institutionalized conduct and as such is an integral facet in socializing future professional hockey players. Because this behavior is institutionalized, it becomes an integral facet of the role obligation of young hockey players and is learned by subsequent novices via formal and informal socialization.

In a study concerned with female college athletes, Malumphy (1970) reported that family influence is a major factor in college women competing in sport. She noted that the typical female athlete has family approval for her participation and competition and is encouraged by at least 50% of her significant others.

Finally, in a recent consideration of socialization into primary involvement (Kenyon, 1970b),* the path analysis technique† was utilized to attempt to explain the factors precipitating primary sport involvement (Fig. 6). According to this system, and for data of these kinds, a sizable portion of the variance was accounted for. Upon examination of the path coefficients, it appears that in order of importance the most influential factors were encouragement from nonschool personnel (e.g., neighbors and relatives outside the nuclear family) to participate in sport (.361), encouragement from school personnel (i.e., teachers, coaches) to participate in sport (.217), primary sport aptitude (.200), the type or size of the community (.198), and the sex (i.e., male) of the subject (.182). Thus, if an individual is a male, has a high level of sport aptitude, lives in a large city or town which has adequate facilities and instructors (i.e., a good opportunity set), and receives encouragement from significant others within and outside the school, the likelihood of his being socialized into a primary sport role is increased considerably.

In summary, it appears that college athletes and Olympic aspirants become interested and involved in sport by age 8 or 9; that they participate, usually with a great deal of success, in a number of sports before they begin to specialize in one sport; that they receive positive sanctions to become involved and to compete from a number of significant others, of which the family, peer group, and coaches appear to be the most influential; and, that although the general socialization process has a number of common ele-

---

* These data were collected in 1970 by members of a graduate class in the sociology of sport at the University of Wisconsin and analyzed by Mr. Tom Grogg. Their assistance is gratefully acknowledged.

† The basic purpose of path analysis is to estimate ". . . the paths which may account for a set of observed correlations on the assumption of a particular formal or causal ordering of the variables involved" (Duncan, 1966). In developing such a system the researcher is greatly aided by the construction of a path diagram which represents graphically the causal theory underlying the system. In the diagram causal relations are represented by unidirectional arrows, while noncausal relations, i.e., correlational, are represented by two-headed curvilinear arrows. Residual variables, also depicted by undirectional arrows, represent the unmeasured variation in the system which may consist of unknown variables, error, or both. Given the path model, it is possible to set up a system of regression equations which in turn supply standardized regression coefficients (betas) which are taken as the path coefficients. Thus, the path coefficient is the proportion of the standard deviation in the dependent variable for which the antecedent variable is directly responsible, with all other variables, including residuals, held constant.

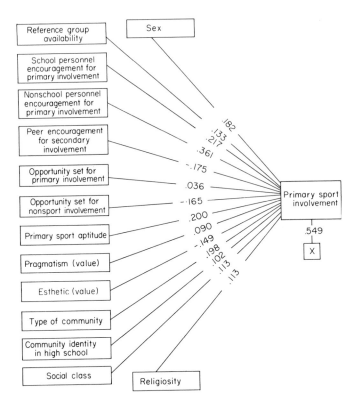

FIG. 6. Factors influencing primary sport involvement.

ments, noticeable between-sport, between-level, and between-life cycle stages occur.

B. SOCIALIZATION INTO ROLES ASSOCIATED WITH SECONDARY INVOLVEMENT

In addition to the study of primary involvement roles, a number of studies have been addressed to the problem of socialization into secondary sport roles. These are described here in terms of consumer roles and leader roles.

1. Socialization of the Sport Consumer

Recent attempts to understand the sport socialization process have been concerned with the role of sport consumer, that is, the viewer, listener, reader, or discussant. Since these studies have attempted to be somewhat more theoretical than some earlier attempts to describe socialization into primary

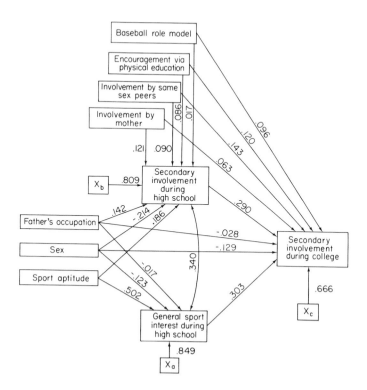

FIG. 7. Test of causal model accounting for socialization into secondary involvement in San Francisco Bay Area major league baseball.

roles, they have lent themselves to the use of path models and path analysis. For example, in a study designed to explain the degree of present enactment of two consumer sport roles, data were collected from 177 college juniors and seniors (Kenyon, 1970a). The observations for 13 variables were placed into the path model shown in Fig. 7, which represents the factors accounting for consumption of major league baseball, and into the model depicted in Fig. 8, which represents the factors influencing the consumption of the Mexico Olympic Games. The causal functioning of each variable is represented in the multistage (high school and college), multivariate (13 variables) models. Upon an analysis of the path coefficients and the residuals it is apparent that much of the variation in each system remains unexplained. Nevertheless, Fig. 7 suggests that the most influential factors, in order of importance, leading to baseball consumption are sport aptitude, general sport interest in high school (.303), involvement by the same sex peers in sport consumption (.143), and secondary involvement in baseball during high

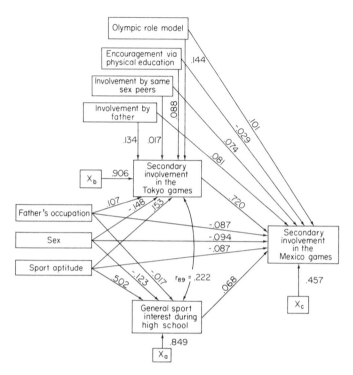

FIG. 8. Test of causal model accounting for socialization into secondary involvement in Mexico Olympic Games.

school (.290). Similarly, Fig. 8 indicates that the most important factors accounting for the consumption of the Mexico Olympic Games include involvement in the Tokyo Olympics (.720) and familiarity with athletes who participated in the Tokyo Olympic Games (.144).

It would appear that in comparing the baseball with the Olympic data, some differences emerge. For example, general sport interest in high school appears to contribute to secondary involvement in baseball in college but not to one's later interest in the Olympic games. In short, the factors which generate interest in one form of sport may not be the same factors which generate interest in other forms. Thus, as in primary sport role socialization, it may be necessary to account for considerable sport differences. Moreover, enactment of secondary sport roles as a young adult may not be very heavily dependent upon enactment as an adolescent. As shown in Fig. 7, to be interested in baseball during high school is only weakly related to interest in college.

In a secondary analysis of some factors hypothesized to be important in the sport socialization of male adolescents in Canada, the United States, and England, Kelly (1970) found that frequency of attendance at sport events was directly related to family size and indirectly related to age. For winter sports only, attendance at sporting events was positively related to social class background in Canada and the United States but negatively related in England.

In a study conducted at the University of Wisconsin in 1970 (Kenyon, 1970b), 96 college sophomores and juniors were interviewed to determine the factors accounting for socialization into the role of sport consumer. Although considerable unexplained variance remained, it was found that the major factor accounting for the learning of the role was primary sport involvement (Fig. 9); that is, the more they enacted primary sport roles, the greater their interest in other facets of sport, and, therefore, the greater their consumption of sport.

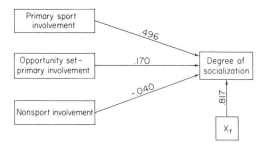

FIG. 9. Factors accounting for socialization into the role of sport consumer.

In a recent study of the sport consumer, Toyama (1971) investigated the influence of the mass media on the learning of sport language—specifically, football terminology. She found that 65% of her sample spent more than 3 hr/week consuming televised sport, and, of these, 36% spent 5 hr or more. It was not surprising then to find that 65% of the sample reported that they learned football terms by watching television. Another 16% learned the terms while actually playing the game. Thus, it appears that the mass media, especially television, is an important agent in cognitive sport socialization.

2. Socialization of the Sport Leader

Although many leadership roles can be found in a sport system, few of these have been the subject of empirical inquiry. Two exceptions, however, are the studies by Bratton (1970) and Pooley (1971).

Bratton (1970) investigated the demographic characteristics of executive members of two amateur sport associations. Although this study was limited to a descriptive analysis, it does suggest how and when one becomes a sport executive. Again, it should be noted that there are sport differences. For example, volleyball executives were 10 years younger than the swimming executives (29.5 vs 40.6 years), with the majority still active as players. It was also observed that fewer than 10% had occupations that fell below 65 (range 34–96) on the NORC occupational status scale, that Catholics were underrepresented, and that people of British origin held a larger proportion of the positions. Thus, it appears that to become an executive for an amateur sport association it is almost essential that one be a member of the higher socioeconomic class.

In a more detailed analysis, Pooley (1971) investigated the sociopsychological elements involved in the professional socialization of physical education students in the United States and England. He found a number of significant national and sex differences in the socialization experiences of the role recruits.

With respect to sport leader socialization in general, the Pooley findings suggest that the process differs cross-nationally as a function of psychological and sociocultural factors operating in the particular social situation, while Bratton's work suggests that the possession of social characteristics and the occupancy of certain roles are prerequisites for gaining access to executive positions in the sport system.

C. SUMMARY

Although the study of sport socialization has only just begun, the findings to date already suggest a number of generalizations. For example, the elite athlete emerges from an environment which was highly supportive; that is, he was exposed from an early age to an abundant opportunity set (middle-class values and ample facilities, equipment, and leadership) and much encouragement, reward, and reinforcement from a variety of meaningful others. The consumer of sport is likely to be or have been active as a contestant, but his consumption is dependent upon the social situation in which he finds himself. The sport leader, in addition to considerable primary involvement, tends to have emerged from a background laden with middle-class values. There are important differences in the socialization process for all forms of involvement. For example, the process varies somewhat from sport to sport and also differs for each stage in the life cycle.

III. A Basis for Future Research

Since most of the studies cited in Section II were not based upon a social role–social system approach, it has not been possible to test the general model presented in Section I. However, by considering the evidence already reported, and through a degree of conceptual extrapolation, we provide below a combination summary of present knowledge and a partial list of hypotheses requiring further empirical testing, using our general model (Figs. 3 and 4) as a frame of reference. In each case, it will be assumed that some minimal degree of motoric and cognitive aptitude is present; that is, the greater the motoric aptitude (sport aptitude), the greater the degree of primary role socialization, and the greater the cognitive aptitude (e.g., ability to comprehend and retain rules, strategy, statistics, and names), the greater the degree of primary or secondary role socialization. The major intervening variable in the model is "system-induced propensity for sport involvement." Thus, we have grouped the propositions around each of the four social systems: the family, the school, the peer group, and the community. Primary and secondary involvement socialization are treated separately. In each case only a few of the many possible propositions are given.

PROPOSITIONS RELATED TO SOCIALIZATION INTO A PRIMARY SPORT ROLE

A. *Family-Induced Factors*

1. The greater parental primary involvement in sport, the greater the degree of sport role socialization among offspring.
2. The higher sport is placed in the parental value hierarchy, the greater the degree of sport role socialization among offspring.
3. The earlier the involvement in sport, the greater the family support, and thus the greater the level of attainment.
4. The higher the ordinal position, the greater the degree of sport role socialization.
5. The higher the social class background, the more favorable the sport socialization situation, and thus the greater the degree of socialization.

B. *School-Induced Factors*

1. The higher sport is placed in the school's hierarchy of values, the greater the degree of sport role socialization.
2. The less sport that is presented in the mass media, the greater the socializing function of teachers and coaches.
3. The greater the frequency of positive sanctions received from school personnel for participation in school sport, the greater the propensity for sport involvement.
4. The greater the success of school athletic teams, the greater the propensity for sport involvement.

C. *Peer-Induced Factors*

1. The greater the peer involvement in sport, the greater the propensity for sport involvement.
2. The greater the positive sanctions from peers, the greater the propensity for sport involvement.

3. The greater the amount of sport-oriented face-to-face interaction with peers, the greater the propensity for sport involvement.
4. The higher sport is placed in the peer group's hierarchy of values, the greater the degree of sport role socialization.

D. *Community-Induced Factors*

1. The greater the role player's success, the greater the reinforcement from significant others.
2. The greater the role player's publicity, the greater the propensity for sport involvement.
3. The more urban the place of residence, the greater the degree of sport role socialization.
4. The greater the opportunity for direct and indirect consumption of sport, the greater the propensity for sport involvement.
5. The larger the community, the greater the propensity for sport involvement.

PROPOSITIONS RELATED TO SOCIALIZATION INTO A SECONDARY SPORT ROLE

A. *Family-Induced Factors*

1. The greater the parental identification with individuals or groups in the sport system, the greater the novice's degree of consumer role socialization.
2. The greater the size of the family, the greater the degree of consumer role socialization.
3. The higher the social class, the greater the degree of consumer role socialization.
4. The greater the number of sport-consuming significant others in the nuclear family, the greater the degree of consumer role socialization.
5. The greater the amount of direct and indirect consumption of sport by significant others in the nuclear family, the greater the degree of consumer role socialization.
6. The greater the number of males in the nuclear family, the greater the degree of consumer role socialization.

B. *School-Induced Factors*

1. Among their hierarchy of values, the higher sport is ranked by the school's instructional and administrative personnel, the greater the degree of consumer role socialization.
2. The greater the number of interscholastic sport teams, the greater the degree of consumer role socialization.
3. The greater the success of interscholastic sport teams, the greater the degree of consumer role socialization.
4. The greater the consumption of interscholastic sport by school personnel, the greater the degree of consumer role socialization.

C. *Peer-Induced Factors*

1. The greater the peer consumption of sport, the greater the degree of consumer role socialization.
2. The greater the peer propensity for sport involvement (e.g., a loyalty for, or an identity with, an entity or group in the sport system), the greater the degree of consumer role socialization.

3. Among their hierarchy of values, the higher sport is ranked by members of the peer group, the greater the degree of consumer role socialization.

D. Community-Induced Factors

1. The greater the primary involvement in sport, the greater the degree of consumer role socialization.
2. The greater the access to the mass media, especially television, the greater the degree of cognitive sport socialization.
3. The greater the identification with one sport team, the greater the degree of consumer role socialization.

IV. In Conclusion

The reader will have concluded by now that the "state of the art" explaining socialization into the many roles associated with sport involvement is not well advanced. Indeed, in 10 years' time, much of that which has been reported in this chapter will seem elementary, at best. The success of future efforts, however, will depend upon a judicious combination of the best theoretical and empirical methods, the willingness to make many observations over time, the concentration upon smaller more manageable social systems (but always within the context of a larger model),* and an abiding pursuit of parsimony (such as Heise's "theory trimming" approach, 1969). Finally, in view of the evidence suggesting sport and national differences, much comparative work will be needed to test the generality of explanatory models.†

References

Bandura, A. (1969). In "Handbook of Socialization Theory and Research" (D. A. Goslin, ed.), pp. 213–262. Rand McNally, Chicago, Illinois.

Bandura, A., and Walters, R. H. (1963). "Social Learning and Personality Development." Holt, New York.

Blalock, H. M., Jr., (1969). "Theory Construction: From Verbal to Mathematical Formulations." Prentice-Hall, Englewood Cliffs, New Jersey.

Blau, P. M., and Duncan, O. D. (1967). "The American Occupational Structure." Wiley, New York.

Bratton, R. (1970). Can. Ass. Health, Phys. Educa., Recreation 37, 26–28.

Brim, O. G. (1958). Sociometry 21, 343–364.

Brim, O. G., and Wheeler, S. (1966). "Socialization After Childhood: Two Essays." Wiley, New York.

Campbell, E. O. (1969). In "Handbook of Socialization Theory and Research" (D. A. Goslin, ed.), pp. 821–859. Rand McNally, Chicago, Illinois.

Clausen, J. A., ed. (1968). "Socialization and Society." Little, Brown, Boston.

Duncan, O. D. (1966). Amer. J. Sociol. 72, 1–16.

Heise, D. R. (1969). In "Sociological Methodology" (E. F. Borgatta, ed.), pp. 38–73. Jossey-Bass, San Francisco, California.

Helanko, R. (1957). Acta Sociol. 2, 229–240.

* See B. D. McPherson, "Socialization into the Role of Sport Consumer: A Theory and Causal Model." Unpublished Ph.D. Dissertation, University of Wisconsin, 1972.

† A cross-national study on leisure role socialization was initiated by the International Committee for Sociology of Sport in 1972.

Homans, G. C. (1950). "The Human Group." Harcourt, New York.
Kelly, C. (1970). Unpublished M. Sc. Thesis, University of Wisconsin, Madison.
Kemper, T. D. (1968). *Amer. Sociol. Rev.* **33**, 31–45.
Kenyon, G. S. (1968). Unpublished study, University of Wisconsin, Madison.
Kenyon, G. S. (1969). *In* "Aspects of Contemporary Sport Sociology" (G. S. Kenyon, ed.), pp. 77–100. Athletic Institute, Chicago, Illinois.
Kenyon, G. S., and Grogg, T. (1969). Unpublished study, University of Wisconsin, Madison.
Kenyon, G. S. (1970a). *Int. Rev. Sport Sociol.* **5**, 191–203.
Kenyon, G. S. (1970b). Unpublished study, University of Wisconsin, Madison.
LeVine, R. A. (1969). *In* "Handbook of Socialization Theory and Research" (D. A. Goslin, ed.), pp. 503–542. Rand McNally, Chicago, Illinois.
Lüschen, G. (1969). *In* "Aspects of Contemporary Sport Sociology" (G. S. Kenyon, ed.), pp. 57–76. Athletic Institute, Chicago, Illinois.
McCandless, B. R. (1969). *In* "Handbook of Socialization Theory and Research" D. A. Goslin, (ed), pp. 791–819. Rand McNally, Chicago, Illinois.
McPhee, W. N. (1963). "Formal Theories of Mass Behavior." Free Press, New York.
McPherson, B. D. (1968). Unpublished study, University of Wisconsin, Madison.
McPherson, B. D. (1971). *Pap., Int. Sympo. Sociol. Sport, 3rd, 1971.*
Malumphy, T. M. (1970). *Quest* **14**, 18–27.
Moore, O. K., and Anderson, A. R. (1969). *In* "Handbook of Socialization Theory and Research" (D. A. Goslin, ed.), pp. 571–614. Rand McNally, Chicago, Illinois.
Parsons, T. (1961). *In* "Theories of Society" (T. Parsons *et al.,* eds.), pp. 30–79. Free Press, New York.
Pooley, J. C. (1971). Unpublished Ph.D. Dissertation, University of Wisconsin, Madison.
Pudelkiewicz, E. (1970). *Int. Rev. Sport Sociol.* **5**, 73–103.
Rehberg, R. A., Sinclair, J., and Schafer, W. E. (1970). *Amer. J. Sociol.* **75**, 1012–1034.
Reiss, A. J., Jr. (1966). Unpublished manuscript, University of Michigan, Ann Arbor (cited by Bandura, 1969).
Roethlisberger, F. A. (1970). Unpublished M.Sc. Thesis, University of Wisconsin, Madison.
Rosen, B. C. (1956). *Ameri. Sociol. Rev.* **21**, 203–211.
Sewell, W. H. (1963). *Anna. Amer. Acad. Political Social Sci.* **394**, 163–181.
Sewell, W. H., Haller, A. O., and Portes, A. (1969). *Amer. Sociol. Rev.* **34**, 82–92.
Sewell, W. H., Haller, A. O., and Ohlendorf, G. (1970). *Amer. Sociol. Rev.* **35**, 1014–1027.
Smelser, N. J. (1962) "Theory of Collective Behavior." Free Press, New York.
Spady, W. G. (1970). *Amer. J. Sociol.* **75**, 680–702.
Stone, G. P. (1969). *In* "Aspects of Contemporary Sport Sociology" (G. S. Kenyon, ed.), pp. 5–16. Athletic Institute, Chicago, Illinois.
Toyama, J. S. (1971). Unpublished M.Sc. Thesis, University of Wisconsin, Madison.
Turner, J. H. (1970). *Sociometry* **33**, 147–165.
Vaz, E. (1970). *Pap. World Congr. Int. Sociol. Ass., 7th, 1970.*
Webb, H. (1969). *In* "Aspects of Contemporary Sport Sociology" (G. S. Kenyon, ed.), pp. 161–187. Athletic Institute, Chicago, Illinois.
Woelfel, J., and Haller, A. O. (1971). *Amer. Sociol. Rev.* **36**, 74–87.
Zetterberg, H. L. (1966). "On Theory and Verification in Sociology." Bedminister Press, Totowa, New Jersey.
Zigler, E., and Child, I. L. (1969). *In* "The Handbook of Social Psychology" (G. Lindzey and E. Aronson, eds.), 2nd ed., Vol. 3, pp. 450–589. Addison-Wesley, Reading, Massachusetts.

Ethnic and Cultural Factors in the Development of Motor Abilities and Strength in American Children

Robert M. Malina

Populations differ in a variety of biological and cultural characteristics, and motor performance and strength are no exception. This section deals with comparative studies of motor development, performance, and strength in the three major ethnic and/or racial populations within the vast American culture complex—American whites, American Negroes, and Americans with Spanish surnames, i.e., Mexican-Americans and Puerto Ricans. Although the present sociopolitical climate has resulted in a variety of ethnic and/or racial labels, those indicated above will be used throughout the review. Interpretations of developmental, performance, and strength differences and/or similarities are also included. These, however, are primarily speculative since motor variables are rarely included in systematic analyses of child socialization, personality development, and cognitive development. Further, motor development and performance are only cursorily mentioned in descriptions of childhood and adolescence in American society. This review is not concerned with the specialized, highly trained performer, the athlete, nor is it concerned with those factors that might motivate individuals to high levels of performance under the stress of competitive athletics. It is concerned, rather, with the average or normal range of variation in motor development, performance, and strength.

I. Infancy and Early Childhood

A. MOTOR DEVELOPMENT

The sequence of motor development during the first 5 years of life has been described in detail in many early studies. These in general are concerned with the description of stages and timetables for the attainment of postural, locomotor, and prehensile control during the first 2 years of life, and then of various motor achievements during the preschool years. The approaches used in these studies are basically descriptive and are "age and stage" oriented, i.e., children are expected to demonstrate certain levels of motor competence at specified chronological ages.

Specific skills such as jumping, throwing, and running have also received detailed developmental analysis. These observations, however, are concerned with the sequential changes in the development of mature patterns of motor skills. In most instances they are limited to relatively small samples of children.

The reported motor developmental data for infancy and early childhood are with few exceptions derived primarily from samples of middle-class American white children. These observations serve as the standards of comparison against which the motor development and early performance of children from other ethnic and/or racial groups are assessed. Nevertheless, the early motor development of children is generally uniform in sequence with considerable variation in developmental rate. Variations in the motor development sequence which might be related to different child rearing practices, opportunity for practice, availability of toys and equipment, and extent of infantile stimulation are apparent. Hence, the range of normal variation, though very wide in itself, is perhaps further extended by various cultural and subcultural practices, demands, and sanctions. It is within this framework that the subsequent discussion of motor development and performance of American white, American Negro, and Mexican-American or Puerto Rican children is set. The available data for the last-mentioned ethnic groups, however, are extremely limited and almost totally lacking for infancy and early childhood.

TABLE I

MEAN POINT SCORES ON THE BAYLEY INFANT SCALE OF MOTOR DEVELOPMENT FOR
NEGRO AND WHITE INFANTS[a]

Age	White		Negro	
(months)	n	X̄	n	X̄
1	41	6.34	41	6.39
2	45	9.31	37	9.89
3	42	12.12	41	13.39
4	47	14.57	31	16.29
5	41	18.83	40	21.25
6	44	25.73	42	25.76
7	47	28.47	41	30.46
8	61	34.41	51	35.67
9	54	37.13	44	38.95
10	53	40.11	40	41.32
11	43	42.84	26	44.00
12	49	44.22	43	45.88
13	44	46.45	36	47.08
14	48	48.33	38	48.68
15	46	49.35	36	48.39

[a] From Bayley (1965, p. 405).

There is a considerable body of literature comparing the early motor development of Negro and white children reared under a variety of circumstances. These studies compare samples of Negro and white children directly, or they compare Negro children to standards established on white children. The available data indicate, in general, motor advancement of the Negro compared to the white infant. The advancement, sometimes termed "motor precocity," is apparent in the newborn and persists during the first 2 or 3 years of life (Bayley, 1965; Knobloch and Pasamanick, 1953; Graham et al., 1956; McGraw, 1931; Pasamanick, 1946; Scott et al., 1955; Sessoms, 1942; Williams and Scott, 1953). These observations, however, are not without exception. Several reports have indicated minor or no consistent differences in the early motor development of American Negro and white children (Knobloch and Pasamanick, 1958, 1960; Solomons and Solomons, 1964; Walters, 1967).

Results of Bayley's (1965) comprehensive study of early development of Negro and white infants,* 1 through 15 months of age, from 12 cities

* Bayley's sample also included Puerto Rican infants; however, the sample sizes were too small for detailed comparisons.

throughout the United States, are presented in Table I. The mean scores for Negro infants are higher at each age except 15 months. The Negro advancement in motor development includes most of the motor tasks in the Bayley scales: "Of the 60 items in the scale, no item for which there is a complete age distribution on percentages passing favors the white babies by as much as half a month" (Bayley, 1965, p. 406). Further, superiority in any one class of motor behavior was not responsible for the Negro motor advancement.

Reports considering the development of Negro and white children between 2 and 6 years of age are limited and inconclusive. In an early study, Van Alstyne and Osborne (1937) noted better rhythmic patterns among

TABLE II

PERCENTAGE OF NEGRO AND PUERTO RICAN PRESCHOOL CHILDREN PASSING GROSS MOTOR ITEMS RELATIVE TO THE DENVER STANDARDS[a]

Item	n	Age group	% Passing	Denver sample norms: age (years) at which given % of population pass items[b]
Balance on 1 foot,	36	4.1–4.5	82.9	75% at 3.9
5 sec.	35	4.6–5.0	97.0	90% at 4.3
	33	5.1–6.0	100.0	
Balance on 1 foot,		4.1–4.5	68.6	50% at 4.5
10 sec.		4.6–5.0	87.9	75% at 5.0
		5.1–6.0	97.1	90% at 5.9
Hops on 1 foot		4.1–4.5	89.1	75% at 4.0
		4.6–5.0	97.1	90% at 4.9
		5.1–6.0	97.1	
Catches bounced ball		4.1–4.5	68.6	50% at 3.9
		4.6–5.0	81.8	75% at 4.9
		5.1–6.0	97.1	90% at 5.5
Heel-to-toe walk		4.1–4.5	65.7	75% at 4.2
		4.6–5.0	93.9	90% at 5.0
		5.1–6.0	97.1	
Backward heel–toe		4.1–4.5	28.6	25% at 3.9
		4.6–5.0	64.1	50% at 4.7
		5.1–6.0	74.1	75% at 5.6

[a] From Sandler *et al.* (1970, p. 777).

[b] Denver norms are from Frankenberg and Dodds (1967, p. 186). The Denver norms are derived from a primarily white sample, although small samples of Negro children and children with Spanish surnames are included.

Negro children 2–6 years of age in regulated and free rhythm situations. Differences between the Negro and white children, however, were more marked for the simple than for the complicated rhythm patterns. Sessoms (1942) compared a sample of low-income Negro preschool children with a sample of Iowa white children on a group of fine and gross motor tasks. Negro boys and girls at 3 years of age were in advance of the white children in hopping, skipping, walking, and step and ladder climbing. At 4 years of age, differences were negligible or in favor of the Iowa white series. In fine motor tasks, Negro children had better scores at both ages 3 and 4. In fact, the manual achievement scores* for Negro children were about 1 year in advance of the scores for white children. The data of Rhodes (1937), on the other hand, suggested little, if any, difference between Negro and white children 2½–5½ years of age in speed of walking a path and three fine motor tasks (needle threading, three-hole test, and stylus tapping). Similarly, Knobloch and Pasamanick (1960) found no significant differences in the Gesell fine and gross motor developmental quotients in Negro and white children 3 years of age. More recently, Sandler *et al.* (1970) compared a small sample of lower-class urban Negro and Puerto Rican children, 4 through 6 years of age, to the Denver Developmental Screening Test norms. In general, the children performed consistently with, if not slightly in advance of, the norms (Table II) on six measures of gross motor behavior. Performance on the fine motor adaptive items of the developmental test showed appropriate functioning on visual motor items involving a model to copy or imitate. Those fine motor adaptive items requiring "cognitive operations" (draws a man and picks longer line) were below the Denver norms.

B. INTERPRETATIONS

The foregoing seems to suggest advanced motor development in Negro children during the first 2 or 3 years of life, with the differences being more marked at the younger ages and less apparent at the later ages of early childhood. Explanations for the observed differences vary. Bayley (1965, pp. 408–409) suggested that ". . .a genetic factor may be operating. That is, Negroes may be inherently more precocious than whites in their motor coordinations." It should also be noted that Negro infants are also advanced in the appearance of the dentition and skeletal ossification centers compared to standards derived from white children (Malina, 1969; Roberts, 1969; Tanner, 1962). Knobloch and Pasamanick (1958), on the other hand, did not find any differences in motor development in a large sample of Negro

---

* The manual achievement test items included the following: pouring sand, screwing lids, turning pages, piling blocks, cutting paper, stringing beads, modeling clay, carrying water, hammering nails, and pulling out nails.

and white infants, leading them to conclude that previous differences, including their own earlier observations on an independent sample (Knobloch and Pasamanick, 1953; Pasamanick, 1946), ". . . could not be attributed to innate racial characteristics" (p. 132). Knobloch and Pasamanick (1958) appeared to favor an explanation related to changes in child rearing practices.

Studies of child rearing practices across social class and ethnic groups, however, have been concerned primarily with socialization and personality development, with emphasis on feeding practices, toilet training, age of weaning, independence, and aggression. More recently, attention has been given to cognitive development and academic motivation relative to rearing practices especially in minority and/or culturally, disadvantaged children. Most child rearing studies have emphasized only the attitudes of the mother, to the neglect of the father and the child's spontaneous peer groups. Further, child rearing studies have not been concerned with motor development and behavior as the central focus of study except for the age of onset of walking.

In a review of child care practices over a 25-year period between 1930 and 1955, Bronfenbrenner (1958) reported a shift in the rearing attitudes of middle-class and working-class mothers. From about 1930 to the end of World War II, working class mothers were generally more permissive than middle-class mothers, but after World War II the middle-class mothers became more permissive in practices related to eating, sleeping, and toilet training. Similarly, middle-class mothers have become more permissive toward the child's expressed desires, but they have greater expectations for the child. Working-class mothers tended to demand obedience and discipline, while middle-class mothers emphasized the importance of self-direction. Waters and Crandall (1964), evaluating maternal behavior between 1940 and 1960, noted that mothers, regardless of social class, showed a tendency to be less coercive and more permissive than before. While child rearing practices differ among social classes, there are some practices that cut across social class lines. The extent to which the observed rearing practice differences and/or similarities affect the child's motor development, however, remains to be answered.

Implicit in the comparisons of child rearing is the assumption of a generalized American family pattern, both white and Negro, an assumption that can be legitimately questioned. Race and poverty are closely related in this country, i.e., many Negro families are living at or slightly above the poverty level. Hence, the number of Negro families that can be classified as middle class is relatively small but steadily increasing. Finally, it would be myopic to view the child rearing situation as static; rather, practices are subject to change, gradual or abrupt.

TABLE III

GROSS MOTOR DEVELOPMENTAL QUOTIENTS (GESELL) OF NEGRO INFANTS 4–18
MONTHS OF AGE GROUPED ACCORDING TO SOCIOECONOMIC STATUS AND REARING
ATMOSPHERE OF THE HOME[a]

Age	Socioeconomic status	
	Lower	Higher
Below 37 weeks	110.7	103.9
38 weeks plus	118.5	111.0
Socioeconomic status	Rearing atmosphere	
	Permissive	Rigid
Lower	114.6	101.6
Higher	113.4	100.5

[a] From Williams and Scott (1953, pp. 112 and 116).

How might child care practices affect motor development and perhaps account for the observed differences between Negro and white infants? Williams and Scott (1953) and Scott *et al.* (1955) have indicated socioeconomic variables as significant in explaining the Negro–white difference. More specifically, they indicate an important role for the child rearing correlates of socioeconomic status—the more permissive atmosphere characteristic of the lower social classes enhances and facilitates motor development. Results of the Williams and Scott (1953) study are summarized in Table III. In the words of Williams and Scott (1953, p. 118):

> . . . the picture for the subjects from the low socio-economic group is a much more active, free and uninhibited one; they have a closer, much more direct and manipulating relationship to their environment than the babies in the high socio-economic group.

In an earlier study, Bayley and Jones (1937) reported no relationship between socioeconomic variables and age of first walking in a sample of white children. They did, however, note a tendency toward an increased number of negative correlations between socioeconomic variables and motor scores, suggesting more rapid development in children from the lower socioeconomic group.

More recently, however, Knobloch and Pasamanick (1958) and Walters (1967) reported social class differences among Negro infants in motor de-

TABLE IV

GROSS AND FINE MOTOR DEVELOPMENTAL QUOTIENTS (GESELL) FOR NEGRO AND WHITE INFANTS APPROXIMATELY 40 WEEKS OF AGE (RANGE 34–69 WEEKS) GROUPED ACCORDING TO ECONOMIC STATUS AND EDUCATION OF THE MOTHER[a]

	White			Negro		
Economic fifth	Gross	Fine	Economic fifth	Gross	Fine
1 and 2	114.8	95.7	1	108.1	97.9
3	115.5	98.9	2	115.1	99.5
4 and 5	114.0	98.6	3, 4, and 5	116.8	100.1
Total	114.7	97.6	Total	113.4	99.2
Education of mother	Gross	Fine		Gross	Fine
Eighth grade or less	114.8	98.2		110.4	97.6
Ninth to eleventh grades	116.9	96.3		117.2	99.5
Twelfth grade or more	112.3	99.4		112.0	102.8
College	113.7	98.3		[b]	—

[a] From Knobloch and Pasamanick (1958, p. 129).
[b] None of the Negro mothers in the study sample had completed college.

velopment but not among white infants. Further, the results were in the opposite direction from that suggested by Williams and Scott (1953), Scott et al. (1955), and Bayley and Jones (1937), i.e., the higher-class Negro infants were more advanced in gross motor development than the lower social class infants (Table IV). Note also that children of Negro mothers with higher levels of education also had higher motor developmental quotients. On the other hand, Bayley's (1965) extensive analysis reported earlier did not show any differences in motor scores on the basis of educational level of either the child's mother or father. Geographic location was likewise systematically unrelated to motor development scores in the Bayley series.

Comparing Negro and white infants (12, 24, and 36 weeks of age) equated for socioeconomic status and other variables that might affect early development (e.g., prenatal care and birth weight), Walters (1967) found the high socioeconomic Negro group advanced on the Gesell developmental scales over all other groups, especially the low socioeconomic Negro group. Willerman et al. (1970) reported similar results for white infants, 8 months of age. Infants from high socioeconomic backgrounds had significantly higher Bayley motor scores than those from low socioeconomic backgrounds.

The socioeconomic hypothesis, though appealing, is by no means clear-cut and is not consistent with the child rearing studies mentioned earlier. Bronfenbrenner's (1958) review specifically indicated a shift toward greater permissiveness in the middle-class mother (and presumably higher social class than the working-class mother) after World War II. This appears consistent with the observations of Knobloch and Pasamanick (1958) and Walters (1967), but it is contrary to Williams and Scott (1953) and Scott et al. (1955). Yet all samples are post-World War II, with the last two mentioned attributing advanced motor behavior among lower-class Negro infants to more permissive child rearing practices. Perhaps the trends in child care practices derived from studies of eating, sleeping, toilet training, and independence training are not applicable to motor development. Or perhaps child rearing trends established on white American families do not apply to American Negro families. Or social class criteria for Negro and white samples might be different. As indicated earlier, the number of middle-class Negro families is probably very small. Although aspirations of Negro and white parents for their children are similar, the Negro family life styles, values, and patterns of socialization are somewhat different from the white cultural tradition (Schulz, 1969; Frazier, 1948; Baughman and Dahlstrom, 1968; Young, 1970).

Nevertheless, the fact that differences in early motor development between samples of American Negro and white children are apparent in most, but not all, studies is significant in itself. Why do findings of separate studies differ? Sampling problems and test items used, of course, must be recognized. However, does the socioeconomic explanation, particularly the permissiveness theme, operate by permitting full development of the infant's genotype, which might differ between populations? Knobloch and Pasamanick (1958, p. 131) seemed to imply something along this line in the following comment: ". . .child-rearing makes no difference in gross motor development in infancy, except when certain biological or cultural thresholds are exceeded."

The correlates of permissiveness and acceptance which might enhance motor development need further exploration. More detailed study of child rearing practices within the context of specific social and/or subcultural units within the vast American culture complex is likewise needed so that the full cycle of social class influence will be recognized. There are, for example, very few systematic long-term studies of child rearing of American Negro and white children in general, and none, to the writer's knowledge, that have the child's motor development as the central focus of study. Much that can and has been said about child rearing and motor development must be inferred from general rearing studies, which have traditionally concerned themselves with personality development and socialization within a psychoanalytic framework.

The amount and extent of stimulation in infancy is recognized as an important element affecting the developmental progress of the child. For example, infants in institutions, lacking the warmth of human contact, are generally delayed or retarded in the attainment of motor development milestones. How much stimulation, on the other hand, is necessary to support normal development? Does excessive infantile stimulation enhance early motor development?

Several reports suggest a great deal of indulgence in and stimulation of infants in Negro households. Schulz (1969, p. 63), describing babyhood in a ghetto, indicated that:

> A great deal of affection and attention is given them until they begin to achieve some degree of autonomy. Then their siblings begin to lose interest in them. Babies are like dolls, things to play with and cuddle, but they soon become more trouble than they are worth. Only one child, the last child, can play the role of "family pet" for any length of time; the rest must quickly relinquish babyhood and the center of attention in the family world.

Young (1970), in a detailed description of childhood in a Southern Negro community, similarly indicated the considerable handling of infants to the extent that most of their first year of life is spent in direct human contact.

> Everyone with the slightest claim to do so holds and plays with the baby. . . . The baby finds that its environment is almost wholly human. Cribs, baby carriages, and highchairs are almost never seen. The baby is held and carried most of the time, and when it is laid down it is seldom without company. . . . the baby experiences many different types of people of all ages, all of whom hold the baby and play with it (Young, 1970, pp. 275–276).

In contrast to the high level stimulation received from the human environment, the infant's

> . . . explorations of the inanimate environment are limited. Few objects are given to babies or allowed them when they do get hold of them. Plastic toys are almost the only objects ever seen in the hands of a prewalker. . . . Babies' reachings to feel objects or surfaces are often redirected to feeling the holder's face, or the game of rubbing faces is begun as a substitute (Young, 1970, p. 279).

Concerning motor developmental milestones and sequences, constant holding prevents infants from crawling on the floor, yet the infants are encouraged to be precocious in talking and walking.

> Babies learn to stand on people's laps. They are jumped generously when they want to flex and stiffen their leg muscles. They are stood momentarily

on the floor between the mother's or father's legs and jumped. But they are seldom allowed to stand alone, sink down on the floor or pull themselves up. All their standing is done on a lap. Babies are observed to walk at ten months, and claims were recorded of walking at nine months, while those learning to walk after twelve months seem relatively few. The baby who is ready to walk is stood on the floor; in the midst of great excitement it lunges and steps to waiting arms and is stood up when it falls either to try again or to be put in the lap (Young, 1970, p. 280).

Changes introduced by walking are many, including new expectations. The 1-year-old child is called the ". . .knee-baby. . .the ambulatory child who leaves the mother and returns to her, who stands at her knee whether there is a new baby in her lap or not" (Young, 1970, p. 280). In this context, the child plays fetching games with his mother. The child likewise has many experiences with older children of the neighborhood once the child can walk: "The youngest child is the toy of all children who are six or older and who can claim the right to carry it" (Young, 1970, p. 281).

Hence, the first year of the Negro infant's life is spent in an essentially human environment to the neglect of the inanimate environment. The child is likewise accustomed to a high degree of stimulation resulting from the almost constant human contact. This stimulation persists through the second year, though perhaps not as intense. The foregoing rearing atmosphere would seem to offer a fertile substrate for motor development with opportunity to practice most newly developing motor powers. However, after 2–3 years of age when the "knee-baby" is in competition with a younger sibling, there is a change in the mother–child relationship. The child

. . . is expected to transfer his dependence to the gang of older children in the family and to the oldest child in it who is in charge of the gang. . . . parents no longer play with children after the knee-baby position. The stimulation that was so intense ends almost entirely (Young, 1970, p. 282).

Interestingly, it is at this time, ages 2–3 and thereafter, that there are inconsistencies and a lack of clear-cut differences in motor development comparisons of Negro and white children.

It is possible within limits, to infer from these observations of a Southern rural community to the crowded conditions of poverty stricken urban dwellers, most of whom are Negro. Poverty and race in the United States are practically coterminous, and crowded living conditions are characteristic of poverty areas. Crowded conditions, with many people around, perhaps provides more stimulation for the developing infant, at least initially. However, after the child has learned to walk, the crowded conditions may hamper motor development since the child's locomotor and manipulative activities may interfere with the adults' activities. This may result in a lessened positive interaction with the parent's removing this as a source of stimulation; hence, further locomotor and manipulative development may also be curbed. On the other hand, the crowded conditions, at least by the white middle-class standards, might be only a sleeping-crowding. Otherwise, little time is spent in the apartment or house by anyone except the youngest children and mother.

In contrast to the intense human stimulation and relative lack of contact with toys and other inanimate objects experienced by Negro infants (Young, 1970), the middle-class white infant appears to get a balance of both kinds of infantile experience. The white infant is reared in an environment relatively free from physical restraint, many toys, and perhaps not nearly as much person-to-person contact. Cribs, playpens, high chairs, and walkers, in addition to an abundance of toys, afford the developing child numerous opportunities to exercise his rapidly developing motor powers.

White middle-class mothers apparently pay close attention to the child's needs and readiness in contrast to the rearing atmosphere in the Negro family described by Young (1970) and Schulz (1969). Fischer and Fischer (1963, p. 942), in their discussion of early socialization in "Orchard Town, U.S.A.," indicate this concern of white middle-class mothers:

> . . . a positive concern with training and controlling the infant in certain respects emerges at an early age. . . . Mothers recognize various signs of "readiness": when a child can hold a cup, when it can sit up in a high chair . . .

Fischer and Fischer also noted considerable anxiety in mothers concerning feeding and weaning. Young (1970, p. 277) did not notice such anxiety in Negro mothers; rather emphasis was on

> . . . the interpersonal exchange aspects of eating. . . . Need, nourishment, succoring, routine—all possible aspects of eating and emphasized in White culture—are not usual concepts.

A high degree of motor stimulation was also characteristic in Fischer and Fischer's white series. Parents, especially mothers, were concerned with the infant's physical maturity, as evidenced in emphasis on comparative ages for the attainment of developmental milestones. The rearing atmosphere indicated not nearly as much carrying and handling as noted earlier for Negro infants.

> A mother also may hold the baby while it practices walking. Much of the day the baby may be in a crib or playpen with sides of convenient height for the baby to grasp and practice standing or walking. At times the mother may

stimulate motor activity by jiggling the baby on her knee or by other physical play. Some adults, especially men, may lift the baby over their heads or swing it around. The baby comes to like this eventually, and this sort of roughhousing continues in later childhood between men and boys, though not so long for girls.

The baby is given a number of toys which encourage motor activity, such as rattles, and later bouncing horses and wheeled toys which jingle or make other pleasant noises when pulled or pushed. Some of these are advertised as aiding the baby's physical development by encouraging exercise (Fischer and Fischer, 1963, pp. 945–946).

Interestingly, though the parents are concerned with their child's social maturity, interactions with other persons are limited and spaced [which is in marked contrast to the constant handling and face-to-face contact described by Young (1970) for Negro infants]:

> . . . prolonged social life is not considered desirable for babies and is felt to tire them. Most of them spend a good part of each day alone in a crib or playpen or in a fenced-in yard. Children learn early in Orchard Town that interaction with others is spaced, separated by periods of withdrawal. . . . Such contacts as the baby has with other human beings are not marked by close bodily contacts as in many societies. There are two opposing needs considered here—one the early need for warmth supplied by close bodily contact, and the other the pleasure in the free movement of limbs. In this society the second is highly satisfied at the expense of the first (Fischer and Fischer, 1963, p. 947).

Limitations inherent in making comparisons of rearing practices as done in the preceding paragraphs are many and should be kept in view. First of all, there is undoubtedly considerable variability in child rearing practices within and across subcultures. Then, most of our rearing data for white infants are derived from middle-class families; hence, direct comparisons of lower-class Negro infants to such observations are difficult. Also, since the number of middle-class Negro families is relatively small, we should strive, if possible, to compare lower-class white and Negro families. Nevertheless, although there are differences in rearing practices such as infantile stimulation during infancy, it is difficult to pinpoint specific factors that might be responsible for the observed differences between Negro and white infants in their motor developmental status.

In another detailed study of the rural South, Baughman and Dahlstrom (1968) reported child rearing differences between Negro and white families. Based upon detailed interviews of mothers of kindergarten-age children, different family life styles were noted for Negro and white families living in close proximity to each other. The differences related to family structure, resources, and activity patterns. Negro mothers favored whippings as a means of punishment, while white mothers used banishment. White mothers expressed greater use of praise than Negro mothers, while the latter reported more frequent use of material rewards. When interviewed about the child's developmental history, including motor behavior, the mothers generally evaluated their children's motor skills favorably. Negro mothers reported early patterns of motor development more frequently than whites, especially for girls. White mothers, on the other hand, reported more disturbing incidents in the lives of their children. Two specific racial differences appeared in the pattern of interview answers concerning the child's current motor skills:

> First, Negro children received fewer judgments of good for their skill in coloring pictures than white children did. (The Negro boys were judged especially poor in this regard.) Second, when mothers were asked if their children had motor talents in addition to those about which inquiry had been made, white mothers responded affirmatively more often than Negro mothers (Baughman and Dahlstrom, 1968, p. 374).

In general, white mothers gave more judgments of good concerning their child's motor skills. When mothers were asked to rank which kinds of behavior (motor, social, or intellective) of the child created "pleasurable feelings" in them, both Negro and white mothers ranked motor behavior third. Among white mothers, the ranking was social, intellective, and motor; among Negro mothers the ranking was intellective, social, and motor.

The following account of dependency and guilt from the Baughman and Dahlstrom study might have some implications for the children's motor development:

> . . . both Negro and white mothers more frequently perceive their sons rather than their daughters as needing encouragement. This sex difference is accentuated among the white children. With respect to racial differences, there is a suggestion that Negro mothers may, in a certain sense, attempt to keep a tighter rein on their children or be more severe with them than white mothers. Thus, there seems to be more certainty that they will punish suspected rule violations; also, that they will keep close track of the child's whereabouts. As to the children themselves, there is an indication that the behavior of white children may be relatively mature compared to that of Negro children. For example, white children in general are described as playing better alone, as being more ready to admit rule violations, and as being less likely to become upset when mother leaves than is true of Negro children. On the other hand, it is also true that Negro children appear to have greater freedom of movement about their neighborhoods than white children do. Why this is so is not clear, however, since Negro mothers also claim to keep closer track of their children than white mothers do (Baughman and Dahlstrom, 1968, p. 391).

Abrahams (1972) suggested that Negro mothers rely upon a reputation for having control over their children to maintain their sense of responsibility. They are apparently judged as mothers on the orderliness of their household, which means how stern they are with children who step out of line. But they expect their older children to maintain the working of this order, and one can thus observe Negro children apparently wandering away from the maternal area.

Freedom of movement about the neighborhood possibly provides the child with a range of situations in which he can try his developing skills, whereas close supervision perhaps functions to inhibit explorations via motor channels. Within this freedom-restriction perspective, the significance of sex differences in encouragement among white children more so than among Negro children is not clear. It is probably related to the generally more favorable position of girls in the Negro family setting. Fischer and Fischer (1963) also noted the tendency for their Orchard Town mothers to "push" their sons more than their daughters. The foregoing observations point to the need of careful study of maternal control strategies and regulatory techniques as they might affect the child's developing motor skills.

The preceding has been concerned primarily with early motor development and possible relationships to the variables associated with ethnicity, socioeconomic status, and child rearing practices. When we attempt to look at subsequent motor development and performance during the preschool years, another concept related to socioeconomic status and child rearing practices must be brought up—that of the "disadvantaged child" (Gordon, 1965; Grotberg, 1969; Richmond, 1970). It must be recognized at the outset that there is probably no such thing as a typical "socially disadvantaged child"; rather, there are probably a great variety of such children (Gordon, 1965). Reviews describing these children, however, generally do not include data on their motor characteristics or the possible effects of their environment on their motor behavior and pursuits. Pavenstedt (1967), for example, noted generally superior gross motor coordination but a lack of "motoric caution" in a small sample of preschool children from lower-class, "disorganized, maximally deprived families" both Negro and white. The motor characteristics of these children are simply described, with little, if any, supporting quantitative data. For example:

> Their motor coordination showed developmental anomalies. In some ways the children evidenced superior gross motor coordination for their ages. They were sure-footed, quick, and capable of many advanced motor feats. They had an astonishingly good sense of rhythm. Motoric activity was preferred and would produce an occasional expression of fleeting happiness on a child's face. But most of the time it appeared to be primarily a vehicle for

tension discharge. . . . They showed a lack of motoric caution which resulted in frequent falls and injuries. Thus we had 4-year-old children who would pump themselves on the swing like 8-year-olds, but fell off backwards like 2-year-olds. . . . In general, their motor activity tended to be rather diffuse rather than focused. . . . the children often used their bodies for diffuse discharge and avoidance, with little focus on the pleasures of attaining mastery (Pavenstedt, 1967, pp. 56–57).

The rearing atmosphere of these children appears to be a mixture of stimulation and deprivation. In contrast to the close personal contract reported by Young (1970), the children of the lower-class, disorganized families are

Growing up in a noisy, hectic apartment with a mother whose physical and emotional presence is unreliable, and who is frequently overwhelmed by the manifold burdens of the household and the care of many children, the young children are frequently left alone for hours; . . . in general, there is little holding, comforting, and handling involved in the mother's infant and toddler care. A quiet baby is considered a "good baby" and self-activation and motor capacity is fine if this answers the mother's need and expectation to have the child grow up fast, and does not place any demands upon her (Pavenstedt, 1967, p. 111).

Further, these children are often left alone in the streets at a very early age, or at times under the guidance of slightly older siblings. Contrast the lower social class child's relative freedom on the streets to that reported by Fischer and Fischer (1963, p. 961) for middle-class preschool children:

The area in which the child is free to move at will increases gradually as he grows older. For a child to the age of 3 or 4, this may be only a few yards adjoining his own yard. Playing away from home is somewhat controlled by the environment. Some children live in more dangerous environments than others. Also, some children live great distances from places where they might play with other children. Most of the 4-year-olds are allowed to go across a quiet street to play. Until they go to school, however, most children stay at home or in neighbors' yards to play. One 4-year-old girl goes to the store, across a busy street, about two blocks from home. She seems to be entirely reliable and capable. Crossing streets which have been forbidden, playing with knives and matches, and going long distances from home without permission are all severely punished.

Evaluating the effects of nursery school experiences on these children, Pavenstedt (1967) noted beneficial effects on the children's motor development. Fine motor skills showed age-appropriate levels of attainment, while

Gross motor skills became functionally excellent. While motor behavior remained a dominant avenue of tension release for some children, they no longer used their bodies in random discharge (Pavenstedt, 1967, p. 207).

Similar results were reported by Hodges and Spicker (1967). Severely disadvantaged preschool children showed deficiencies in motor skills compared to middle-class age peers. These deficiencies, however, were improved by a special instructional program. Lillie (1966, as cited by Hodges and Spicker, 1967) exposed a group of disadvantaged preschool children to a motor development instructional and practice program (65 lessons) and then compared their motor status to children receiving a traditional preschool program and children remaining at home. All three groups improved significantly in gross motor items, while only the experimental and traditional preschool curriculum groups improved in fine motor tasks. The foregoing are of importance for the two school groups had daily, organized physical education classes, leading Hodges and Spicker (1967, pp. 282–283) to comment that ". . . the running, jumping, balancing, and climbing opportunities available in the home and neighborhood are sufficient for developing gross motor skills."

The preceding discussion of motor development during the preschool years indicates a general lack of comparative data for Negro and white children between 2 and 6 years of age. By and large, studies of infancy using the Bayley and Gesell items are concerned with the prediction of later intelligence and mental development. After the ages of 2 and 3 years, motor items are seldom used and even less frequently reported since these items as well as other measures are not predictive of later intelligence. Advanced motor behavior in infancy apparently does not represent a superior intellectual potential (Knobloch and Pasamanick, 1963).

The need for studies relating early motor development to later developmental status and performance is clearly indicated. Such studies should give attention to a host of factors including the sociocultural milieu, family environment and size, sibling status, maternal behavior and control—all of which relate to the motor development of children. Such analyses should likewise consider physical and physiological characteristics affecting motor development and motor performance early in life (see Malina, 1971).

II. School Ages

A. MOTOR PERFORMANCE

It is generally agreed that after the fifth or sixth year, on the average, no new basic skills appear in the child's motor repertoire; rather, the quality of performance continues to improve through refinements in movement patterns (Espenschade, 1960). Motor performance and strength gradually increase both qualitatively and quantitatively during the years of middle childhood. Except for the classic studies of Jones (1949) and Espenschade (1940), longitudinal studies of motor performance and strength during middle childhood and adolescence are strikingly lacking. The work of these two investigators clearly shows that the adolescent years bring about marked gains in strength and performance of males and a lesser improvement or rather stable levels of performance in females.

The majority of motor performance data for the school ages are product-oriented, concerned with the end result of a particular motor act, i.e., the time elapsed and the distance covered. Studies of the movement processes underlying the motor acts, the quality of movement, the specific patterns or styles of movement, and the specific components of the total motor act are for the most part lacking. Clearly, the processes of movement are just as important as the products. Variations in movement styles and patterns are probably more subject to local adaptations and cultural and/or subcultural influences than are the products of the motor acts.

The few reports comparing the performance of American Negro and white children of elementary school age indicate generally higher levels of performance among Negro children compared to their white age and sex peers. Temple (1952), for example, noted a tendency toward superior performance levels in Negro children 7 through 9 years of age in five motor tasks.* Of the five tasks only the jump and reach difference for both sexes and the difference in the throw for distance among girls attained statistical significance. Espenschade (1958) noted a greater percentage of Negro boys passing all items of the Kraus–Weber test than white boys at the fourth grade level. Differences between Negro and white girls at the same grade level, however, were not apparent. Hutinger (1959) found fourth, fifth, and sixth grade Negro boys and girls superior to their white grade and sex peers in the 35-yard dash. Ponthieux and Barker (1965a) found fifth and sixth grade Negro boys superior to white boys of the same grade levels in five of the seven items of the AAHPER fitness test. Negro girls surpassed white girls on four of the seven items.

In a longitudinal study of 1 year's progress in running, jumping, and throwing performance of Negro and white children 6 through 12 years of age, Malina (1968) found Negro children of both sexes faster in the 35-yard dash. No consistent differences between the Negro and white children were apparent in the softball throw for distance. The standing broad jump, however, illustrated an interesting pattern. At the first measurement period, white boys and girls were on the average, slightly in advance of the Negro children; however, at the second observation period, Negro children performed better.

Two studies considered the performance of elementary school age, urban, disadvantaged children with conflicting results. Bartholomew (1966) found

* Jump and reach, broad jump, wall test, distance throw, and target throw.

fourth grade Negro children from a disadvantaged urban area, on the average, in advance of the norms for the New York State Fitness Test. On the other hand, Safrit (1969, no date) reported the performance of inner city, disadvantaged Negro children, 9–11 years of age, to be slightly lower than the AAHPER standards for all items except pull-ups and sit-ups. Interestingly, Bartholomew's Negro children were markedly better in the standing broad jump, while Safrit's Negro children were deficient in jumping performance. It should be noted that disadvantaged children tend to be somewhat shorter and lighter in height and weight, respectively, and size at the elementary school ages is significantly related to performance in gross motor tasks (see Rarick and Oyster, 1964; Malina, 1972).

Finally, Muzzey (1933) reported significantly better motor rhythm in Negro compared to white children of elementary school age. The motor rhythm of Negro children was better both at the beginning and at the end of a rhythm training period. The relationship of motor rhythm to motor performance as generally tested is not clear. Using Negro girls in grades nine through twelve, Bond (1959) found essentially no relationship between rhythmic perception tests and performance in several motor items.

Detailed comparative studies at the junior high school and senior high school ages are less available than those at the earlier ages. In an early study, Walker (1938) compared the performances of Negro and white boys, 10 through 17 years of age, on five events. With the boys classified on the basis of age, height, and weight into eight categories (after Neilson and Cozen's classification index), differences across categories consistently favored the white boys in the broad jump and dash. No consistent racial differences were evident in pull-ups, push-ups, and the ball-put. Codwell (1949) attempted to assess the effects of American Negro hybridity on performance by using skin color as the primary classification criterion. Generally greater levels of motor ability (McCloy's battery) were noted in high school boys classified as "dominantly negroid" compared to those classified as "intermediate" and as "strong evidence of white" admixture. The "intermediate" boys resembled the "dominantly negroid" in performance more closely than they did the "strong evidence of white" group. The limited validity of skin color (and the extent of variation characteristic of this anthroposcopic variable) as an indicator of the degree of Negro–white admixture should be carefully noted.

Comparing the performances of tenth grade high school boys of varied ethnic groups on five events,* Ahern (1969) found the performance of the Negro boys significantly superior to that of white and Oriental boys. Using

* The five items were the 100-yard dash, running broad jump, high jump, 440-yard run, and shot put.

high school samples, Laeding (1964), Marino (1966), and R. M. Martin (1966) reported significantly better vertical jumping ability among Negro as compared to white males.

Motor performance comparisons of Negro and white girls of high school age are for the most part lacking. Espenschade (1946), in one of the few studies published, found no differences between Negro and white tenth grade girls in total Brace motor ability test scores. Analysis of specific items, however, indicated superior levels of balance in white girls and greater arm strength in Negro girls (see below).

The preceding studies involved comparisons of American Negro and white children. Few, if any, have considered the performance of Mexican-American or Puerto Rican children. Thompson and Dove (1942) and Thompson (1944) reported significantly better levels of performance on a six-item battery* for 12-year-old Mexican-American boys compared to Anglo (white) boys of the same age. More recently, Miller (1968) reported different results. Comparing the performances of seventh and eighth grade Mexican-American and white males on the California Physical Performance Test, Miller found better performance levels among white boys on six of the eight test items at an initial observation session and on four of the eight items at the second measurement session 6 months later. The foregoing would seem to suggest that the Mexican-American boys benefited more from the physical education program during the intervening 6 months. It should be pointed out that the seventh and eighth grade levels include the age range 13–15 years, the time of male adolescence with its characteristic variability. Also, Mexican-American boys in general tend to be somewhat smaller in height and weight than white boys. Size–performance relationships during the adolescent years need no further discussion here.

B. STRENGTH

Comparative studies of strength performance among different ethnic and/or racial groups are few. Hrdlička (1900) reported greater grip and traction strength in Negro compared to white children of both sexes from 6 to 16 years. Malina (1968) also found greater grip, pushing and pulling strength values in Negro compared to whilte children of both sexes, 6 through 12 years of age. The latter observations were longitudinal over a 1-year period. Velocities, i.e., annual increments, however, did not consistently differ between the Negro and white samples. Montpetit et al. (1967) found greater grip strength in Negro girls, 9 through 17 years of age, compared to white girls. Among boys, however. Montpetit et al. noted greater

---

* The six items included baseball throw, base running, chinning, 60-yard dash, jump and reach, and shot put.

grip strength in Negroes 9 through 11 years old. White boys, on the other hand, were stronger thereafter through 14 years of age.

Comparing Negro and white high school boys on isometric knee extension strength, R. M. Martin (1966) found no racial difference. Also, comparing Negro and white boys of high school age (subjects were Job Corps candidates, 16–18 years old), Thomas (1967) found only one significant strength difference (right shoulder extension) out of six tests used. The difference favored the Negro subjects.

Goss (1968) assessed grip strength in small samples *(n = 8 per cell)* of white, Negro, and Latin children at the third, sixth, ninth, and twelfth grade levels. Latin children of both sexes had the lowest dynamometric strength scores. Negro girls had greater grip strength at all grade levels compared to white girls, while grip strength differences between Negro and white boys were slight and inconsistent across the grade levels studied.

C. INTERPRETATIONS

The foregoing discussion of motor performance and strength of children of different ethnic and/or racial groups indicates generally higher performance levels among Negro children during the elementary school years. During the high school years, the data for males show mixed results except for consistently better performance levels for Negro boys in the vertical jump. It should be remembered that male adolescence, with its characteristic variability and effect on strength and performance (see Tanner, 1962), results in a wide range of variation in strength and performance per se. Also, the role of physical activity pursuits, prowess, and strength in the adolescent male peer group culture (see below) is of considerable importance.

Strength and performance data on high school girls are generally lacking. Negro girls appear to be stronger than white girls in dynamometric strength tests at the elementary and high school ages. The meager amount of data available on Mexican-American and Puerto Rican children is insufficient to indicate any suggestive trends.

Racial differences in body proportions and composition, and possible implications for performance have been discussed at length elsewhere (Malina, 1969, 1972). The subsequent discussion will concern factors in the sociocultural setting that might be of importance in affecting the activity pursuits and performance levels of children.

Socioeconomic status or background is frequently implicated as a significant factor affecting the activity pursuits and performance levels of children. This was pointed out earlier in this chapter where data were presented which suggested that lower socioeconomic and/or disadvantaged children had greater freedom to move about the neighborhood than children of more affluent families. Such an atmosphere might be conducive to greater free-dom of activity and opportunity for practice in movement experiences. Jones (1949), however, found no relationship between socioeconomic fac-tors and strength in his longitudinal series of white adolescents. Though av-erage values for the socioeconomic extremes were not markedly different, Jones (1949, pp. 30–31) noted a suggestive negative relationship between socioeconomic status and strength:

> . . . the averages actually obtained are to the disadvantage of the "high" rather than of the "low" selection, for the high cases averaged slightly below the mean of the total sample . . . while the low cases averaged almost exactly at the group mean. . . . This is perhaps surprising, for many studies of physical size have shown a positive relationship between height or weight and socioeconomic status.

For a more detailed discussion of socioeconomic effects upon physical growth and maturation, see Tanner (1962).

Ponthieux and Barker (1965b) reported significant correlations between socioeconomic status and performance on the AAHPER test in a mixed ra-cial sample of fifth and sixth grade children. The significant relationships, however, did not consistently favor either the high or low socioeconomic groups. Three of the more commonly measured performance items (the broad jump, 50-yard dash, and throw for distance) showed significant cor-relations with socioeconomic status. The dash and throw were significantly related to lower socioeconomic status for both sexes, while the broad jump was significantly related to upper socioeconomic status for girls only. Pull-ups and sit-ups, on the other hand, had significant relationships with higher socioeconomic status. More recently, Barker and Ponthieux (1968) con-trolled for the effects of socioeconomic status and noted higher levels of performance among the Negro children of this fifth and sixth grade sample compared to the white children. Controlling for the effects of the socioecon-omic status variable apparently intensified the performance differences fa-voring Negro children reported in their earlier study (Ponthieux and Bar-ker, 1965a). In a similar situation, Berger and Paradis (1969) reported higher motor performance levels among Negro compared to white seventh grade boys of the same socioeconomic level. Thus, the foregoing would seem to suggest that socioeconomic status per se is not a significant variable un-derlying the performance differences noted between Negro and white chil-dren.

In a mixed racial sample of high school girls, Freytag (1967) found little relationship between socioeconomic status and rhythmic ability. The rela-tionship of rhythmic ability to music and dance background, however, var-ied to some extent with socioeconomic level. These variables were substantially related in the "middle," slightly related in the "low," and not related in the "high" socioeconomic groups.

As indicated above, socioeconomic status per se does not seem to be a critical factor affecting strength and motor performance, although some relationship might be suggested. The concomitants of high or low socioeconomic status which might inhibit or enhance strength and motor performance thus need further examination. Child rearing practices, for example, have been implicated. The generally less restrictive atmosphere of the lower socioeconomic neighborhoods involving more time spent in the streets with the local peer groups might be conducive to a greater freedom of physical activity and exploration.

In their early study of Mexican-American boys, Thompson and Dove (1942) and Thompson (1944), for example, suggested that Mexican-American boys tended to lead physically vigorous lives throughout childhood because of relatively poor home conditions and lesser parental supervision. Such a home atmosphere encourages play outside the home and few restrictions of movements. In his analysis of ghetto socialization, Schulz (1969) also indicated these children had considerable freedom to move about (or alternatively less parental attention and/or restriction). Elementary school age children were frequently out of the house for a major portion of the day. Discussing children 6–10 years of age, Schulz (1969, p. 42) offered the following about their care: "They attract comparatively little parental attention as they go about the business of growing up, largely under the influence of their peers and siblings."

Young (1970) indicated active participation of Negro children in the family-based gang from preschool ages on:

> The core of any gang is the children of one family. . . . Boys from about nine years leave the mixed-sex gang if they themselves are not the nurse-child. They range farther from home than the gangs with young children, playing games of athletic skill . . . The gang with young children sits and gossips, runs, climbs, swings, lines up for inspection, teases. . . . There are almost no toys, almost no use of natural materials, no "projects" such as the middle class child, whether in the sandbox or supplied with paper and crayons or dress-ups, is usually involved in. Equipment such as swings and slides are used where available, . . . but simple equipment, such as ropes, sticks, balls, and stones, is almost never seen in use. . . . *For the most part children are the playthings and the players.* (Italics mine) . . . There are few displays of proficiency of the kind learned at school and little use of school as a play model (Young, 1970, pp. 282–283).

The preceding observations appear in accord with an early study of play activities of Southern Negro and white children. G. Martin (1931) noted a lack of interest among Negro children in games involving apparatus and equipment, perhaps reflecting their lower socioeconomic status and availability of equipment and facilities. These observations might also possibly reflect a long-term conditioning in infancy—things being taken away from children and their attention directed toward people (as discussed earlier in this report). G. Martin's (1931) study also noted no differences in the total number of activities of Negro and white children. Kinds of activities, however, varied. In general, activities involving teamwork and specific skills seemed more characteristic of the play pursuits of white children, while rhythmic and social activities appeared more characteristic of the play pursuits of Negro children. It should be noted that the observations of G. Martin (1931) are almost two generations old and are perhaps of limited applicability today. Needless to say, play analyses of different ethnic and/or racial groups are warranted both in terms of activities and possible effects of these activities on motor performance.

Abrahams (1966) has described the singing games and rhythmic hand-clapping activities of Negro girls in Philadelphia. The girls are initiated into these games early in life and actively participate in them at least until puberty. The activities are noncompetitive with emphasis on skill. For example, in hand-clapping games

> . . . there is no isolation of a central person, but to a degree the performance of the individual is all-important, and emphasis is once again on dexterity. Most commonly, the girls would form a ring and set up a complex hand-clapping pattern, involving not only the hitting of the hands but also the shoulders, knees, etc., and the striking of the hands of the players on each side (Abrahams, 1966, p. 131).

A recent film of playground singing games involving fourth grade Negro girls illustrates the role of these activities in American Negro childhood. Hawes (1969) indicated that Negro girls begin to learn the games at approximately 6 or 7 years of age. The games persist through middle child-hood only to disappear or be transmuted into social dance forms with the onset of the adolescent years. Interestingly, Hawes (1969, p. 2) also noted that boys of the same age know the singing games, but ". . . do not perform them in public situations such as the school yard. In backyards or alleys, however, the games may be played by mixed groups." The physical involvement in the games is a very distinctive stylistic feature:

> Motor expressiveness is elaborated; musical expressiveness is not. Though the children clap, their clapping style seems to stress tactile rather than tonal values. Their hands are quite relaxed; they stroke instead of making an impact. This effect is emphasized by the degree of body empathy the children share; they move over, make room, spread out, close together, move in tandem and

adjust to each other's physical presence in a thousand subtle ways. Physically speaking, they enjoy group blend to a degree that white society only seems to achieve under the strictest imposed discipline. . . . It is important to bear in mind that this tradition is a child-initiated and child-directed activity; there is a minimum of adult intervention (Hawes, 1969, p. 2).

The descriptions of Abrahams (1966) and Hawes (1969) imply a markedly social aspect of the Negro children's involvement in their games and physical activities. Note also that G. Martin's (1931) study indicated extensive participation in social games among Negro children. Hawes' observations further imply more independence in the play activities of Negro children than of whites. This might be related to the lack of adult supervision as previously indicated above as well as by Hannerz (1969, p. 128) in his description of play activities of Negro ghetto children:

. . . most of them seemed very independent and self-confident, and handled their interaction without much of the parental mediation which has a relatively large place in small middle-class white children's play.

The foregoing might be related to the highly competitive nature of our predominantly white, success-oriented American culture. Early in the life of American white children, the spirit of competitive self-enhancement and success styles are well established. Boys 8 through 12 years of age, for example, are already able to distinguish various success styles, i.e., success via strategy, power, or a combination of the two. Girls, on the other hand, distinguish between those who succeed through good fortune and those who fail (Sutton-Smith and Roberts, 1964). These success styles are related both to child rearing patterns and games played, particularly among boys.

Hawes (1969, p. 2) also noted a degree of "social tolerance and mutual support" among the Negro children participating in singing games from circle and line formations, which is ". . .inconceivable to those used to supervising white children's play." Hawes suggested that the high social empathy of Negro children is perhaps inhibited in "less formal relationships" or in structured outside institutions.

Although the preceding discussion provides some insight into the rearing and activity patterns of Negro children and by inference those of white children, specific factors in the children's environments which might hinder or enhance strength and motor performance need to be identified. Data in general, however, are lacking, and much must be inferred from discussions of life styles and socialization patterns.

Among Negroes, it is generally assumed that boys must learn their masculine identity and roles outside the home. Schulz (1969, p. 64), for example, indicates that a Negro boy

. . . must develop his sense of being a person and a man largely outside of the home and under the negative evaluation of his mother. He must further do this when the father is often missing or, if present, is likely to be preoccupied with the attempt to cover his own deep sense of failure with a facade of competence and rough hewn masculinity . . . Quite likely the male model provided by his father and older brothers is totally inadequate as a model for achievement mobility in the larger society.

Abrahams (1970, pp. 17–18), on the other hand, viewed the absence of a permanent father in a mother-dominated family differently:

For boys, many older males in the community—if not in the home—provide a strong male image and patterns of life the boys are able to emulate, thereby developing their masculine identities. Since so much of the life of the young, especially the male children, is spent playing and talking and running errands on the streets, the absence of a permanent man in the house does not create the emotional deprivation so often imputed to their lives.

Middle class America predicates adequate emotional strength on the ability to develop depth relationships that grow out of the model parent-child involvement. Lower-class Afro-American culture offers an alternative that approaches the ideals and attitudes of the extended family . . . While the white middle class seeks to keep its children off the streets and in the homes, playing with a limited number of other children and developing some meaningful and continuing play relationship . . ., black mothers teach their children to get out of the home and onto the streets. There they learn to develop relationships of a conventional nature with a potentially large number of peers.

Hannerz (1969) suggested a similar role for adult male neighbors in the socialization of ghetto boys, particularly in the context of providing occasional instruction (especially in boxing) and in serving as an audience for their games and activities, occasionally breaking up fights.

A similar father-absence situation can be inferred to white middle American families relative to the male role models. Citing the work of Talcott Parsons, Hannerz (1969, p. 120) indicated that:

. . . girls can be initiated into a female role from an early age because their mothers are usually continuously at home doing things which are tangible and meaningful to the children, while fathers do not work at home so that their role enactment remains to a large extent unobserved, inaccessible and relatively unknown. Girls can help their mothers with many domestic activities and thus get sex role training; the boys have little chance to emulate their fathers in action, partly because of the abstract and intangible nature of many middle class male tasks.

Thus, both in the lives of Negro and white boys there appears to be, to some extent, a lack of a constant male model. This perhaps functions to reinforce the strength of male peer groups, gangs, and so on, in the young male's search for his masculine identity. The role of physical prowess, including strength and performance feats, in male gangs needs little further elaboration. The role of strength and power in the male cultural ideal, a sex role learned early in life, would seem to suggest the importance of extraneous factors affecting the performance of boys. Thus, the foregoing in conjunction with the physical and maturational variability of adolescence might contribute to the lack of clear-cut differences in strength and performance among adolescent Negro and white males. Both are more or less subjected to similar pressures in American society at large, although role models and directives are different for Negroes and whites, leading to what Hannerz (1969) called "ghetto specific masculinity." Both Hannerz (1969) and Abrahams (1970), for example, emphasized the role of athletes, their physical accomplishments and life styles, as strong models for ghetto Negro youth. The significance of socialization in and through sports and related activities in Negro and white youth, however, needs further study before specific effects on performance can be indicated.

An observation common to descriptions of socialization among Negro children is the constant downgrading of the Negro male relative to his female siblings. Schulz (1969, p. 64) noted that "The small boy in the project is made aware of his mother's preference for his sisters in numerous ways." Abrahams (1970, p. 28) indicated that "Men are regarded by the women as untrustworthy for the most part." This attitude is transmitted to the children early in life and persists into adulthood. Baughman and Dahlstrom (1968) documented the inferior role of the Negro male child via peer popularity estimates. Negro girls were consistently rated higher in popularity than Negro boys beginning early in the elementary school years. Interestingly, such a sex difference in popularity was not noted among white children, the available data thus spotlighting

> . . . the unfortunate position of the Negro boys. At each of eight age levels studied, Negro girls were favored over Negro boys. The fact that the gap appears so early and persists into adolescence (in fact, it tends to grow larger with advancing age) indicates that there is a pervasive devaluation of the Negro male within his own culture, and well before he can be expected to demonstrate attributes of manhood (Baughman and Dahlstrom, 1968, p. 463).

The significance of this constant downgrading as a motivating device to seek excellence via physical performance feats and sports activities needs further elaboration.

The Negro female dominance is demonstrated in Goss' (1968) study of actual and estimated dynamometric strength in Latin, white, and Negro children. Negro girls were stronger than white and Latin girls, while white boys were only slightly and inconsistently stronger than Negro boys, leading Goss (1968, p. 289) to conjecture:

> Whether the Negro female actually needs to be stronger due to social differences in the Negro culture (due to need-achievement) or whether the differences are purely physiological is unknown.

Strength estimates indicated a pattern identical to that for demonstrated dynamometric strength. Negro girls had the highest strength estimates among the female samples, while white boys had the highest strength estimates among males. Estimates for Latin children were the lowest. Opposite-sexmate strength estimates indicated the dominant role of females in the Negro social setting:

> . . . the Anglo population estimated both boys and girls as strongest and . . . the Latin female population estimated the male as the strongest . . . the female Negro estimates placed the Negro male lower than both the Anglo and the Latin male estimates (Goss, 1968, p. 289).

III. A Suggested Hypothesis

This discussion of the performance of Negro and white school age children is closed with a hypothesis based upon data reported herein and personal observations from work with Negro and white children in Brooklyn and Philadelphia. During the elementary school years, Negro boys tend to perform better than white boys because of a greater freedom to move about in and explore their neighborhoods. Equipment and facilities are generally minimal and not readily accessible. Hence, in the course of moving about their home neighborhoods, basic motor skills are refined under a variety of situations ranging from street games to street fights. Adult supervision as well as adult pressures to conform or to win are minimal. In contrast, the white elementary school age boys receive most of their experiences with physical activities through organized programs—Little League sports competition, Cub Scouts, Boy Scouts, etc. Although white children usually live in better neighborhoods and have more equipment and facilities, they are generally limited in the duration and extent of unsupervised mobility in their neighborhoods. Adult supervision, organization, and pressures are maximal. This hypothesis, in general, also applies to young girls, although in the American culture complex girls are not supposed to be as physically aggressive and proficient in basic motor and athletic skills as boys. Rather, girls, both Negro and white, are supposed to be more oriented toward maternaltype activities. This sex difference in activity preferences is learned early in life.

Performance comparisons of Negro and white adolescent youth of both sexes cannot at this point be made because of insufficient data. The data suggest little, if any, differences in strength and performance between Negro and white males except for consistently better vertical jumping performance in Negro males. Perhaps the characteristic variability of male adolescence in conjunction with the teen-age male concern with physical prowess and activities in the form of heightened sports competition might offset the performance advantages noted in elementary school Negro boys. Too often, however, emphasis during the adolescent years is on interscholastic competition in which the better performers are selected and trained to the neglect of the average and poorer performers.

In the case of Negro and white adolescent females, extensive motor performance data are insufficient to make reliable comparisons and interpretations. The cultural socialization of the female, however, reduces her interest in physical skills and athletic pursuits, although this is a cultural value that is perhaps undergoing change today.

References

Abrahams, R. D. (1966). In "Two Penny Ballads and Four Dollar Whiskey: A Pennsylvania Folklore Miscellany" (K. Goldstein and R. Byington, eds.), pp. 121-135. Folklore Associates, Hatboro, Pennsylvania.

Abrahams, R. D. (1970). "Deep Down in the Jungle: Negro Narrative Folklore from the Streets of Philadelphia," 1st rev. ed. Aldine, Chicago, Illinois.

Abrahams, R. D. (1972). Personal communication.

Ahern, F. J. (1969). Unpublished Master's Thesis, University of Washington, Seattle, (abstr.).

Barker, D. G., and Ponthieux, N. A. (1968). Res. Quart. 39, 773.

Bartholomew, W. M. (1966). Pa. J. HPER 36, 13 (Dec.).

Baughman, E. E., and Dahlstrom, W. G. (1968). "Negro and White Children: A Psychological Study in the Rural South." Academic Press, New York.

Bayley, N. (1965). Child Develop. 36, 379.

Bayley, N., and Jones, H. E. (1937). Child Develop. 8, 329.

Berger, R. A., and Paradis, R. L. (1969). Res. Quart. 40, 666.

Bond. M. H. (1959). Res. Quart. 30, 259.

Bronfenbrenner, U. (1958). In "Readings in Social Psychology" (E. E. Maccoby, T. M. Newcomb, and E. L. Hartley, eds.), 3rd ed., pp. 400-425. Holt, New York.

Codwell, J. E. (1949). J. Negro Educ. 18, 452.

Espenschade, A. S. (1940). Monogr. Soc. Res. Child Develop. 5, No. 1.

Espenschade, A. S. (1946). Child Develop. 17, 245.

Espenschade, A. S. (1958). Res. Quart. 29, 274.

Espenschade, A. S. (1960). In "Science and Medicine of Exercise and Sports" (W. R. Johnson, ed.), pp. 419-439. Harper, New York.

Fischer, J. L., and Fischer, A. (1963). In "Six Cultures: Studies of Child Rearing" (B. B. Whiting, ed.), pp. 869-1010. Wiley, New York.

Frankenberg, W. K., and Dodds, J. B. (1967). *J. Pediat.* **71**, 181.

Frazier, E. F. (1948). "The Negro Family in the United States," rev. abridged ed. Dreyden Press, New York.

Freytag, J. G. (1967). Unpublished Master's Thesis, University of Washington, Seattle, (abstr.).

Gordon, E. W. (1965). *Rev. Educ. Res.* **35**, 377.

Goss, A. M. (1968). *Child Develop.* **39**, 283.

Graham, F. G., Matarazzo, R. G., and Caldwell, B. M. (1956). *Psychol. Monogr.* **70**, 427.

Grotberg, E., ed. (1969). "Critical Issues in Research Related to Disadvantaged Children." Educ. Test. Serv., Princeton, New Jersey.

Hannerz, U. (1969). "Soulside: Inquiries into Ghetto Culture and Community." Columbia Univ. Press, New York.

Hawes, B. L. (1969). "Study Guide for Film *Pizza Pizza Daddy-O.*" Univ. of California Media Center, Berkeley.

Hodges, W. L., and Spicker, H. H. (1967). *In* "The Young Child" (W. W. Hartup and N. L. Smothergill, eds.), pp. 262–289. Nat. Ass. Educ. Young Children, Washington, D. C.

Hrdlička, A. (1900). "Anthropological Report." N. Y. Juvenile Asylum, New York.

Hutinger, P. A. (1959). *Res. Quart.* **30**, 366.

Jones, H. E. (1949). "Motor Performance and Growth." Univ. of California Press, Berkeley.

Knobloch, H., and Pasamanick, B. (1953). *J. Genet. Psychol.* **83**, 137.

Knobloch, H., and Pasamanick, B. (1958). *Psychiat. Res. Rep. Amer. Psychiat. Ass.* **10**, **123**.

Knobloch, H., and Pasamanick, B. (1960). *Pediatrics* **26**, 210.

Knobloch, H., and Pasamanick, B. (1963). *Amer. J. Dis. Child.* **106**, 43.

Laeding, L. (1964). Unpublished Master's Thesis, Michigan State University, East Lansing, (abstr.).

McGraw, M. B. (1931). *Genet. Psychol Monogr.* **10**, No. 1.

Malina, R. M. (1968). Doctoral Dissertation, University of Pennsylvania, Philadelphia.

Malina, R. M. (1969). *Clin. Pediat.* **8**, 476.

Malina, R. M. (1971). *In* "Motor Development Symposium" (H. Eckert, ed.), pp. 22–46. Dept. Phys. Educ., Univ. of California, Berkeley.

Malina, R. M. (1972). *In* "Physical Education: An Interdisciplinary Approach," pp. 237–309. Macmillan, New York.

Marino, A. (1966). Unpublished Master's Thesis, University of Oklahoma, Norman.

Martin, G. (1931). Unpublished Master's Thesis, George Peabody College for Teachers, Nashville, Tennessee.

Martin, R. M. (1966). Unpublished Master's Thesis, University of Toledo, Toledo.

Miller, F. S. (1968). Unpublished Master's Thesis, California State College, Long Beach (abstr.).

Montpetit, R. R., Montoye, H. J., and Laeding, L. (1967). *Res Quart.* **38**, 231.

Muzzy, D. M. (1933). *Res. Quart.* **4**, 62.

Pasamanick, B. (1946). *J. Genet. Psychol.* **69**, 3.

Pavenstedt, E., ed. (1967). "The Drifters: Children of Disorganized Lower-Class Families." Little, Brown, Boston, Massachusetts.

Ponthieux, N. A., and Barker, D. G. (1965a). *Res. Quart.* **36**, 468.

Ponthieux, N. A., and Barker, D. G. (1965b). *Res. Quart.* **36**, 464.

Rarick, G. L., and Oyster, N. (1964). *Res. Quart.* **35**, 523.

Rhodes, A. (1937). *Child Develop.* **8**, 369.

Richmond, J. (1970). *Yale J. Biol. Med.* **43**, 127.

Roberts, D. F. (1969). *J. Biosoc. Sci.* **1**, 43.

Safrit, M. J. (1969). "The Physical Performance of the Disadvantaged Child." Paper presented at the National Conference of the AAHPER, Boston. mimeo

Safrit, M. J. (no date). "The Physical Performance of Inner City Children in Milwaukee, Wisconsin," mimeo.

Sandler, L., Van Campen, J., Ratner, G., Stafford, C., and Weismar, R. (1970). *J. Pediat.* **77**, 775.

Schulz, D. A. (1969). "Coming Up Black: Patterns of Ghetto Socialization." Prentice-Hall, Englewood Cliffs, New Jersey.

Scott, R. B., Ferguson, A. D., Jenkins, M. E., and Cutter, F. F. (1955). *Pediatrics* **16**, 24.

Sessoms, J. E. (1942). Unpublished Master's Thesis, State University of Iowa, Iowa City.

Solomons, G., and Solomons, H. C. (1964). *Child Develop.* **35**, 1283.

Sutton-Smith, B., and Roberts, J. M. (1964). *J. Genet. Psychol.* **105**, 13.

Tanner, J. M. (1962). "Growth at Adolescence," 2nd ed. Blackwell, Oxford.

Temple, A. L. (1952). Unpublished Master's Thesis, University of California, Berkeley.

Thomas, L. (1967). Unpublished Master's Thesis, San Jose State College, San Jose, California (abstr.).

Thompson, M. E. (1944). *J. Educ. Psychol.* **35**, 49.

Thompson, M. E., and Dove, C. C. (1942). *Res. Quart.* **13**, 341.

Van Alstyne, D., and Osborne, E. (1937). *Monogr. Soc. Res. Child Develop.* **2**, No. 4.

Walker, L. (1938). Unpublished Master's Thesis, George Peabody College for Teachers, Nashville, Tennessee.

Walters, C. E. (1967). *J. Genet. Psychol.* **110**, 243.

Waters, E., and Crandall, V. J. (1964). *Child Develop.* **35**, 1021.

Williams, J. R., and Scott, R. B. (1953). *Child Develop.* **24**, 103.

Willerman, L., Broman, S. H., and Fiedler, M. (1970). *Child Develop.* **41**, 69.

Young, H. V. (1970). *Amer. Anthropol.* **72**, 269.

Competitive Sports in Childhood and Early Adolescence

G. Lawrence Rarick

I. Introduction

Programs of competitive sports of the varsity type* for children and young adolescents sponsored by school and community groups have in-

---

* Competitive sports of the varsity type as used in this chapter follows the definition given by the Joint Committee of The American Association for Health, Psysical Education and Recreation and the Society of State Directors of Health, Physical Education, and Recreation (1968). This means interschool or interagency competition of children who have been selected because of their special athletic skill to compete individually or on teams in formally organized leagues and tournaments on a within city and/or between city basis.

creased rapidly in the last 20 years. The expansion of these programs has generated considerable controversy among educators, physicians, lay leaders, and parents. Parents and lay groups for the most part have tended to react favorably to this development, whereas educational and medical organizations have over the years been conservative in their point of view. The conservative policies of the latter organizations reflect the responsibility which the school and the medical profession assumes for the physical and mental health of children and the belief that such programs are not appropriate for children. More specifically educators have felt that these programs are peripheral to the central purposes of education. There has also been the concern that certain of the undesirable features of athletic programs in the secondary schools and colleges may find their way into varsity-type sports programs for children and youth. On the other hand, parents and lay groups who strongly support community- and agency-sponsored competitive sports programs do not see this as a problem, and they would like to see the schools assume a more vigorous role in the conduct of such programs.

Any discussion of the place of competitive sports in our society cannot rightly ignore the basic premise that the competitive spirit is deeply ingrained in the nature of man. It is an expression of man's biological inheritance, for man in his own evolution has been forced to compete for survival in an environment which has often been hostile and dangerous. It is not surprising then that we see signs of the competitive nature of man appearing early in life as the infant competes for the affection of a parent or for a place in the family. The competitive drive is present as a major factor in the developing ego. As the child grows his self-image is enhanced or placed in jeopardy by his successes and failures. Regardless of how the child is sheltered he lives in a competitive world. The basic philosophical question revolves around how this basic drive can be most effectively channeled. Clearly, it cannot be ignored nor should it be completely suppressed. But the kind of experiences which the developing child should have in order to build a balanced personality is not at this point clear.

It is in the domain of physical activity that children give their fullest expression to the competitive drive. In the competitive world of children's games the youngster experiences the thrill of winning and the harsh reality of losing. But he soon learns that competition extends beyond his play life and becomes directly or indirectly a potent factor in many of his daily activities. The child soon finds that competition is a significant force in our social order and that from the time he enters nursery school until he completes his education he will be competing for grades, promotion, and for recognition among his peers in music, art, drama, and sports. It is not surprising then that parents urge their children to develop their talents so that they can hold their own among their peers. All would agree that the ability to react in acceptable ways to success and failure is an important aspect of development. Experiences in both are essential in developing a well-balanced personality. Physical activities of a competitive nature provide children with perhaps the most effective medium for this kind of experience.

Few informed people would argue against competitive sports for children and early adolescents when conducted within the framework of the school's physical education classes or in the after-school intramural program. The desirability of varsity-type athletic competition for children of this age on an interschool or interagency basis is the issue to which this chapter addresses itself. To this end the remainder of the chapter will consider the following: (1) past policies pertaining to this question, (2) the current status of competitive sports for children and early adolescents, (3) research findings on competitive sports of the varsity type for children and early adolescents, and (4) present policies and recommendations on competitive sports for children.

II. Past Policies on Competitive Sports for Children

Prior to 1968 the report of the Joint Committee of the American Association for Health, Physical Education, and Recreation and the Society of State Directors of Health, Physical Education, and Recreation (1952) served as the general policy for programs of competitive sports for children of elementary and junior high school age. The recommendations in the report were based upon (1) questionnaires returned by a sample of cardiologists and orthopedic surgeons; (2) the responses of a group of 31 psychiatrists, psychologists, and child development specialists to a set of questions pertaining to the psychological effects of these programs on children; and (3) data obtained by a questionnaire from a sample of 20 state directors of physical education which provided information on the status and conditions of athletic competition for children and early adolescents.

The results coming from the Committee's study served as the basis for a conservative policy statement. The vast majority of the physicians polled recognized the potential values of a well-supervised sports program in the schools. However, the possibility of continuing activity to the point of harmful exhaustion was regarded by a high proportion of the physicians as the most important single factor in determining the type of activity suitable for this age group. Orthopedists rated the vulnerability of the joints and the danger of injury to the epiphyseal structures as hazards of the first order. Tackle football was the only activity singled out by the physicians as being of sufficient danger to be banned.

The psychiatrists and child development specialists were generally of the opinion that the "high pressure" aspects of competition for children in this age group were highly undesirable. The comments generally stressed a belief that highly organized forms of competition involving championship play and newspaper and radio publicity added not only undue psychological stress on the participants but also tended to distort the child's sense of values.

The results of the survey of the state directors indicated that the general public does not understand the objectives of the total educational experience and as a result pressure is frequently exerted on the school to include programs of interschool competition rather than to provide support for a broad program of physical education for all. The group pointed out that many athletic programs are sponsored by community agencies and that those in charge have little knowledge of the risks involved. It was the opinion of this group that the interests of community agencies and of the persons in charge were often placed ahead of the welfare of the participants, this occasionally bordering on exploitation.

Based on the above findings the Committee recommended that interschool competition of the varsity type not be sponsored either by school or nonschool agencies for children below the ninth grade. Strong recommendations were included against all pressure practices such as interschool or interagency leagues, championship tournaments, commercial programs, or the grooming of players for high school teams. The Committee stressed the importance of providing all children with a sound program of instruction in physical education and a varied intramural program. It did, however, recognize the value of play days and occasional invitational games involving children in two or more schools.

Two years after the above report was in print the Educational Policies Commission (1954) released a report on school athletics stating that "athletics can and do serve valuable purposes in school programs." The report warned that too much of the educational potential in these programs is not used and in some instances misused. The report emphasized that the curriculum at all levels should include broad programs of physical education, serving all pupils, and that these programs should be under the supervision of teachers on the regular school staff and under the jurisdiction of school authorities.

More recently, the White House Conference on Children and Youth (1960) and the American Academy of Pediatric's Committee on School Health (1966) made similar recommendations. The latter group was particularly concerned about the susceptibility of children at this age to bone and joint injury and the possibility that such injuries might interfer with normal bone growth. This report also pointed out the susceptibility of children at this age level to chronic fatigue and the resulting interference with optimal body functions and predisposition to illness. Attention was also directed to the tendency of the adult leadership to produce strong and unhealthy emotional responses in the young athlete by stress on winning, undue adulation of the skilled performer, and the coercion of some children beyond their interest or ability.

III. Extent of Participation in Competitive Sports Programs

A. ELEMENTARY SCHOOLS

Prior to 1958 little, if any, information was available on the extent to which children participated in organized athletic programs sponsored by schools or community agencies. Schneider's questionnaire study (1959) covering 532 school systems in cities with populations of over 10,000 found that in 44% of these systems either schools or agencies outside the school sponsored athletic competition during the school year for boys beginning in grades three, four, five, or six. Of the urban areas sponsoring such programs approximately half indicated that organized athletic competition for children of this age level was sponsored exclusively by the school, and a third had organized athletic competition sponsored exclusively by outside agencies or organizations. Fifteen percent of the group stated that some of the activities

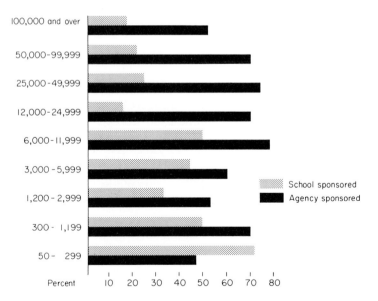

FIG. 1. Percentage participation by schools in school- and agency-sponsored athletic competition according to school district enrollment. Data from AAHPER Committee Report (1968).

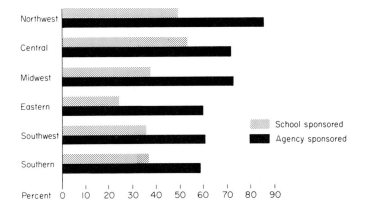

FIG. 2. Percentage participation by schools in school- and agency-sponsored athletic competition according to geographical region. Data from AAHPER Committee Report (1968).

were sponsored by the school and some by agencies other than the school during the same school year. Only 11% of the school systems reported interschool, inter-playground, or interagency athletic competition for girls of this age level during the school year. It should be noted that at this time school-sponsored athletic programs were more prevalent than those sponsored by outside agencies.

A recent national survey (AAHPER Committee on Desirable Athletic Competition for Children of Elementary School Age, 1968) of 786 elementary school principals disclosed that 37% of the schools with an organizational structure ranging from K–4 to K–8 had some form of organized interschool athletic competition. As one would expect, the percentage of schools sponsoring organized athletic programs varied according to the grade level organization of the schools. For example, 63% of the schools with an organizational structure K–8 sponsored varsity-type athletic programs, whereas this was the case in only 26% of those with a K–6 organization.

The percentage participation by school- and agency-sponsored programs according to school district enrollment and geographical location is shown in Figs. 1 and 2, respectively. The greatest proportion of schools sponsoring organized athletic programs were in the districts with low enrollments and in the central and northwest sections of the country, and those with smallest proportion were in school districts with the largest enrollments and those in the eastern part of the United States. There is a logical reason for this since the smaller schools in the less populous regions would have too few

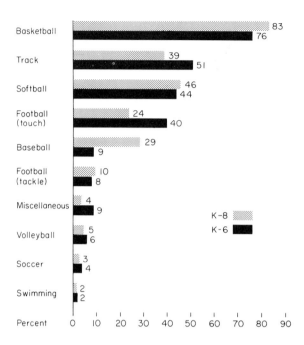

FIG. 3. Percent of elementary schools engaged in interschool athletic competition according to sports. By permission from AAHPER Committee Report (1968).

children to field competing teams on an interclass or intramural basis. The percentage of schools with students on agency-sponsored teams did not vary appreciably according to school size but followed the same pattern as that for school-sponsored programs when viewed in the light of geographical regions. It should be noted that the percentage of schools with children on agency-sponsored teams was generally substantially greater than that for schools sponsoring interschool teams. This is a reversal of conditions as given in Schneider's 1958 survey, representing a significant change in parental and community attitude toward highly competitive sports for children.

The most popular sport in schools with competitive athletic programs was basketball (Fig. 3). Nearly half of these schools sponsored interschool athletic programs in track and softball. There was surprisingly little difference between elementary schools organized on the 8-year basis as compared to those with a K–6 organizational structure in regard to the percentage sponsoring competitive programs in particular sports. This was even true for tackle football in which 10% of the K–8 schools sponsored programs in this sport in comparison to 8% of the K–6 schools.

B. JUNIOR HIGH SCHOOLS

Information on competitive sports in junior high schools was accumulated by Thompkins and Rowe (1958) in a nationwide survey of interscholastic athletic programs in separately organized junior high schools, a project of a committee of the National Association of Secondary School Principals. A breakdown of the status of interscholastic athletic programs and the attitudes of principals toward such programs is given in Table I. It is evident from this survey that a large proportion of the schools surveyed sponsored interscholastic athletic programs at this time. Approximately 85% of the schools had some kind of program of interscholastic athletics. Of the principals responding 78% indicated that they were in favor of interscholastic athletics at the junior high school level, 15.4% indicated their opposition, and 6.6% abstained. Approximately one-fourth of the principals favoring such programs favored competition only among junior high schools within the same school system. It should be pointed out that at the time of this study a greater number of schools had interscholastic athletics than the number favoring such programs. Basketball was the most frequently reported sport followed in descending order by track, tackle football, baseball, softball, swimming, wrestling, soccer, tennis, and volleyball. A relatively small proportion of the schools (5%) anticipated a change in policy on in-

TABLE I

ATTITUDES OF JUNIOR HIGH SCHOOL PRINCIPALS TOWARD
INTERSCHOLASTIC ATHLETICS[a]

Attitude	Number	Percent
Have interscholastic athletics and favor such programs	1211	52.4
Have interscholastic athletics and favor them in local school system	498	21.5
Have interscholastic athletics but oppose them	143	6.2
Do not have interscholastic athletics and favor them	38	1.6
Do not have interscholastic athletics and favor them only among schools in local school system	68	2.9
Do not have interscholastic athletics and oppose them	213	9.2
Have interscholastic athletics but made no comment	116	5.0
Do not have interscholastic athletics but made no comment	25	1.1

[a] Based on 2313 responses.

terscholastic athletics, slightly more than half anticipating an extension and approximately one-third planning to discontinue or curtail the program. In the above survey 80% of the schools did not have interscholastic playdays for girls. Of the 20% that had interscholastic athletics for girls basketball was most frequently mentioned followed in order by softball, volleyball, tennis, hockey, track, and swimming. Current data on a nationwide basis on athletic competion for boys and girls of junior high school age are not available. There is little reason to believe that the extent of such programs is less than it was in 1958.

IV. Effects of Athletic Competition on Children

A. GROWTH AND DEVELOPMENT

Few would question the positive effects of physical activity on the growth and development of children. Summaries of research (Malina, 1969; Rarick, 1960) on the effects of exercise on animals and humans clearly show that physical activity is necessary to support the normal growth of bone and muscle tissue and serves as a stimulant to the development of the heart, lungs, and other internal organs. Only when the physical activity is sufficiently strenuous and repeated often enough to bring about chronic fatigue or when stressful enough to induce trauma to a body part is there danger of adverse effects on normal growth. We are not concerned here with elaborating on the benefits or the hazards of physical activity per se, but rather with consideration of the dangers to physical growth and development of children which some authorities have attributed to competitive sports of the varsity type. These hazards can be conveniently grouped into three categories, namely: (1) the physical demands and the psychological stresses of highly competitive athletics may be great enough to affect adversely growth and development; (2) repeated stress resulting from a particular movement peculiar to a sport may in some instances be sufficiently great to introduce trauma to a body part, thus impairing its normal growth; and (3) a blow or forceful impact may be sufficiently great to be fatal or to do permanent damage to a growing structure.

At the outset it should be recognized that there is not a sufficient body of reliable information to make complete and valid assessments of the effects of competitive athletic programs on children. Clearly, any evaluation of the hazards and benefits which may occur from such programs must for the most part be based on an assessment of the growth of participants over a period of many years. Such observations must give attention to a multitude of factors such as the type of athletic activity, the intensity and duration of the training regimen which is followed, the nature and frequency of the contest, the number of years of participation, and the age at which the activity is started. The problem is further complicated by the difficulty of reaching an agreement on valid and reliable assessment procedures. While it might appear that we are still completely in the dark on this matter, considerable data are at hand from which we can begin to make reasonably valid judgements about the values and dangers of competitive sports for children.

1. Physical Growth

Several short-term studies have been conducted on the effects of competitive sports on physical growth with controversial results. Rowe (1933) and Fait (1951) reported that the physical growth of boys in the early pubertal years was slowed down by a 6-months program of interscholastic sports. On the other hand, Schuck (1962) found that the growth trends of a large sample of seventh, eighth, and ninth grade boys who were involved in a season of interscholastic sports did not differ from the growth trends of nonparticipants. There was, however, a somewhat greater lag in the growth of those who participated in the entire season (17 games) in comparison to those who were involved in a shorter season (12 games).

The research of Astrand et al. (1963) in which longitudinal data on height and weight were obtained on 30 girl swimmers has yielded perhaps the most complete data to date on the effects of rigorous training and highly competitive athletics upon the growth of humans. Periodic checks of the growth of the girls during the years of competition indicated that their growth curves were normal, in fact somewhat accelerated above the average for their age group. The majority of the girls had started heavy training prior to 13 years of age, some as early as 10 years, so that the observations were made during the period of accelerated growth. Two years after the investigation had been completed, when the girls were 18 years of age, medical examinations revealed no harmful effects of the competition or of the training regimen which had been followed.

Another aspect of Astrand's investigation (Astrand et al., 1963) involved the administration of a questionnaire to the 84 women finalists in the Swedish swimming championships for the years of 1946–55. With few exceptions these women, some 6–15 years after competing, indicated that they were enjoying good health, and their responses indicated that they were socially and psychologically well adjusted. Three out of four of the group were married and two-thirds had children. While they still had a positive attitude toward swimming, many took a dim view of present strenuous training programs, some considering them to be detrimental.

Selective factors, either native or acquired are obviously important in determining boys who will gravitate to competitive athletics. The evidence is strong that young boys who join varsity-type athletic games are physiologically and skeletally more mature than their nonathletically oriented counterparts. For example, Hale (1956) and Krogman (1959) reported that the boys who participated in the Little League World Series in 1955 and 1957 were skeletally and physiologically advanced. Further evidence that young athletes are stronger and structurally more mature than nonathletes of a similar chronological age was reported by Clarke and Peterson (1961). There would seem to be little doubt that the bigger, stronger, and more mature boy is more nearly ready psychologically and is better equipped physically for this type of activity than his less mature counterpart.

As indicated in this section there is little reason to believe that the demands of competitive sports in any way adversely affect the physical growth of the young athlete. The critical question, of course, is how much physical activity is optimal and how much places undue stress on the growing organism. This undoubtedly varies with the maturational level of the child, with his constitutional makeup, and with the state of his health. Clearly, the critical question is not so much a matter of whether the activity is competitive in nature but rather the care that is taken to insure that undue stresses are not placed on the participant. While there may be isolated instances where undue stress has been placed on young participants, well-documented descriptions of these instances have not been published.

2. Physiological Effects

Concern has been voiced by some that the stresses imposed by certain competitive sports, particularly those of an endurance nature, may throw excessive demands on the cardiovascular system of children and early adolescents. There is little, if any, published data on the physiological responses of children either during or immediately following participation in competitive sports. It should be pointed out that the distances in the running events in track and in swimming and the time duration of team contests are materially less than for adults, the reduction theoretically being commensurate with the age and maturity level of the child.

There is now general agreement that the normal heart is not damaged by heavy exercise. The early research of Seham and Egerer-Seham (1923) indicated that children in the age range 6–15 years who exercised on a bicycle ergometer to the point of exhaustion had completely recovered within a few hours. The exercise was so severe that many in the group were nauseated at the end of the exercise period, but none showed signs of acute dilatation of the heart. Seldom, if ever, are demands of this kind placed on children in competitive contests. Even so, more evidence is needed regarding the exercise tolerance of children and early adolescents.

Common sense would dictate that the age and the maturity level of the child be considered in the selection of athletic activities. It is well known that the working capacity of young boys is substantially less than that of older boys. Such differences, however, result more from variations in body size than from organic differences in the capacity of the heart to respond to exercise (F. H. Adams *et al.,* 1961; Morris *et al.,* 1949).

Unfortunately, there have been cases of cardiac arrest and sudden death in young athletes (James *et al.,* 1967). Autopsies of such cases have generally disclosed a history of cardiac problems in the family or earlier cardiac difficulties that have either been ignored or faultily diagnosed. Complete medical histories and physical examinations of children prior to admission to programs of competitive sports should provide reasonable assurance that problems of this kind will not occur.

3. Psychological and Emotional Effects

One of the most frequently expressed concerns about competitive sports for children is that the psychological stresses of league and tournament competition endanger the emotional stability of some children. In view of the difficulties encountered in assessing the effects of psychological and emotional stress, it is not surprising that the data bearing on this problem are at best meager and contradictory. The research of Hanson (1967) and Scubic (1955), in which physiological assessments were made of the affective state of Little League baseball players during and after the contest, indicated that the state of emotional tension was not materially affected by Little League play. The tension which a few exhibited subsided shortly after the termination of the contest.

Questionnaires issued by the Research Division of Little League Baseball (1962–1963) to some 1300 physician-fathers of Little League players indicated that 97% of the fathers were of the opinion that the contests did not have sufficient emotional impact to adversely effect the health of their sons. On the other hand, Scubic (1955) reported that one-third of the parents of the participants in Little League play stated that their sons were too excited to eat normally after the games. The excitement in some instances was of sufficient duration to delay the normal onset of sleep. Sleep disturbances in young children following competitive athletic events have also been reported by Giddings (1956). Based on the evidence to date, one could not with confidence conclude that competitive sports constitute an emotional danger to the young participant.

B. INJURY

Perhaps the greatest hazard in competitive sports, particularly those involving body contact, is that of physical injury. Yet this in the minds of many is a consideration of limited significance. While careful records of injuries are kept in most schools, we have very little large-scale data on the frequency and severity of injuries occurring in varsity-type athletic programs for children. Knutson (1960), in a 5-year study in Elgin, Illinois, compared the frequency of reported injuries in junior and senior high school physical education classes with the injury rate in intramural and interscholastic sports. He found that the number of injuries per student hour of participation was highest at the senior high school level in interscholastic sports and was substantially lower among junior high school participants, even though the athletic programs were practically the same. Differences were negligible when comparing the injury ratios in intramural athletics, interschool sports, and physical education classes for seventh and eighth grade boys. This raises questions regarding the relative seriousness of the injury problem in interschool sports at the junior high school level. Clearly definitive conclusions on this cannot be drawn until we have considerably more data.

The vulnerability of the epiphyseal structures in early adolescence to the stresses and strains of competitive sports has been a matter of concern to many physicians over the years (Krogman, 1959). In considering the kind and extent of athletic activity appropriate for youth in the 12–15-year age bracket, some 70% of the orthopedic surgeons polled by the AAHPER Committee (1952) stated that they would give particular attention to the hazard of the activity in respect to possible fractures of the epiphyseal area of the long bones. In support of this, J. E. Adams (1965), in an examination of the roentgenograms of the elbows of the throwing arms of 162 Little League players 9–14 years of age, revealed there were varying degrees of epiphysitis, osteochondritis, or accelerated growth and separation of the medial epichondylar epiphysis in 76 of the 80 pitchers but in only 7 of the 47 nonpitchers. This raises a serious question regarding the tolerance limits of the epiphyseal zones during the pubertal years, particularly where the mechanical stresses of contact sports and the peculiar torsions characteristic of some sports may be excessive. As a result of his research, J. E. Adams (1965) recommended that Little League pitchers be limited to two innings per game and that curve ball pitching be prohibited for those under 14 years of age.

There are many in the medical profession who have a less conservative point of view regarding epiphyseal injury and do not see this as a major problem in competitive sports for children. Shaffer (1964), for example, believed that although the epiphysis is vulnerable to injury, damage to the growing bones in the early adolescent years comes from many causes other than sports injuries.

In a study of 1338 consecutive athletic injuries seen by four orthopedists in Eugene, Oregon, Larson and McMahon (1966) reported that only 20% of these occurred in participants 14 years of age and under. The injury incidence was 40% in the age group 15–18 years. In the group under 15 years of age only 1.7% of the injuries were epiphyseal. Most epiphyseal displacements, according to Larson and McMahon (1966), can be reduced with minimal likelihood of permanent damage. Sigmond (1960) agreed that if the physician is on the alert for symptoms of bone or joint disease and acts promptly to prevent further damage most epiphyseal injuries can be handled without adverse long-range effects. However, in view of the division of opinion among medical authorities, and the conflicting evidence, it would seem wise to err on the side of caution until we have more conclusive findings on this question.

Tackle football which has a great appeal for the developing adolescent is a contact sport in which the potential for injury is relatively high. Recent data on 16,500 junior high school football players in New York showed that the injury incidents was 11.2% (Emerson, 1964). Other reports give lower injury figures (Lane, 1966, Allman, 1966). While there are scattered statistics of this kind at the local level, data on a board scale are not available in which records of the nature and kind of injuries have been accurately kept over a period of many years. While fatalities have occurred in tackle football, it should be noted that in the Pop Warner Tackle Football Program for young boys there has not been a fatality during the 34 years it has been in operation (Allman, 1966). That contact sports do not have complete approval is indicated by the results of a recent national poll showing that 43.5% of the physicians contacted expressed unqualified opposition to these sports (Allman, 1966).

An extensive study (Hale, 1967) of the injuries incurred in Little League baseball based on more than 5 million participants indicates that less than 2% of the boys sustained injuries of sufficient severity to necessitate medical attention. Most of the injuries were of a minor nature, well over half being abrasions, contusions, and lacerations. The batter was most frequently injured, more than 40% of the injuries being contusions. With the development of the new protective helmet and the adjustment of the pitching distance, concussions from the pitched ball were markedly reduced. Of the 2% injured in Little League baseball there were 41% contusions, 19% fractures (of which 43% were fractures of the fingers), 18% sprains, 10% lacerations, 5% dental injuries, 3% abrasions, 2% concussions, and the remaining 2% injuries of various kinds. The older boys (9–12 year olds) were more frequently injured than the younger, less skilled, minor league players. The 13–15-year-old group sustained more fractures, sprains, dislocations, and epiphyseal injuries than the 9–12-year-old boys. The steel spiked shoe used in the senior division was apparently a major cause of injuries. In this group there were approximately twice as many fractures, sprains, and dislocations of the lower extremity than among the 8–12-year-old players who did not wear steel spikes. The report indicates that with the prohibition of steel spikes, fractures, sprains, and dislocations to the lower extremities have been markedly reduced.

Little League has over the years modified its program and instituted the use of protective equipment in an effort to safeguard the welfare of the participants (Hale, 1967). The changes put into effect have come largely as a result of an analysis of data which have been kept on Little League injuries.

Data of the above kind are currently not available on injury rates of elementary school children involved in interschool or interagency athletic competition. The AAHPER Committee Report (1968) indicates that of the elementary school principals surveyed, only 2% had observed serious injuries in school-sponsored interschool athletics. On the other hand, 8% of the principals noted that children in their schools had injuries serious enough to require medical attention as a result of agency-sponsored athletics.

Provisions for safeguarding the health and safety of the young athlete is of utmost concern to school and health authorities. The universal recommendation is that children should have a physical examination (by a physician) before being permitted to join an athletic team. That this has not been the general practice is evident by the fact that the 1968 national survey showed that only 58% of the school-sponsored programs and only 17% of the agency sponsored programs had such a requirement (AAHPER Committee Report, 1968). Furthermore, only 28% of the coaches in agency-sponsored programs were professionally trained, whereas 81% of those in school athletic programs were regular teachers with training or experience in the sport. It is clearly evident that the conditions under which the school programs functioned were decidedly superior from the health standpoint to the conditions which prevailed in agency-sponsored programs.

The above is further supported by the expressed reactions of the principals in the survey. While in general the vast majority of the principals had not noticed any harmful effects of the athletic programs in their schools, a small proportion had noticed evidence of physical strain (8%) and a lesser number (4%) had seen children seriously injured. In agency-sponsored programs these figures were increased fourfold.

In view of the above it is not surprising that a larger proportion of principals opposed agency-sponsored programs as compared to school-sponsored programs (Fig. 4). Obviously, many of the principals were not in the position to make competent judgments about the agency-sponsored programs. This was evident by the fact that more than one-fourth of them were undecided in their reaction to these programs. Clearly, the principals were divid-

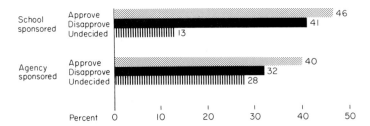

FIG. 4. Reactions of elementary school principals to agency-sponsored and school-sponsored athletic competition. Data from AAHPER Committee Report (1968).

ed in their reactions, the percentage favoring interschool and interagency programs being only slightly greater than the percentage voicing disapproval. The figures were in part colored by the school practices, for the number favoring such programs were greater in the small rural schools where such programs existed and less in the larger school districts where the programs were less evident.

Although information on sports injuries in this age bracket is limited, it is clear that the problem in this age range is sufficiently great that it cannot be ignored. It is also evident that youth of this age level will engage in unsupervised sports, including contact sports, unless challenging supervised activities are provided. The desire for team competition and in some instances the desire for sports with the element of danger become great. Part of this need can be met through well-designed physical education and intramural programs. In many communities schools have simply not faced up to the responsibility of providing well-organized physical activity and recreational programs after school hours, on weekends, and during the summer months.

The burden of providing recreational programs for youth has in many localities been assumed by churches and civic groups. This has frequently led to the development of league and tournament competition without adequate provision for trained leadership. As pointed out earlier the requirement of a medical examination prior to participation is the exception rather than the rule. Experience generally has shown that the supervision of such programs is far from adequate, and there is reason to believe that many such organizations fail to provide adequate safeguards to protect the welfare of the participants.

V. Competitive Sports for Girls

It is clearly evident that our culture impresses on children at an early age distinctly different patterns of behavior for the two sexes. This is not only reflected in the early play patterns of boys and girls but also in the kinds of sports programs that our culture feels are appropriate for each sex. It is not surprising, then, that most parents are not as favorably disposed to varsity athletic programs for young girls as for young boys. While this attitude has tended over the years to result in relatively few girls developing a strong sports orientation, there are a number of skilled girls who through their own initiative or on the encouragement of parents and friends have become interested in competitive sports. In recent years this interest has been expanded through the development of local swim clubs and through athletic programs initiated by the YWCA, by civic and community organizations, or in some instances by the schools. No reliable figures are available on the percentage of girls involved in such programs, but it is much smaller than that of the boys. The purpose of this section is to consider the effects of organized competitive athletic programs on the well-being of girls.

A. PHYSIOLOGICAL EFFECTS

The young female responds physiologically to physical training in much the same way as the young male. While the physiological effects of training regimens for girls are ordinarily not as dramatic as for boys, substantial differences have been demonstrated in the working capacity of trained as compared to untrained girls in the age range 12–18 years. For example, Astrand et al. (1963) reported that the functional dimensions of 30 girl swimmers (ages 12–16 years) involved in rigorous competitive swimming were substantially greater than healthy girls of similar body size who had not been involved in a special training program. The greatest differences were in heart and lung volumes and in their functional working capacity as expressed by maximal oxygen uptake. The intensity of the training program is indicated by the fact that the average training volume ranged between 6,000 and 65,000 meters/week and 6–28 hr/week in the water. When the girls were classified according to the number of years in training, and the hours and the meters of swimming per week, it was evident that the increases in heart volume and functional capacity were largely a function of the training volume. There seemed to be no doubt that this intensive swimming program had resulted in positive adaptive changes of a significant nature. Similar results have been noted by Drinkwater and Horvath (1971) in tests of the aerobic working capacity of girls 12–18 years of age in training for running events. The research of Wilmore and Sigerseth (1967) with untrained subjects indicates that the working capacity of boys and girls of similar age and body size in the age range 7–13 years is not materially different. In view of the above there is little reason to believe that the physiological stress imposed on young girls who are participating in well-supervised sports programs is likely to be physiologically harmful.

B. Psychological and Emotional Effects

The statement is frequently made that girls and women are not by temperament suited for competitive athletics. Although data on this are meager, the research of Astrand et al. (1963) indicates that girls who have participated over a period of several years in competitive swimming are as well balanced psychologically and emotionally as nonparticipants. The commonly held belief, nurtured chiefly by males, that women are poor losers and that they cannot psychologically stand up to the stresses of sports, has not been substantiated. There is every reason to believe that responses to frustrations are learned, and there is no well-founded psychological research to indicate that there are differences between the sexes in this type of learning.

C. Sports Participation during the Menstrual Period

The literature generally agrees that for most healthy girls the level of physical performance is not materially affected by the menstrual period. Where such differences do exist the performance tends to be somewhat poorer in the premenstrual and early menstrual period, with improving performance late in the period and the best performance in the postmenstrual period (Erdelyi, 1962).

While there is no firm evidence to indicate that participation in sports during the menstrual period is likely to do permanent damage to the female reproductive organs, sports physicians are not in complete agreement regarding the advisability of strenuous competitive sports for women at this time. The majority tend to be conservative in their point of view, although there are many who believe that a healthy girl may safely continue participation at this time. The latter group maintains that the disposition of the athlete may serve to indicate how much stress she can safely tolerate. Conservative physicians do not favor competition during menstruation in swimming and in events requiring long periods of exertion such as skiing, gymnastics, tennis, and rowing. This is predicated on the greater incidence of menstrual disorders associated with participation in these activities. Erdelyi (1962) reported that he has observed an increase in the frequency of injuries and accidents in gymnastics during the menstrual period. A drop in the performance level of the girl during menstruation may indicate that the stress of the sport is imposing demands sufficient to make it inadvisable from the standpoint of her health to participate at this time.

Astrand (Astrand et al., 1963) reported that in his group of 30 highly trained girl swimmers half of the group swam least well during menstruation and one-third indicated that swimming caused pain in the lower abdomen at this time. In one-third of the girls infectious organisms were recovered from the vagina. Gynecological investigation of the 30-girl swimmers did not, however, provide evidence of any detrimental effects of the intensive swimming training. In the light of the above observations, Astrand advised against swimming, both training and competition, during the menstrual period.

The possibility of damage to the uterus or displacement of this organ from excessive jumping, particularly during the early part of the menstrual period, has been frequently mentioned. Evidence in support of this is meager. McCloy (1931), in a detailed study of the landing shock in jumping for women, demonstrated that the shock is negligible. The study included an analysis of the shock imposed on the uterus and the pelvic floor resulting from the standing and running broad jump, the running broad leap, the running high jump, and the basketball jump. The findings indicated that there is no reason why girls whose pelvic floors are uninjured and show no signs of pelvic infection should not jump. It should be noted that the intent of this investigation was not directed to answering the question of the advisability of these activities during menstruation.

The use of hormones to alter the time of the menstrual period to meet the convenience of scheduled athletic events has been proposed. Erdelyi (1962) does not recommend this practice because postponement of the onset of menstruation may only lengthen the unfavorable premenstrual phase.

The age of menarche of the young girl athlete is essentially the same as that of her nonathletic counterpart. There is some evidence to indicate that girls who start intensive athletic training before menarche tend to have a higher percentage of menstrual disorders than those who begin training at a later time (Bausenwein, 1954). This problem tends to subside with a reduction in the work load.

Gynecological problems tend to occur more frequently among competitive girl swimmers and skiers than for participants in other sports. Overtraining and excessive stress in the female athlete have been cited as a cause of menstrual disorders (Erdelyi, 1962).

D. MASCULINIZATION

It is held by some that heavy athletic training in the female brings on signs of masculinization. While girls of masculine build tend to become involved in competitive sports more frequently than those with feminine characteristics, there is no evidence to indicate that athletic participation in and of itself brings on masculinization. Erdelyi (1962) stated that while some of those in his sample of 729 Hungarian women athletes showed male characteristics in respect to body build and musculature, they came originally from the less feminine, intersexual type. The majority in the sample, according to Erdelyi, showed no signs of masculinization after many years of heavy competition.

Few would categorize the female dancer or swimmer as having masculine characteristics. On the other hand, girls who fall on the masculine end of a somatic androgeny scale (masculinity–femininity scale of body build) are stronger per pound of body weight than girls who are at the feminine end of the continuum (Bayley, 1951). The demands of the sport probably serve as a selective factor attracting or repelling girls of a particular body build.

E. CHILD BEARING FUNCTIONS

While some girls have expressed concern that their child bearing functions may be adversely affected by participation in competitive sports during adolescence, the facts do not bear this out. In fact evidence to date indicates that the young women who have engaged in athletic training are less likely to have complications in pregnancy than nonathletic women. Erdelyi (1962) reported that the duration of labor was shorter in 87.5% of his female athletes than in control subjects, and they delivered their babies faster than the general average. The second stage of labor was approximately half that found in the average nonathlete. Astrand (Astrand *et al.,* 1963) reported that in his group of postactive female swimmers there was no evidence of increased incidence of gynecological disturbances or morbidity which could be attributed to the vigorous swimming program. Many authorities believe that the musculature of the abdomen and pelvic floor of the female athlete is better equipped to handle the problems of labor than that of the nonathlete.

VI. Policies Governing Competitive Sports for Young Girls

There is at present no formal national policy governing competitive sports for girls below the junior high school level. The general assumption is that schools will not sponsor such activities in the elementary schools. The Division for Girls and Women's Sports (D.G.W.S.) of the American Association for Health and Physical Education (1966) released guidelines for the administration of athletic programs for junior high school girls. The rationale for such programs is based on the belief that junior high school girls should have the opportunity to explore a wide variety of sports in many kinds of sports situations. Some girls with high skill potential may even wish to extend their training and competitive experiences beyond the jurisdiction of the school. While the school physical education program should provide good instructional and recreational experiences for all, opportunities may be provided for interschool competition in the form of a limited number of sports days at the conclusion of the intramural season. Such competitive sports opportunities should be separate and distinct from the program of competitive athletics for senior high school girls and should be planned for the good of the participant rather than for spectators or for training personnel for senior high school teams. Extramural sports programs should be provided only after provision has been made for a well-rounded instructional and intramural program adequate to meet the needs of all girls.

The leadership and administration of these programs must be under the direct control of the school. The program should be directed, coached, and officiated by qualified school personnel. Qualified medical assistance should be available for all contests. Particulars regarding the extent and type of participation are spelled out in the D.G.W.S. guidelines.

VII. Guidelines for Competitive Sports for Children of Elementary School Age

In view of the trend toward greater participation of children of elementary school age in competitive athletics a new and somewhat liberal joint policy statement on this issue was recently released by the American Association for Health, Physical Education, and Recreation, the American Academy of Pediatrics, the American Medical Association Committee on Medical Aspects of Sports, and the Society State Directors of Health, Physical Education, and Recreation (1968). The document places particular emphasis on (1) proper physical conditioning of all participants prior to entering competition; (2) grouping of participants according to body size, sex, skill, and physical maturation; (3) periodic health appraisals of participants; and (4) the immediate availability of a physician during all games and practices.

The report emphasizes that each athletic activity be supervised by personnel competent to provide instruction in the sport. The game equipment, facilities, and rules should be apropriate to the maturity level of the participant and commensurate with the hazards of the sport. The authority and responsibility of school administrators, parents, physician, and coach must be clearly defined.

The statement specifies that prior to instituting an interschool athletic program provisions should be made for daily physical education instruction for every child. Opportunities should also be provided for all children in the upper elementary grades to participate in an organized and supervised intramural program with assurance that the interschool athletic program will in no way curtail either the time allotment or budget of the instructional and intramural physical educational programs.

Interschool athletic participation should be restricted to children in the upper elementary grades, and the contests should be limited to a neighborhood or community basis. Play-offs and all-star games should be prohibited and every effort should be made to avoid excess publicity, paid admissions, victory celebrations, commercial promoting, and high pressure public contests. Particular care should be taken that there is no exploitation of children in any form.

The statement recognizes that such sports as baseball, basketball, football, hockey, soccer, softball, and wrestling entail varying degrees of collision risk. While the hazards of such competition are debatable, the risks are ordinarily associated with the conditions under which these sports are conducted and the quality of the supervision which is provided. Boxing is not an appropriate sport for children of this age because its prime purpose is to inflict injury, and the benefits attributed to it can be attained through other less hazardous sports.

The report emphasizes the importance of making children aware of the physiological values of sports and their lifelong recreational value. Such competitive sports as archery, bowling, golf, skating, swimming, tennis, and track are appropriate for children of elementary school age, most of which can serve as a recreational outlet in the years ahead.

It is clear from the above that the policy statement, while more liberal than the earlier policy statement (AAHPER Committee Report, 1952), places well-defined restrictions on the extent and type of competitive athletics for children of elementary school age. Throughout the report stress is placed upon the need for exemplary medical and educational supervision, and unless this can be assured schools and communities should not become involved in programs of competitive sports for children of this age level.

References

AAHPER Committee on Athletic Competition for Children of Elementary and Junior High School Age. (1952). "Desirable Athletic Competition for Children." American Association for Health, Physical Education, and Recreation, Washington, D.C.

AAHPER Committee on Desirable Athletic Competition for Children of Elementary School Age. (1968). "Desirable Athletic Competition for Children of Elementary School Age." American Association for Health, Physical Education, and Recreation, Washington, D.C.

AAHPER Division for Girls and Women's Sports. (1966). *J. Health, Phys. Educ. Recreat.* **37**, 36.

Adams, F. H., Linde, L. M., and Miyake, H. (1961). *Pediatrics* **28**, 55.

Adams, J. E. (1965). *Calif. Med.* **102**, 127.

Allman, F. L. (1966). *J. Ga. Med. Ass.* **55**, 464.

American Academy of Pediatric's Committee on School Health. (1956). *Pediatrics* **18**, 672.

American Academy of Pediatric's Committee on School Health. (1966). "Report of Committee on School Health," pp. 71–75. Amer. Acad. Pediat., Evanston, Illinois.

Astrand, P. O., Eriksson, B. O., Nylander, I., Engstrom, L., Karlberg, P., Saltin, B., and Thoren, C. (1963). *Acta Paediat. (Stockholm), Suppl.* **147**, 1–75.

Bausenwein, D. D. (1954). *Deut. Med. Wochenschr.* **79**, 1526.

Bayley, N. (1951). *Child Develop.* **22**, 47.

Clarke, H. H., and Peterson, K. H. (1961). *Res. Quart.* **32**, 163.

Drinkwater, B. L., and Horvath, S. M. (1971). *Med. Sci. Sports* **3**, 56.

Educational Policies Commission. (1954). "School Athletic Problems and Policies." National Education Association and American Association of School Administrators, Washington, D.C.

Emerson, R. S. (1964). *Med. Trib. (Sports Rep.)* **5**, 36–37.

Erdelyi, G. J. (1962). *J. Sports Med. Phys. Fitness* **2**, 174.

Fait, H. (1951). Ph.D. Dissertation, University of Iowa, Iowa City.

Giddings, G. A. (1956). Cited in "Sleep in Children." A Report of the Joint Committee on Health Problems in Education of NEA and AMA.

Hale, C. J. (1956). *Res. Quart.* **27**, 276.

Hale, C. J. (1967). *Pap., Amer. Acad. Orthop. Surg., 1967* (unpublished).

Hanson, D. L. (1967). *Res. Quart.* **38**, 384.

James, T. N., Frogatt, P., and Marshall, T. K. (1967). *Ann. Intern. Med.* **67**, 1013.

Knutson, J. A. (1960). Master's Thesis, University of Wisconsin, Madison.

Krogman, W. M. (1959). *J. Health, Phys. Educ. Recreat.* **26**, 12.

Lane, R. G. (1966). "Midget Tackle Football," Proceedings of the Athletic Workshop. Georgia Recreation Commission.

Larson, R. L., and McMahon, R. O. (1966). *J. Amer. Med. Ass.* **196**, 607.

McCloy, C. H. (1931). *Arbeits physiologie* **5**, 100.

Malina, R. M. (1969). *Clin. Pediat.* **8**, 16.

Morris, M., Schultz, F., and Cassels, D. (1949). *J. Appl. Physiol.* **1**, 683.

Rarick, G. L. (1960). *In* "Science and Medicine of Exercise and Sports" (W. A. Johnson, ed.), pp. 440–465. Harper, New York.

Research Division of Little League Baseball. (1962–1963). "1300 Doctors Evaluate Little League Baseball." Little League Baseball, Inc., Williamsport, Pennsylvania.

Rowe, F. A. (1933). *Res. Quart.* **4**, 108.

Schneider, E. (1959). "Physical Education in Urban Elementary Schools." U.S. Department of Health, Education, and Welfare, Washington, D.C.

Schuck, G. A. (1962). *Res. Quart.* **33**, 288.

Scubic, E. (1955). *Res. Quart.* **27**, 97.

Seham, M., and Egerer-Seham, G. (1923). *Amer. J. Dis. Child.* **25**, 1; **26**, 254.

Shaffer, T. E. (1964). *Theory Into Practice* **3**, 95.

Sigmond, H. (1960). *Pediat. Clin. N. Amer.* **7**, 165.

Thompkins, E., and Rowe, V. (1958). "A Survey of Interscholastic Athletic Programs in Separately Organized Junior High Schools." National Association of Secondary-School Principals, Washington, D.C.

White House Conference on Children and Youth. (1960). "Children and Youth in the 1960's." Committee on Studies for the Golden Anniversary White House Conference on Children and Youth, Inc., Washington, D.C.

Wilmore, J. H., and Sigerseth, P. O. (1967). *J. Appl. Physiol.* **22**, 923.

Author Index

Numbers in italics refer to the pages on which the complete references are listed.

Subject Index